Sustainable Supply Chains and Digital Transformation

This book focuses on the impact of digitalization on supply chains and how it is affecting every aspect of people's lives. It covers the operational changes that are happening in organizations due to the use of technologies such as IoT, cloud computing, smart sensors, electric vehicles, blockchain, AI, drones, smart factories, smart logistics, and smart warehouses. Additionally, it explores how digitalization is helping organizations achieve sustainability through methods like life cycle costing, carbon emission reduction, green supply chains, and recycling technologies.

Sustainable Supply Chains and Digital Transformation includes case studies, and exploratory studies utilizing quantitative analysis, scientific and qualitative studies to demonstrate how innovation and technology in supply chains contribute to business sustainability in emerging economies and the global economy and discusses the impact of digitalization on supply chains from both operational and sustainability perspectives. It also highlights the role of various stakeholders in achieving digital transformation through sustainability.

Corporate executives, entrepreneurs, government officials, professionals, academics, postgraduate students, and research associates will undoubtedly find this book to be a valuable addition to their libraries.

Emerging Trends in Technology in Management and Commerce

Series Editors: Vikas Garg and Rashmi Agrawal

In the present competitive scenario the recent changes in hardware, software, and telecommunication technologies has transformed society and is continuously evolving. In the past few decades, there has been tremendous change witnessed in information technology. We have entered the phase where the way we communicate and perform day to day activities is dependent upon telecommunication technology combined with increasing the power of computer use. This change leads to the profound changes in technology that supports our organizations. These changes influence the way organizations function and paves the way for new business opportunities. It also creates the need to understand these technologies in the area of management.

Due to dynamics technology is influencing each and every aspect of an organization. In this regard, the main aim of this new series is to understand the latest technologies that are emerging and changing the entire outlook of management. How the emerging technologies are influencing the communication and day to day activities of an organization. There are a number of books that address different technologies such as AI, blockchain technology, data analytics, multimedia etc. These trends are visible, but at times too complex to be of use to the organization and management team. There is a need for books that can provide the business solutions to many management problems. It is crucial for readers to be exposed not only to different emerging technologies and their applications, but also their impact on management and the role they play in an organization. Strategic managerial issues arising from and dealing with management of information technology and systems are also included in this series.

If you are interested in writing or editing a book for the series or would like more information, please contact Cindy Carelli, cindy.carelli@taylorandfrancis.com

Sustainable Technology for Society 5.0
Case Studies, Examples, and Advanced Research Findings
Edited by Tilottama Singh, Richa Goel, and Jan Alexa Sotto

Sustainable Supply Chains and Digital Transformation
Edited by Saurabh Tiwari and Richa Goel

Sustainable Supply Chains and Digital Transformation

Edited by
Saurabh Tiwari and Richa Goel

CRC Press
Taylor & Francis Group
Boca Raton London New York

CRC Press is an imprint of the
Taylor & Francis Group, an **informa** business

Designed cover image: Shutterstock

First edition published 2025
by CRC Press
2385 NW Executive Center Drive, Suite 320, Boca Raton FL 33431

and by CRC Press
4 Park Square, Milton Park, Abingdon, Oxon, OX14 4RN

CRC Press is an imprint of Taylor & Francis Group, LLC

ISBN: 978-1-032-76048-3 (hbk)
ISBN: 978-1-032-76133-6 (pbk)
ISBN: 978-1-003-47724-2 (ebk)

DOI: 10.1201/9781003477242

Typeset in Times
by SPi Technologies India Pvt Ltd (Straive)

Contents

Preface

Over the past few decades, the business environment of the global has changed significantly due to globalization, climate change, and technological innovation. At the heart of these changes is the shift in the perception of supply chains from a mere channel to get the product through, to a key supply chain driver. The 21st century is a different ball game—first, the COVID-19 virus, second, a volatile global environment, and third a push toward sustainability. We've come a long way from blaring out the fact that the environment is something that can and should be pushed to the side; the word "sustainability" is no longer simply the buzzword of the decade. The consumers, investors, and other stakeholders expect organizations to avoid polluting the environment, violating people's rights, and affecting them negatively. Global industries and businesses across all sectors play a crucial role in this transformation, and this is through supply chains. Sustainability is not an escapade in business anymore, but a vital management strategy as influenced by the prevailing societal, regulatory, and market demands. They include the deliberate seeking of sustainable business value by activities that respect environmental, social, and economic consequences. Sustainability is material management, waste minimization, and carbon footprint in businesses for organizations and in the supply chain for companies. This also covers issues of ethical employment standards, a commitment to fair trade, and support for the community. Managers can manage risks that threaten their companies and simultaneously take advantage of opportunities to develop new sources of corporate value, strengthen their position in the eyes of consumers and investors, and meet their new expectations related to sustainability. But getting to sustainable supply chain management is not without its complications. Carbon emissions, all sorts of resources, and ethical supply chain concerns cannot be solved in isolation. Just like the sustainability penchant, the wheel of digital innovations wheel that is proactively transforming the supply chain management landscape. Digital transformation cannot be interpreted as just implementing new technologies, but changes the way how company works. AI, blockchain, the IoT, and Big Data are technologies that some businesses are using to achieve great insight and flexibility in operations and supply chains. Digital transformation entails the use of digital technology to improve how business is performed, how value is created for consumers, and how competitive advantage is achieved. It has to do with rethinking work processes, value propositions, and markets from scratch. AI, cloud computing, the IoT, and data analytics create opportunities for companies to optimize processes, gain insights, and promote innovation. Digital also enables companies to be more flexible in responding to changes in the market, and the continuation of the digital offering to the customer base, as well as internal process improvement. In other words, digital transformation is a cultural issue, leadership issue, and a coordination issue across organizational silos. To this, businesses must embrace this new direction in order to remain relevant as the world changes with advancements in technology. Thus, the basic tenet for this book is in harmonizing sustainability and digitalization opportunities for organizations. It is thus possible to powerfully argue that only organizations that combine these two priorities shall be able to increase their competitiveness, while at the same time helping to create a fair and sustainable future on planet Earth. But to achieve such integration, it takes more than vision, cooperation, and willingness to change.

Further, this book has been organized in a reader-friendly way that makes it easy to understand the content. Important information that has been properly analyzed is underlined. The reader's access to materials in the book creates the possibility for more in-depth research. It explores the transformative impact of emerging technologies on supply chain management, emphasizing sustainability, innovation, and leadership. Divided into three thematic blocks, it provides insights into how robotics, blockchain, and digitization are revolutionizing logistics and operations. The reader will be able to quickly grasp the chapters' major ideas and summarizing content.

Chapter 1 Talks About Enhancing Sustainability in Supply Chain Through Robotics

Environmental responsibility is now a crucial aspect of modern industry. Embracing eco-consciousness in every aspect of the supply chain enhances organizational reputation and further legitimizes the organization. To improve supply chain sustainability, companies have also started implementing advanced technologies, including robotics. Various research has been conducted on ways to adopt and integrate robotics in the automobile, electronics, and pharma industries. However, most of them focus on the technological impact of robotics in improving productivity and supply chain performance. In contrast, this study discusses the implications of robotics in making a sustainable supply chain. The study used narrative literature to review methodology and analyzed the application of robotics for sustainable supply chain management across industries. Also, this study presents how robotics technology is ideal for waste management and reverse logistics operations. The trends and gaps identified through this study allow us to develop a reasoned agenda to guide future technology-enabled sustainable supply chain measures. This study would be helpful for practitioners, policymakers, and academia working in robotics and automation of supply chain operations.

Chapter 2 Talks About Digitization in Road Transportation for Logistics Industry

The road transport and logistics industry could benefit significantly from the use of digital technology to enhance operations—particularly in a volatile market with increasingly demanding customers—and this is applicable, even if not to precisely as high an extent. Existing work on digitization and its profound impacts in the area of road transportation and logistics is reviewed throughout this chapter. It describes the traffic conditions, and elucidates artificial intelligence (AI) from a perspective of road transportation as an application that offers multiple advantages in real life. It incorporates the efficient routing of predictive analytics with flexible traffic management.

Applying AI in logistics can make a difference to the market and, finally, result in good sales for a company. It's improving the way logistics and transport are managed by using smart algorithms to make operations better with every passing day. This chapter indicates the use of AI in logistics attached to deep analysis from a corresponding study and signifies its mission-critical side when it comes to performing route optimizations, tracking vehicles under surveillance, and forecasting traffic situations. This study is an overall significant paper on digitization and AI with regard to their intersection with the road transport and logistics industry as well, by illustrating what the opportunities are and the barriers faced throughout undergoing a transformation.

Chapter 3 Talks About The Role of Technology Leadership in the Use of Blockchain in Supply Chain Management

Technology has become an integral part of every business. The use of new age technologies such as blockchain helps in improving supply chain management (SCM). Technology can be best utilized when it is part of strategic planning. Therefore, technology leadership plays an important role in improving the functionality of the distribution channel system through the use of blockchain technology. This chapter aims to propose a research model on the above moderating role of technology leadership. The study is based on a qualitative approach and is conceptual in nature to emphasize the importance of technology leadership in supporting the supply chain with the blockchain technology. The findings of the chapter are valuable for policymakers and organizational leaders to provide direction for blockchain implementation through technology leadership. The model proposed in the conclusion strengthens existing literature and knowledge while adding the perspective of leadership in technology implementation.

Chapter 4 Talks About Optimization of the Virtual Supply Chain

Today, the manufacturing sector has embraced integrating modern information technologies to enhance operational efficiency and facilitate collaboration with geographically dispersed supply chain partners, particularly in expanding physical trade constraints on a global scale. Traditional supply chains encounter numerous challenges, including uncertainty, high expenses, complexity, and susceptibility. Traditional supply chain systems often suffer from poor efficiency, lack of confidence, and reduced trust among stakeholders. These issues can cause inconsistencies in inventory documentation and delays in order processing, ultimately leading to higher costs and unhappy customers. To address these concerns, it is essential for supply chains to evolve into more intelligent systems. Integrating digital technologies into manufacturing is becoming more vital in today's worldwide business environment. Digital technology can be incorporated into the design, implementation, analysis, and business process improvement phases, applied not only within the organization but also in external IT interfaces and collaborations within the supply chain. Digital technologies provide greater visibility into the supply chain by enabling faster and more economical product traceability, improved coordination among partners, and access to flexible financing options. Additionally, these technologies can assist in gathering customer feedback, which offers valuable insights into market trends. In this research, the goal is to optimize the virtual supply chain to maximize the profit and time of information processing while taking into account virtual costs, information security, and energy consumed.

Chapter 5 Talks About Industry 4.0: Driving Sustainability in the Digital Age

Industry 4.0 integrates independent machines, processes, and systems through the Internet of Things (IoT) while incorporating technologies like artificial intelligence (AI), intelligent systems, and data analytics. The interconnectedness of these technologies with the ecosystem helps in simultaneous optimization, monitoring, and enhancing supply chain efficiency and productivity throughout its lifecycle. Advancements in this field hold unmeasurable competency to advance sustainability by minimizing waste and consumption of resource. However, alongside its promise, Industry 4.0 introduces economic and social challenges. Automation raises concerns about job displacement, emphasizing the need for reskilling and upskilling the workforce. The widening digital divide between advanced and developing regions risks exacerbating global inequalities. Despite these inhibitions and challenges, Industry 4.0 offers much to transform opportunities addressing pressing environmental and social concerns. By leveraging advanced technologies, industries can drive the efficiency of resources, reduce ecological impact, and innovate solutions for a sustainable future. Achieving this vision demands collaborative efforts across sectors, ethical governance, and a collective obligation to build an additional equitable and balanced world for future generations.

Chapter 6 Talks About Ensuring Security Across Digital Supply Chain

In this modern digitalization era, we see a complete shift of supply chains into digital technology, creating an intricate maze for consumption. The digitization of supply chains also reveals their weaknesses, making them vulnerable to numerous cybersecurity threats. However, this pays off greatly in terms of the innovative productivity it brings about in the realm of sustainability. This study seeks to assess the challenges and possible solutions that guarantee security within the digital supply chains. Topics explored include: cybersecurity threats and challenges to collaboration, regulatory compliance, and best practices; research methods include document analysis, focus groups, and interviews. Findings prominently support strong cybersecurity measures, stakeholder cooperation, regulatory compliance, and ongoing security posture monitoring. Options include employee training, supplier risk management, and a security-conscious culture, all of which could strengthen digital supply chains against the effects of cyberattacks. The study will elucidate an interesting view on security issues in a digital supply chain in Industry 4.0 across diverse industrial sectors.

Chapter 7 Talks About Ranking of Barriers to Adoption of Blockchain Technology in Supply Chain

Integrating blockchain technology into supply chain management offers a revolutionary approach to enhancing transparency, traceability, and security within the industry. Despite its potential to address inefficiencies and trust issues, widespread adoption faces significant barriers. This chapter uses both qualitative and quantitative data to systematically analyze and rank the challenges to blockchain adoption in supply chains. In this study, utilizing the Fuzzy Making use of the Decision-Making Trial and Evaluation Laboratory (DEMATEL) and the Fuzzy Analytical Hierarchy Process (AHP), key issues are identified with reference to technical integration, regulatory issues, and concerns over scalability, and environmental effects. This research emphasizes the need for understanding, technological propositions, regulation assurance, and business practices for adopting blockchain supply chain integration; the influence of the TOE framework is found to be insufficiently explored across different contexts.

Chapter 8 Talks About Role of Corporate Social Responsibility in Circular Economy: A Bibliometric Mapping Research Trends

With emphasis on sustainability and resource use optimization, the current era has shifted into a greater discourse on the circular economy (CE). It has become one of the critical drivers to advance the engrailment of principles of CE as a part of the wider corporate social responsibility (CSR) agenda, which will enable organizations to blend profit-making with environmental stewardship and social welfare. This research focuses on how CSR and CE relate to each other, using bibliometric mapping of the concepts to analyze thematic concentrations, and identify the existing knowledge gaps in this fragmented and interdisciplinary field. Analyses have been carried out on publications from various sources around the globe concerning advanced bibliometric tools in subject areas such as sustainable business practices, waste reduction, resource efficiency, and stakeholder engagement. CSR initiatives often promote optimization of resource use as encouragement to design production and processes in such a way that waste is minimized and renewable or recycled materials are favored.

Chapter 9 Talks About Sustainable Transformation Through Organizational Ambidexterity: A Supply Chain Imperative

In today's world, where sustainability is super important, and everything is changing so fast because of digital, having the ability to be ambidextrous is a key skill for supply chain companies. This chapter takes a look at how finding the right mix between using what you do best and looking for new ideas can help businesses move toward sustainable practices and digital change. We draw on some exciting ideas from people who studied the exploration-exploitation paradox. They focus on ambidextrous organizations, too. When considering digital transformation and sustainable supply chains, we draw out the meaning of ambidexterity. It's about figuring out how to assist supply chain corporations in developing this crucial expertise, similar to what other researchers have discussed. This chapter explains sustainable transformation through organizational ambidexterity by combining strong theories with real world, results, and recent examples. Having this talent is very critical for thriving in today's complex and ever-changing corporate landscape!

Chapter 10 Talks About Balancing Innovation and Efficiency: The Role of Organizational Ambidexterity in Digital Supply Chain Management

The ever-changing corporate world faces the dual challenge of exploiting existing resources while exploring new opportunities to maintain competitiveness and drive sustainable growth. This balance, organizational ambidexterity, is particularly critical in the digital supply chain management framework, where rapid technical innovations and market dynamism create complex and multifaceted challenges. This research aims to investigate the interaction between organizational ambidexterity and supply chain management, enlightening how the organization uses ambidexterity to optimize operations and strengthen versatility, and establish a competitive edge. Theoretical framework from the existing literature and the empirical proof from the case studies of the learning organizations, the chapter aims to review the intricacy of digital supply chain management via organizational ambidexterity, practical implications are discussed as well as the strategies for balancing exploitation and exploration, and administering the technological advancements are discussed enhancing transparency and creating digital potential. Furthermore, the chapter proposes propositions for forthcoming research to strengthen our knowledge of organizational ambidexterity in the digital supply chain context.

Chapter 11 Talks About Innovation and Governance in the Digital Era: Exploring the Complexities of the Digital Supply Chain

This chapter investigates multiple factors and dynamics related to the digital supply chain and the involvement of national institutions, government authorities, and various stakeholders. In the highly advanced era of technological innovations and globalization, the digital supply chain has become decisive for modern economies. An interdisciplinary focus addresses the implications of digitalization for female workers, industrialization tendencies, global supply chains, and the enforcement of corporate codes of conduct. Given that the modern digital technologies alter traditionally accepted methods of production and supply, it is important to understand the social and economic effects on female workers and the changes in opportunities and conditions for them at the current point. The purpose of the current research is to identify gender gaps and access to working opportunities and investigate the role national institutional frameworks and government authorities play in supporting women's empowerment in the digital supply chain. Primarily, the chapter aims at assessing the implications of digitalization on industrialization in the developed and developing world. Additional focus will be made on the opportunities and obstacles associated with automation, data analytics, or artificial intelligence, and ways of applying them to ensure sustainable development.

Chapter 12 Talks About From Vulnerability to Resilience the Impact of Eliminating Child Labor on Supply Chain Sustainability

Child labor is still very much prevalent in the international supply chain network. It threatens the ethical integrity and long-term sustainability of the modern trade at all levels. A lot of efforts have been put to tackle this issue of child labor in supply chains, but still this problem has not been fully removed, and children are still working in agriculture as well as industry level. This feeds the cycle of poverty and opens the door for exploitation of children. However, working in the correct direction for elimination of child labor has the protentional to bring about a significant change in the society—a cascade effect—in this context. This chapter will investigate the very implications of removing the problem of child labor on supply chain networks. Examining the situation from an economic perspective, the abolition of child labor will help to lower all the possible risks of reputational damage, legal requirements, and operational difficulties to the companies. Its abolition will additionally help in building the trust in the eyes of consumers and will also help in enhancing the reputation of the brand. In this chapter, a detailed analysis will be done of the present literature available in this area, and this chapter will try to examine the benefits that companies and communities will get after ending the practice of child labor. This chapter will also examine the positive effects on environment also by ending child labor.

Chapter 13 Talks About Inbound and Outbound Logistics in Supply Chain Management: Building Efficiency and Resilience

Logistics, the sector leading the way toward global supply chain transformation, is influenced by the forces of technology, sustainability imperatives, and the growing demands of modern businesses. Building on this foundation, the chapter discusses the important roles, the associated challenges, and the emerging trends in logistics, bringing out why logistics is key to efficient, resilient, and responsible operations. In addition to the challenges of globalization, technology, and sustainability issues, it first gives a summary of the essential logistics functions, including inbound and outbound operations, inventory management, and distribution. The chapter examines how advancements are changing logistics to make it more effective, less expensive, and more transparent via the prism of enablers of technology, including blockchain, robotics, IoT, and artificial intelligence. Finally, the development of automation in last-mile deliveries and warehouse management, as well as how drones and driverless cars are transforming the conventional logistics process, are discussed. The chapter tries to cover the aspect of sustainability by discussing green logistics practices, renewable energy adoption, and reverse logistics systems focusing on reducing environmental impact. The chapter also emphasizes logistics' supporting role in building supply chains that are robust, flexible, and sustainable—a forward-looking development. In addition to adding to scholarly discussions, this research gives practitioners practical advice that aids them in resolving today's logistical challenges.

Chapter 14 Talks About Unchaining Humanity: Pioneering Solutions to Eradicate Exploitation in Global Supply Chains

This chapter focuses on the widespread problem of child labor, bonded labor, and human trafficking within global supply chains, drawing on the complexities, difficulties, and solutions to these reprehensible acts. Labor exploitation is discussed first with a detailed analysis of the magnitude of the issue and its gravity. Then, the chapter provides an argumentation of the ethical, legal, and humanitarian necessity of solving those problems. The chapter aims to discuss the density of factors that cause the situation of labor exploitation to persist within such economic inequality, vulnerabilities in social life, as well as legal lacunas. On top of that, it assesses the existing legal and regulatory frameworks on both the international and national levels, looks at whether they are implemented properly, and at what obstacles may be faced by their implementation. Corporate responsibility is analyzed focusing on the necessity and role of business organizations in shifting workplace exploitation through proper assessment, evaluation, and up-to-the-mark remediation of incidents. Innovation and technology become critical instruments for the battle against labor exploitation, as issues like blockchain, artificial intelligence, and emerging technologies are being discussed as ways of improving supply chain transparency and accountability. The case studies describe approaches that have been generated and have helped to tackle these issues, which underscore that tech innovation has transformative power for driving ethical practices.

Therefore, the purpose of this book is to extend and disseminate knowledge about how digital technologies, sustainability programs, and stakeholder engagement shape supply chain management in Industry 4.0. It seeks to offer information on technology trends, sustainability solutions, the barriers faced, key collaborators, and socially responsible commodities, exploitation, and child labor, with the comprehensive intention of setting up a reliant and ethical supply chain models. The book therefore aims at positioning the researcher, practitioners, and policymakers with knowledge and frameworks for the digital supply chain.

Saurabh Tiwari & Richa Goel

Acknowledgments

In the first place, it is our immense pleasure to pay our devotion to our God Almighty. While working on putting together this book titled "Sustainable Supply Chains and Digital Transformation", we realized how intricate this process can be and got the strength to work toward achieving dream with hope and zeal. I know your guidance did help us to find strength and to see that there is a clear picture. We would have been incomplete without your blessings.

We also thank our families for their constant support during the challenging steps of writing this book. Their patience, encouragement, and believing in us has been very great.

The editors would also like to extend our gratitude to all those who worked on this project, with the particular mention of the authors and reviewers who have done a great job in contributing to the successful completion of this work. They have been highly dedicated in ensuring that the chapters meet high quality and are well integrated. Thanks are also due to the contributors and chapter writers who have so kindly given their time and expertise. We are equally thankful to the reviewers for their constructive criticism that was useful in stickling down and focusing on the content. The many roles of the authors included also the referee where recommendations were sound and constructive that prompted us to fetch out concepts, focus on specific core areas, and defend proposals. This brilliant contribution enhanced the general quality of the provided book in numerous ways. Finally, we would like to express our appreciation to the key mentors and peers who have helped us through years to achieve mastery of the networking art as well as the digital supply chain discipline. It has been our source of learning inspirations and upbringing throughout our respective studies.

Editors
Saurabh Tiwari & Richa Goel

About the Editors

Saurabh Tiwari, Ph.D., is an associate professor at Jindal School of Banking & Finance (JSBF) at O.P. Jindal Global University (Institution of Eminence, Ministry of Education, Govt. of India) Sonipat, Haryana, India. As an ardent student of management, Dr. Tiwari brings approximately 18 years of multifaceted experience in the fields of higher education teaching, research, and administration, encompassing a tenure of three years within the corporate sector. He has co-authored several research articles, book chapters, and conference papers, all indexed in the WoS/ SCOPUS, ESCI, and ABDC databases. Throughout his professional journey, he has demonstrated a commitment to excellence, fostering a culture of innovation, and driving growth in various capacities. He has been instrumental in shaping the next generation of professionals, sharing his expertise and insights with students and professionals alike.

Richa Goel, Ph.D., is an associate professor of Economics and International Business, Symbiosis Centre for Management Studies, Noida, Symbiosis International (Deemed University), Pune, India. She has more than 21 years of experience in academics and consistently strives to create a challenging and engaging learning environment where students become life-long scholars and learners. Imparting lectures using different teaching strategies, she is an avid teacher, researcher, and mentor. She has to her credit a number of books and research papers published in reputed national and international journals, and has participated in many professional conferences as well. She is serving as a member of review committees for conferences and journals, and also acts as associate editor of the *Journal of Sustainable Finance and Investment*. She is a gold medalist in her Master of Economics with dual specialization at the master level, with an MBA in HR, and with dual specialization at the graduate level with gold medal in Economics along with a Bachelor of Law. She also holds a Ph.D. in Management and has about six years of working in the area of diversity management.

Contributors

Areena Parveen Ansari
University of Ottawa
Ottawa, Ontario, Canada

Prakash Chandra Bahuguna
School of Business, UPES
Dehradun, India

Balvinder Singh Banga
Swiss School of Business Management
Geneva, Switzerland

Sukanta Kumar Baral
Department of Commerce, Faculty of Commerce & Management, Indira Gandhi National Tribal University (A Central University)
Amarkantak, India

Aditi Bisht
School of Business, UPES
Dehradun, India

Nikolay Bodrov
Department of Forensic Science, Kutafin Moscow State Law University
Moscow, Russia

Saloni Chitkara
Jaipuria School of Business
Ghaziabad, India

Shikha Dimri
School of Law, UPES
Dehradun, India

Salu Dsouza
School of Law, Christ University,
Lavasa, India

Seema Garg
Amity University
Noida, India

Bridget Irene
Business Management, School of Leadership, Management & Marketing
Leicester Castle Business School, De Montfort University
Leicester, United Kingdom

Muskan Khan
Department of Management Studies, Galgotias College of Engineering and Technology
Greater Noida, India

Col Sharad Khattar
Amity University
Noida, India

Chitra Krishnan
Symbiosis Center for Management Studies, Noida, Symbiosis International (Deemed University)
Pune, India

Arpana Kumari
Symbiosis Center for Management Studies, Noida, Symbiosis International (Deemed University)
Pune, India

Anchal Luthra
Amity University
Noida, India

Jasmine Mariappan
University of Technology and Applied Sciences
Ibra, Oman

Nidhi Shridhar Natrajan
Symbiosis Centre for Management Studies, Noida, Symbiosis International (Deemed University)
Pune, India

Saurav Negi
Modern College of Business and Science
Muscat, Oman

Mohsen Shafiei Nikabadi
Faculty of Economic, Management and Administrative Sciences, Semnan University
Semnan, Iran

Md. Zafar Mahfooz Nomani
Faculty of Law, Aligarh Muslim University
Aligarh, India

Akshat Kumar Rai
School of Business, UPES
Dehradun, Uttarakhand, India

T. Bangar Raju
Indian Maritime University
Chennai, India

Rudra Rameshwar
Academic Unit: S.H.S.S., Thapar Institute of Engineering and Technology (Deemed-to-be-University)
Patiala, India

Nataša Rupčić
Department for Organization and Management, Faculty of Economics and Business, University of Rijeka
Rijeka, Croatia

Supriya Lamba Sahdev
Alliance University, Alliance School of Business
Bangalore, India

Abdullah Samdani
School of Law, UPES
Dehradun, India

Rinku Sanjeev
Symbiosis Centre for Management Studies, Noida, Symbiosis International (Deemed University)
Pune, India

Hanieh Shambayati
Industrial Management, Faculty of Economic, Management and Administrative Sciences, Semnan University
Semnan, Iran

Ritu Sharma
Torrens University, Adelaide Campus
Adelaide, Australia

Vinita Sharma
Amity University
Noida, India

Anuja Shukla
Jaipuria Institute of Management
Noida, India

Anamica Singh
Amity University
Noida, India

D. D. Tewari
Department of Economics, Faculty of Commerce, Administration & Law, University of Zululand

Shantanu Trivedi
O.P. Jindal Global University
Sonipat, India

Liudmila Veselova
Department of Management, School of Economics and Management, HSE National Research University
St. Petersburg, Russia

Madhuri Yadav
Department of Commerce, Indira Gandhi National Tribal University (A Central University)
Amarkantak, India

Part 1

Digitalization of Supply Chain with Respect to Industry 4.0

1 Enhancing Sustainability in Supply Chain Through Robotics

Saurav Negi and Shantanu Trivedi

1.1 INTRODUCTION

In the quest for sustainable development, it becomes clear that balancing economic prosperity with environmental care is crucial. Balance is key to making sure today's generations can meet their needs without risking the needs of future generations. Sustainable development requires a holistic view, one that weighs environmental issues and economic growth equally. It stresses the need for progress that doesn't harm our planet, ensuring future generations can meet their needs too. The Brundtland Commission (United Nations, 1987) defines sustainability as "meeting the needs of the present without compromising the ability of future generations to meet their own needs." This underscores the importance of balancing current demands with long-term care for the environment.

This convergence often puts sustainability in pursuit with cost reduction and productivity improvement in most industries. This is because green practices benefit the environment as well as providing major economic benefits to an organization. In all of this, robotics becomes a big player and eventually crucial in replacing older methods and driving manufacturing toward more sustainability. Using sophisticated automation, manufacturers can make the processes smoother, cut down on waste and energy use, and save resources (Folk, 2019). To describe this, robots work with high precision, accuracy, and consistency at work, therefore producing fewer errors and material wastage. Additionally, the efficiency offered by robots is compared to that of a human being, which results in reduced usage of energy and thus a minimum carbon footprint. Adding robotics to manufacturing would not only save efficiency but also be apt for a wider goal of sustainability by saving resources and reducing environmental damage. The dual benefit in itself demonstrates just how well robotics can drive sustainable practice across domains such as manufacturing (Javaid et al., 2021; Puica, 2022). According to Ojo et al. (2023), who especially emphasize sustainability for global supply chains in manufacturing industries and the crucial roles of automation and technology in these endeavors. These technologies will contribute to organizations saving waste, improving productivity, reducing the level of pollution by the organization, and saving costs, among others. During all this development in technology currently happening in the digital world, the efforts that robotics technology has toward

DOI: 10.1201/9781003477242-2

supply chain sustainability and waste management remain grossly underestimated. There is a need to do more research on important economic, environmental, and social impacts of using robots in supply chains and waste management. It is also critical to consider the ethical implications of the use of these technologies in various fields (O'Brolchain & Grau Ruiz, 2021). For instance, Sarc et al. (2019) indicate the need for further studies of environmental factors in connection with the effect of robotics on the supply chain and waste management. Although robotics may make operations more efficient and waste less, a full assessment of their environmental impact, including energy use and carbon emissions, is needed to fully understand the case. Also, most present studies forget how robotics technology can support waste management in supply chains; this is a critical gap that has to be closed with urgency. There is clearly a call for additional research explaining how robotics could help improve the sustainability of supply chains and even better manage waste. Tao (2021) concludes, "digitalization and smart robotics is gaining more importance in waste management; where future studies will have the task of evaluating its efficacy on waste management performance toward cost, quality, and environmental sustainability dimensions."

Considering these research directions in the field of sustainable supply chain leveraging robotics technology, the present study is carried out with the following objectives:

- To understand the supply chain sustainability issues and measures adopted by integrating robotics in supply chain management.
- To analyze the applicability of robotics in enhancing the sustainability of supply chains from the Triple Bottom Line (TBL) perspective.

This study aims to illuminate the integration of robotics in enhancing supply chain sustainability and waste management practices, empowering stakeholders with insights on informed decision-making and strategic planning in the manufacturing sector. With the increasing prominence of digitalization and intelligent robotics, this research aims to enrich the ongoing dialogue surrounding sustainable supply chain practices, leveraging robotics technology. Through a comprehensive analysis of supply chain sustainability issues and a thorough assessment of robotics' applicability from a TBL perspective, this study seeks to equip stakeholders with best practices for realizing sustainable outcomes in manufacturing operations.

The rest of the study is structured as follows: Section 1.2 describes the technique used in this study. Section 1.3 discusses Industry 4.0 and sustainability, encompassing themes such as robots for supply chain sustainability, robotics for reverse logistics operations, and studying supply chain sustainability through robotics from a TBL perspective. In Section 1.4, future research directions in robotics for sustainable supply chains are outlined. Finally, the conclusion and the limitations of the study are presented in Section 1.5.

1.2 METHODOLOGY

The current study used an exploratory research approach, which has the shape of a qualitative narrative literature review. First, the study attempted to explore what kind

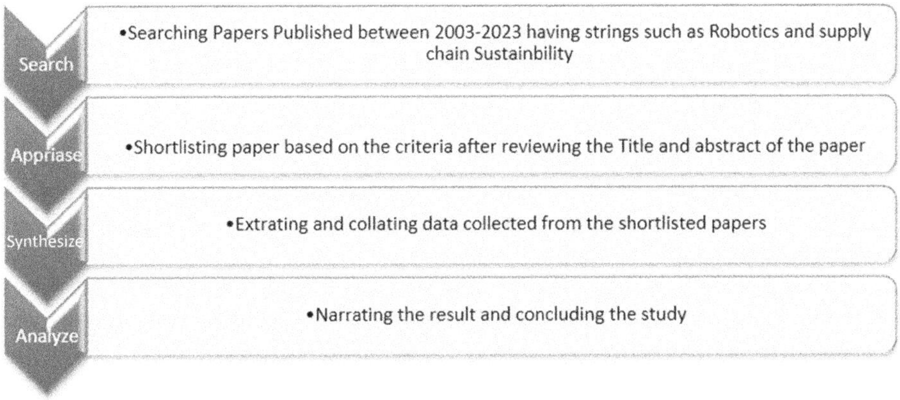

FIGURE 1.1 SALSA process.

Source: Authors.

of challenges come with robotics toward sustainability in supply chains and identify sustainability metrics. For this intent, the authors conducted in-depth review of past literature. For this, the authors relied on qualitative evidence recommended by Tranfield et al. (2003). A narrative literature review has been chosen because it may provide a framework to the area of study and act as a workable resource in future studies. A very broad range of sources was reviewed in these works. In these research papers, articles, conference papers, books, reports, and other publications concerned with supply chain sustainability and waste management were included. Since Scopus is considered one of the largest, entailing a huge collection of scholarly literature (Mengist et al., 2020), there is full recovery of all relevant studies by the authors.

The literature review was based on the SALSA framework, which contains four major stages: (1) Search: keywords/key words and database selection, (2) Appraisal: specifying inclusion and exclusion criteria and quality screening, (3) Synthesis: extracting information and classifying them, and (4) Analysis: writing conclusions and results. This also stated a particular structured approach developed by Kashem et al. (2023), on its way to deliver a structured and logical check along the required predefined criteria. As indicated in Figure 1.1, the SALSA framework is used for this study.

1.3 DISCUSSION

This section critically examines the intersection of Industry 4.0 and sustainability and what that entails for supply chain strategy. An interesting real-world application is the exploration of how robotics can enhance sustainability initiatives within a supply chain, as well as how these have been applied in reverse logistics to ultimately achieve sustainability objectives. Finally, the discussion situates sustainability in a TBL framework, which underscores the need to factor in economic, social, and environmental needs. The sustainability now demands a harmonious balance by organizations for sustained economic growth, social equity, and environmental sustainability through a holistic approach.

1.3.1 Industry 4.0 and Sustainability

With the arrival of Industry 4.0 in the present age of manufacturing, it really is a game changer that is displacing old methods with digital automation technology. Often called Smart Manufacturing or the Internet of Things (IoT), Industry 4.0 introduces a litany of new ideas and protocols that will forever change the way we manufacture. Based on the research of Kamble et al. (2018) and Muller et al. (2018), Industry 4.0 has already had a pivotal impact, particularly its ability to reduce lead times and enhance product quality in organizations.

For example, jobs in the manufacturing sector are likely to change in the future due to the emergence of sophisticated technology roles. According to the report, 40% of respondents expect a high demand for sustainable manufacturing engineers and digital supply chain analysts, while 60% of respondents anticipate considerable development in opportunities for IoT specialists, robotics technicians, and AI professionals (Kaushik, 2024).

However, while Industry 4.0 has many benefits, understanding the full scope of this new phenomenon is complex. Although Kamble et al. as well as Muller et al. shed light on its bright side, we still lack an understanding around both the challenges and opportunities that will arise from IoT for users' implementers, environmental and social effects on people close by. Manavalan and Jayakrishna (2019) look at some immediate gains from deploying Industry 4.0, such as the ability to respond quickly to customer orders. Yet, they also point out the competitive challenges and long-term sustainability issues that organizations face concern echoed by Kamble et al. (2018). In order to overcome these obstacles, Bai et al. (2020) recommend a methodical way to assess Industry 4.0 technologies according to their sustainability performance. To optimize beneficial sustainability effects, they stress the necessity of strategic adoption. The influence of each technology on industry and sustainability should also be carefully examined, according to de Sousa Jabbour et al. (2018), who emphasize the necessity of adopting solutions with consideration.

Eleven critical success elements are identified by Luthra and Mangla (2018) as being necessary for successfully integrating Industry 4.0 with ecologically friendly production methods. They think that in order to fully realize Industry 4.0's promise in advancing environmentally friendly manufacturing processes, this collaboration is essential. Furthermore, there are exciting opportunities to reduce waste in supply chains with creative solutions like the Returnable Transport Items covered by Schneikart et al. (2024). However, as stressed by Kamble et al. (2018), their successful implementation hinges on sound reverse logistics procedures and strategic alignment among all stakeholders. Innovations like the IoT-enabled systems mentioned by Kumar et al. (2024) demonstrate how Industry 4.0 may greatly support sustainability initiatives by encouraging the circular economy and optimizing resource consumption. In a similar vein, Pachouri et al. (2024) present outstanding accomplishments in a variety of Industry 4.0 enabling technologies, demonstrating a wide spectrum of innovations that are expected to build a more sustainable future. The IoT, artificial intelligence, blockchain technology, cloud computing, edge/fog computing, Big Data analytics, machine learning, digital twin technology, the Metaverse concept, virtual reality (VR), augmented reality (AR), mixed reality (MR), and robot deployment in built environments are some of these innovations. According to Pachouri et al. (2024), however, Industry 4.0's full potential in advancing circular economy concepts has not yet been realized and requires further study and research, particularly in areas like robotics for closed-loop production systems to reduce waste and environmental impact.

1.3.2 Robotics for Supply Chain Sustainability

Today, supply chain sustainability is important because stakeholders and customers pressure for organizations to act with more social and environmental responsibility. When implemented, supply chain management with robotics can assist firms in enhancing sustainability by optimizing productivity, minimizing waste, and reducing risks (Fragapane et al., 2021; Goddard et al., 2021). These supply chain sustainability problems will be discussed in detail in this part as well as how the use of robotics might help the organization's supply chain. There is the implication of a comprehensive principle, particularly supply chain sustainability, from the study by Walker et al. (2008). They identify the main business processes and topics in the supply chain sustainability in the body of literature and stress the need to incorporate sustainability into the approaches in supply chain management. This is built on by Govindan et al. (2015) who speak of sustainable supply chain, emphasizing the social and environmental aspects apart from the economical. To address some sustainability concerns, the report warns organizations of the need to do the following: closed loop supply systems and reverse logistics. The role of technology in supporting organizations' net-zero objectives and cutting carbon in logistics is examined by Ashcroft (2022). The report reveals that the usage of automation, robotics, and drones will soon most probably be part of supply chain completion. These technologies eliminate waste, thus enabling businesses to cut on their expenses and consequently lower service costs. In the same study, Gereffi et al. (2005) suggest that there is a major barrier to the development of a more sustainable value chain in that suppliers should be prompted to assume their social and environmental responsibilities. Besides recommending companies to set long-term sustainable business objectives, they also require suppliers to conform to selected industry benchmarks and offer training to enable suppliers to effect significant changes to their production. In this connection, Rowan (2023) pointed out that robotics is indispensable in supply chains linked to such industries as agriculture and fishing. Such works involve mending nets, cleaning fish ponds, feeding fish and monitoring their behavior, administering vaccines, aching diseased fish, assessing general fish health, and possible escapes in underwater nets, among others. Allaoui et al. (2018) opined that through sustainability practices achieved through efficient agri-food supply chain management, a sustainability strategy in the agriculture food supply chain provides satisfaction to food demand. Slowly and steadily, the conventional agriculture food supply chain in India has been facing challenges related to food loss prevention, spoiling and storage, and procurement. The problems require application use of technologies, such as AI, blockchain, robotics, and RFID (Aboah et al., 2019). They aim to help create responsible and explainable supply chain activities at multiple tiers within the supply network (Cohen, 2020). There are many technological advancements possessing the capability of solving some of these problems and or changing the fishing and farming sectors. The above are sensors, blockchain, intelligent drones, cloud-edge computing, and immersive technologies. Robotics is a broad field focused on the construction and administration of robots, which may also include multi-robot or swarm systems – systems of multiple robots synchronized with a set goal. Additionally, industrial or collaborative robots, commonly referred to as cobots, are designed to be used alongside people in zones that are closely or even adjacent. As stated by McClements et al. (2021), border restrictions led migrant workers to employ robotic equipment for crop planting and picking more, as these services are less accessible. Likewise, today's food production and processing industries are hiring robots for the tasks such as sorting, processing, and packing that used to be performed by people. In the current period of lockdowns, the direct interaction was minimized in many areas through the help of robotic delivery vehicles. According to the study, these technologies could also do a lot in improving the efficiency of the given food system through building food quality and helping in controlling food wastage, even if they have some negative impacts as well. Bensassi et al. (2022) explored how robotics and other Industry 4.0 technologies are creating new opportunities to enhance the circular economy, especially in sustainable manufacturing. Their study highlighted that these technologies optimize resource use, cut down waste, manage returns, and enable plants to adopt cleaner industrial practices. Likewise, Peter and Mbohwa (2018) pointed out that cyber-physical systems, robotics, autonomous vehicles, alternative energy technologies, and additive manufacturing are advancing swiftly due to extensive research. These innovations have the potential to replace repetitive tasks within future supply chain operations. Additionally, the adoption of reverse logistics, green supply chains, and re-engineering techniques has promoted more sustainable methods. This has led to the development of new technologies that modernize industrial flow in supply chain cycles.

Sathiya et al. (2021) investigated mobile robots' role in sustainable supply chain management at India's Maruti True Value Company. These robots were implemented to combat pollution. The study found that using mobile robots has resulted in notable environmental benefits. Additionally, Kintscher et al. (2020) proposed a comprehensive approach to information sharing with several participants in the battery supply chain. Their study confirmed that automation strategies and smart robotic systems integrated with architectures like theirs can boost the economic and environmental application of recycling processes.

Arunkumar et al. (2022) examined how robotics enhances the visibility of end-to-end deliveries, automates order processing, and ensures reliable deliveries. The use of supply chain robots decreases errors, provides timely information, fastens up deliveries, and increases efficiency and trust across various operational contexts. Dependable for automating both internal and external supply chain responsibilities, these robots benefit suppliers, customers, and logistics providers. Akandere et al. (2023) noted industries and institutions are progressively using robotics, smart sensors, and Industry 4.0 technologies, such as the IoTs and cyber-physical systems, to achieve a competitive edge and foster better sustainable supply chains for the future. In another study, Kintscher et al. (2020) discovered numerous ways reverse logistics supports the concept of the circular economy. The authors noted that reverse logistics allow organizations to develop circular product designs and integrate forward and reverse product flows to eliminate waste. Moreover, by using innovative tools like autonomous bikes and robots, organizations can enhance the return rate of used items, accelerating recycling. Advancements in technology (like IoT and Big Data) also help with product traceability, cutting down on waste. The study emphasized the usage of courier robots to gather items from hard-to-reach locations, which helps improve product returns.

Gamal et al. (2023) state that disruptive technologies, such as robotics, blockchain, IoT, and 3D printing, are becoming more common and are set to change sustainability practices in supply chains globally. E-Fatima et al. (2023) explored the financial risks and challenges tied to using robotic process automation in the beef industry. They pointed out its potential as a strategy for sustainability. Their research indicates that effectively incorporating robotics can streamline supply chain operations, reducing complexity while enhancing financial and operational efficiency. Consequently, businesses may see better productivity, extended shelf life for products, and improved food quality. In particular, the study underscores the centrality of cost management and increased production efficiency in the creation of sustainable value, robotics. In another study, Thomas and Mishra (2024) presented a smart circular model regarding the e-vehicle battery market at their study. This includes the use of robots to manage the production process as well as other elaborate technologies to recondition, reuse, and properly dispose of EV batteries. They even incorporated the purification of wastewater using the reverse osmosis technique into their simulation model. In their analysis, they noted a phenomenal 9.72% enhancement in profitability margins owing to corporate investment in robotic systems within the industrial manufacturing processes and environmental conservation techniques such as pollution control systems and wastewater treatment systems. In addition to this, the report encourages a further examination regarding the extent to which sustainable practices can be incorporated into the supply chain networks with a view to fostering environmentally sustainable production and delivery of commodities. These two studies demonstrate that robots have a lot of potential to enhance material handling and sorting, decrease packaging waste, and enhance recycling initiatives. In summary, implementing robots' technology has enormous potential to improve waste management and sustainability in a variety of industries. Businesses are recognizing the need to incorporate sustainability into their supply chain strategy as a result of growing stakeholder and consumer pressure to embrace more socially and environmentally responsible practices. Robots are an effective way to boost productivity, reduce waste, and improve safety in the supply chain. As businesses endure innovation, robotics will become a crucial part of shaping the future of sustainable supply chains.

1.3.3 Robotics for Reverse Logistics Operations

Though offering "free and easy" returns sounds straightforward, managing it is quite challenging when over 30% of all online orders are returned. Last year alone, consumers returned products worth nearly $816 billion. This number is expected to top $1 trillion by 2024. To cope with this surge in returns, many warehouses are turning to automation and robotics to ease the process (Dhaliwal, 2020; Musaoglu, 2023).

Within reverse logistics, robots can assist with collecting, transporting, sorting, dismantling, storing, and retrieving returned items at various points in the reverse logistics networks (Sathiya et al., 2021). Krstić et al. (2022) emphasized that adopting advanced robotics in reverse logistics can significantly boost efficiency, flexibility, reliability, and precision. They also proposed that robotic process automation should be crafted to benefit all stakeholders involved.

Karabegović et al. (2020) stress that Industry 4.0 depends on the real-time exchange of data between interconnected robots, computers, sensors, and humans via the internet. But to make Industry 4.0 fit into a sustainable environment, it must deal with legal issues tied to operational processes in organizations. Wijewickrama et al. (2021) point out that having a system full of information – and integrating people, processes, policies, and technology – is vital for quality in the reverse logistics of demolition waste. This idea must be supported by the government, regulatory bodies, industry and academic institutions to better process quality and thus boost reverse logistics performance.

Kumar and Verma (2022) studied a GPS-based, e-smart system for solid waste management. Their study discovered that this system helps cut logistic expenses and reduces the workload for municipal corporations. They also noted that advancements in robotics could transform municipal solid waste systems by enabling better management and fostering a more sustainable future.

According to Williams (2007), implementing automated scheduling for materials recovery in reverse supply networks may require earlier adoption of automated materials identification processes. A significant research gap is seen in reverse logistics design networks, which need to be structured to recover diverse goods efficiently and automatically. The study emphasizes the importance of advances in selective robotic disassembly techniques, highlighting the need to define task precedence relationships and processing times. Proposed innovations include new approaches to integrate imaging and materials identification technologies, along with developing fresh processing methods for material separation. Based on this, Bensassi et al. (2022) presented a sustainable reverse logistics architecture by employing multi-agent and expert systems, where robots' implementation in manufacturing workshops is designed for flexibility and productivity improvement. We note that due to the rising incidence of product returns, facilities have had to shift to the use of automation and robotics in reverse logistics systems. In these networks, robotics has the responsibility of picking, conveying, sorting, disassembling, stowing, and picking up returned products so as to enhance efficiency, variability, reliability, and accuracy. Real-time data sharing of robots, computers, sensors, or humans is quite achievable in Industry 4.0; however, their integration into a sustainable environment addresses legal concerns. The main activities of reverse logistics involve processes, people, policies, and technology that require synergy of government, regulatory agencies, industry, and research organizations. Ideas like GPS-based e-smart solid waste storage systems suggest that there are better ideas than what has been discussed here in this chapter on how to enhance the overall municipal solid waste management and a step toward a sustainable future. However, challenges remain in designing reverse logistics networks optimized for automated recovery processes, highlighting the need for advancements in selective robotic disassembly techniques and material separation methodologies. Proposals for sustainable reverse logistics architectures stress the importance of integrating robots into manufacturing workshops to enhance flexibility and productivity, signaling a transformative shift toward more efficient and sustainable reverse logistics operations.

1.3.4 Supply Chain Sustainability Through Robotics from Triple Bottom Line Perspective

Robotics holds a pivotal role in supply chain management by automating various activities within the process. This leads to increasing efficiency and reducing costs (Brzezinski, 2022). The application of robotics can significantly assist in maintaining the ecosystem for a sustainable and better future by performing various tasks that can minimize environmental pressure. Integrating robotics into supply chain management has the potential to improve sustainability from a triple bottom-line perspective – this addresses economic, environmental, and social impacts. Robotic technology helps in boosting efficiency, reducing waste, and improving safety. Therefore, it has a positive effect on all three aspects of sustainability. This section explores, through the lens of the TBL, how robotics could potentially enhance supply chain sustainability. Following the analysis done by Dinh (2021), sustainable supply chain entails the element of ethical behaviors in the process, upholding human rights, and decreasing the adverse impacts on the environment. Logistics automation and robotics are recognized and encouraged as beneficial tools that enhance safety standards and reduce emissions. Similarly, Duong et al. (2020) identify significant opportunities in the case of robotics, with a focus on the food industry; however, they also have certain reservations concerning some of the employment of these innovations. The chapter also provides an insight into how automation can decrease food waste and improve the efficiency of the supply chain, which shows how they can enhance performance.

Zain (2021) and Lo et al. (2022) explore the sustainable supply chain impact of automation and robotics. To reduce the amount of negative effects that logistics has on the environment, they employ technologies, such as self-driving robots, electric vehicles, and tracking devices to improve efficiency and safety in operations. Javaid et al. (2021) especially, who studied automation and robotics for sustainable supply chain management in the context of Industry 4.0, stated that through the use of autonomous robots the companies can prevent risks to human life, reduce unnecessary expenses, and enhance the effectiveness of business processes.

Companies may establish an effective corporate model that can help strike a harmony between the economic, social, and environmental impact of operations through the use of automation and robotics in their supply chain systems. Such solutions help organizations improve resource usage, waste reduction, and manage their overall carbon footprint. From the perspective of the TBL, automation has huge opportunities to progress sustainable supply chains significantly. These developments can be utilized by businesses to increase general productivity, improve safety, and reduce effects on the environment concerning trash disposal (Sarc et al., 2019). Technology such as digitalization and smart robotics is seen as key to making sustainable changes and improving wastes (Matsuo et al., 2012). These developments are gradually taking organizations toward a less damaging state.

More recently, Ghobakhloo et al. (2023) explored the organizational, environmental, human resource, and technological antecedents of intelligent automation adoption. They show one's corporate responsibility to the TBL and also how intelligent automation, particularly in this case, its a positive impact on the environment. As stated in the report, intelligent automation supports sustainable processes of supply chain management, increases the efficiency of energy use, reduces the amount of waste, and enhances the monitoring of the environment. Perhaps, it is worthy to note that at implementing these technologies, organizational and human factors received highest priority while technological and environmental factors received second and third priorities. The implication of this is that, strategic management and deployment of intelligent automation systems demands a coordinated strategy that considers environmental objectives. This link makes it possible for organizations to be assured that the automation endeavors are also as supportive of broader sustainability purposes.

Kshirsagar et al. (2022) examined the potential for robotic approaches within the AI framework concerning the control of reusable garbage. These research findings show that through the use of autonomous robots, organizations are able to enhance the utilization of resources, decrease the wastage rates to near negligible, and enhance the measures for waste management. Similarly, Gupta et al. (2022) and Sun et al. (2022) have introduced the concept of Reverse Logistics 4.0 that define the reverse logistics as a smarter and green system for managing material and returned products. The study is directed at demonstrating the efficiency of minimizing waste and optimizing resource utilization in reverse logistics by utilizing autonomous robots. In addition, Alvarez-de-los-Mozos and Renteria (2017) supported the idea of collaborative technologies known as cobots in dealing with the challenges concerning e-waste management; the authors explained how cobotization can improve the e-waste treatment process and reduce wastage. An excellent opportunity for supply chain management sustainability regarding the three dimensions of the TBL is the use of robotics. Journals have proved that through the availability of robots wastes can be prevented and resources conserved as well as becoming safer. Again, it might also ensure that processes are quick, costs are slashed, and the impact on the environment is not severe. In addition, intelligent automation is critical in enhancing associated environmentally friendly behavior and minimizing harm to the environment. Robotics systems, and most notably cobots, together with artificial intelligence approaches, suggest new ways of solving specific challenges, such as e-waste management. In recap, therefore, the incorporation of robotics in supply chain management systems is a major step toward developing TBL-compliant SCM. Organizations may meet the changing needs of the stakeholders and customers besides creating a higher and sustainable future through the use of automation and robots.

1.4 FUTURE RESEARCH DIRECTIONS IN ROBOTICS FOR SUSTAINABLE SUPPLY CHAINS

In supply chain and waste management, the application of robotics can provide a more innovative way of boosting the practices of sustainability in modern business operations. Robotics promises efficient solutions to challenges such as waste reduction, efficient energy use, and labor-intensive tasks in supply chains and reverse logistics. Still, there are vast gaps and opportunities to be explored in this area, despite the very extensive research done so far. This section captures some relevant research directions that have been based on the extant literature to improve on the effective use of robotics in sustainable supply chain and waste management. The filling in of such gaps will then open avenues where academics can unlock more competent robots leading to supply chain efficiency, sustainability, and responsibility – all toward a better TBL performance of business and the environment.

- **Robotics to Enhance Waste Reduction and Sorting in Supply Chains**: Future research should consider how robotics technology can more significantly enhance packaging designs and the automation of the waste sorting process for further reductions in waste. For instance, advanced autonomous waste sorting robots with computer vision and machine learning technologies would be underperformed as waste would be sorted more precisely and efficiently.
- **Specialized Robots for Labor-Intensive Jobs in Reverse Logistics**: Research in labor-intensive jobs in product disassembly and material sorting in reverse logistics promises to be promising. Research work should be aimed at designing specialized robots for disassembly processes, which can integrate machine vision and artificial intelligence to automatically identify components during their disassembly operation.
- **Strong ROI Models for Robotic Adoption**: Wider robot adoption in supply chains may need robust return on investment models. Subsequent studies should all-in embrace efficiency gains, labor cost savings, and environmental benefits that accrue after introducing robotics into the system and standardize return on investment models to guide decision-making.
- **Data Security in the Context of Robotic Sorting and Processing Systems**: Data security will be one aspect of future research, given that reverse logistics activities often involve the handling of sensitive consumer information. Secure handling of data and minimization of possible breaches of such data should, therefore, be ensured to safeguard consumer privacy.
- **Consumer Behavior and Robotics-Enabled Supply Chains**: It is necessary to understand how consumer behavior influences the sustainability of robotics-enabled supply chains. Research should be conducted into how robotics can drive sustainable consumption patterns and environmentally friendly product choices for consumers.
- Next, Advanced Autonomous Robots for Sustainable Supply Chain Advanced autonomous robots, including combinations of computer vision and machine learning, will increase the accuracy and efficiency of waste management. Additionally, the use of collaborative robots in supply chain operations and their relation to human workers may create more sustainable logistics practices.
- **Environmental-Friendly Robotic Technologies**: It is essential to investigate the environmental-friendly robotic technologies and applying robotic process automation to make sustainability reporting streamlined. In

addition, holding focus on robotic systems for e-waste recycling and reverse logistics will be highly impactful in the direction of a circular economy.

- **Research TBL Impact of Robotics in Supply Chain Management**: Some areas that must be further researched are the economic, environmental, and social impacts of robotics in supply chain management to totally understand its role in enhancing the sustainability of the chain. Robotics may be seriously transformative for companies when there is a quest to improve sustainability while achieving excellence in operations simultaneously.

1.5 CONCLUSION, IMPLICATIONS, AND STUDY LIMITATIONS

The study investigates the linkage between Industry 4.0 and sustainability, particularly on the enhancement of supply chain sustainability through robotics. It examines the applications of robotics in reverse logistics and demonstrates how Industry 4.0 technologies support sustaining industrial environments in regard to the goals set for sustainability. The move stresses that holistic sustains economic, social, and environmental dimensions in relations to the TBL framework of achieving comprehensive sustainability with a balance in economic success, social equity, and environmental responsibility.

The results also have practical implications for firms eager to exploit robotics to make their supply chain more sustainable. They provide actionable insights into how automation and robotics can potentially improve business operations while mitigating harmful impacts on the environment, thereby creating incentives to innovate. They can inform strategic choices and investments that organizations make in robotics technologies to foster sustainable growth and competitiveness.

Despite revealing useful inferences regarding the relation of Industry 4.0 and sustainability, this study has limitations. Certainly, it does not cover all areas of an interactive nexus between robotics, supply chain management, and sustainability. Other areas, therefore, remain somewhat unexplored within the scope of this study. The specific contexts and sectors lead to further limitations in the scope toward generalization of the findings. Though the study identifies future research directions, the feasibility of such routes may vary from one organization and context to another. Nonetheless, this study provides good grounds for analyzing the role of robotics in promoting sustainability within supply chains and paves the way for further research into this critical area.

REFERENCES

Aboah, J., Wilson, M., Bicknell, K., & Rich, K. (2019). Identifying the precursors of vulnerability in agricultural value chains: A system dynamics approach. *International Journal of Production Research*, 59, 1–19.

Akandere, G., Khajeh, E., & Paksoy, T. (2023). The internet of things and cyber-physical systems. In *Smart and Sustainable Operations and Supply Chain Management in Industry 4.0* (pp. 277–306). https://doi.org/10.1201/9781003180302-14

Allaoui, H., Guo, Y., Choudhary, A., & Bloemhof, J. (2018). Sustainable agro-food supply chain design using a two-stage hybrid multi-objective decision-making approach. *Computers & Operations Research*, 89, 369–384.

Alvarez-de-los-Mozos, E., & Renteria, A. (2017). Collaborative robots in E-waste management. *Procedia Manufacturing*, 11, 55–562.

Arunkumar, A., Ramabalan, S., & Elayaraja, D. (2022). Optimum design of stair-climbing robots using Taguchi method. *Intelligent Automation & Soft Computing*, 35, 1230–1244.

Ashcroft, S. (2022, December 01). *Article Supply Chain Digital*. Retrieved from Supply Chain Digital: https://supplychaindigital.com/logistics/automation-robotics-the-future-of-sustainable-logistics

Bai, C., Dallasega, P., Orzes, G., & Sarkis, J. (2020). Industry 4.0 technologies assessment: A sustainability perspective. *International Journal of Production Economics*, 229, 1–15.

Bensassi, N., Rezzai, M., Wafaa, D., & Medromi, H. (2022). Sustainable manufacturing in Industry 4.0 context: Theoretical background and multi-agent architecture. *International Journal of Engineering Trends and Technology*, 70, 179–193.

Brzezinski, L. (2022). Robotic process automation in logistics – A case study of a production company. *European Research Studies Journal*, 15(2B), 307–315.

Cohen, M. (2020). Does the COVID-19 outbreak mark the onset of a sustainable consumption transition? *Sustainability*, 16, 1–3.

de Sousa Jabbour, A. B., Jabbour, C. J., Foropon, C., & Filho, M. G. (2018). When titans meet – Can industry 4.0 revolutionize the environmentally-sustainable manufacturing wave? The role of critical success factors. *Technological Forecasting and Social Change*, 132, 18–25.

Dhaliwal, A. (2020). The rise of automation and robotics in warehouse management. In *Transforming Management Using Artificial Intelligence Techniques* (pp. 63–72). https://doi.org/10.1201/9781003032410-5

Dinh, L. (2021, December 29). *Article: Robotics Tomorrow*. Retrieved from Robotics Tomorrow: https://www.roboticstomorrow.com/article/2021/03/2021-top-article-sustainable-supply-chains-in-the-era-of-industry-40/16420

Duong, L. N., Al-Fadhli, M., Jagtap, S., Bader, F., Martindale, W., Mark, S., & Paoli, A. (2020). A review of robotics and autonomous systems in the food industry: From the supply chains perspective. *Trends in Food Science & Technology*, 106, 355–364.

E-Fatima, K., Khandan, R., Hosseinian-Far, A., & Sarwar, D. (2023). The adoption of robotic process automation considering financial aspects in beef supply chains: An approach towards sustainability. *Sustainability*, 15, 1–23.

Folk, E. (2019, August 20). *Blog: Clean Green Guide*. Retrieved from Clean Green Guide: https://greencleanguide.com/how-robotics-is-revolutionizing-sustainability/

Fragapane, G., Koster, R. D., Sgarbossa, F. and Strandhagen, J. O. (2021), Planning and control of autonomous mobile robots for intralogistics: Literature review and research agenda. *European Journal of Operational Research*, 294(2), 405–426. https://doi.org/10.1016/j.ejor.2021.01.019

Gamal, A., Mohamed, R., Abdel-Basset, M., Hezam, I. M., & Smarandache, F. (2023). Consideration of disruptive technologies and supply chain sustainability through α-discounting AHP–VIKOR: Calibration, validation, analysis, and methods. *Soft Computing*, 1–21. https://doi.org/10.1007/s00500-023-08819-0

Gereffi, G., Humphrey, J., & Sturgeon, T. (2005). The governance of global value chains. *Review of International Political Economy*, 12, 78–104.

Ghobakhloo, M., Asadi, S., Iranmanesh, M., Foroughi, B., Mubarak, M. F., & Yadegaridehkordi, E. (2023). Intelligent automation implementation and corporate sustainability performance: The enabling role of corporate social responsibility strategy. *Technology in Society*, 74, 1–20.

Goddard, M. A., Davies, Z. G., Guenat, S., et al. (2021), A global horizon scan of the future impacts of robotics and autonomous systems on urban ecosystems. *Nature Ecology & Evolution*, 5, 219–230. https://doi.org/10.1038/s41559-020-01358-z

Govindan, K., Khodaverdi, R., & Vafadarnikjoo, A. (2015). Opportunities and challenges in sustainable supply chain: An operations research perspective. *European Journal of Operational Research*, 556–557.

Gupta, A., Schoor, M. J., Bräutigam, J., Justo, V. B., Umland, T. F., & Göhlich, D. (2022). Autonomous service robots for urban waste management – Multiagent route planning and cooperative operation. *IEEE Robotics and Automation Letters*, 7, 8972–8979.

Javaid, M., Haleem, A., Singh, R. P., & Suman, R. (2021). Substantial capabilities of robotics in enhancing Industry 4.0 implementation. *Cognitive Robotics*, 1, 58–75.

Kamble, S. S., Gunasekaran, A., & Gawankar, S. A. (2018). Sustainable Industry 4.0 framework: A systematic literature review identifying the current trends and future perspectives. *Process Safety and Environmental Protection*, 117, 408–425.

Karabegović, I., Karabegović, E., Mahmić, M., & Husak, E. (2020). Implementation of Industry 4.0 and Industrial Robots in the Manufacturing Processes. In *Lecture Notes in Networks and Systems* (pp. 3–14). https://doi.org/10.1007/978-3-030-18072-0_1

Kashem, M. A., Shamsuddoha, M., Nasir, T., & Chowdhury, A. A. (2023). Supply chain disruption versus optimization: A review on artificial intelligence and blockchain. *Knowledge*, 3(1), 80–96.

Kaushik (2024). *Financial Times*. https://www.financialexpress.com/jobs-career/workforce-resistance-to-ai-could-be-resolved-through-upskilling-ey-ficci-report-3622019/

Kintscher, L., Lawrenz, S., Poschmann, H., & Sharma, P. (2020). Recycling 4.0 – Digitalization as a key for the advanced in circular economy. *Journal of Communications*, 15, 252–660.

Krstić, M., Agnusdei, G. P., Miglietta, P. P., & Tadić, S. (2022). Evaluation of the smart reverse logistics development scenarios using a novel MCDM model. *Cleaner Environmental Systems*, 7, 100099.

Kshirsagar, P. R., Kumar, N., Almulihi, A., Alassery, F., Khan, A. I., Islam, S., … Debeka, K. (2022). Artificial intelligence-based robotic technique for reusable waste materials. *Computational Intelligence and Neuroscience*, 2022, 1–9.

Kumar, A., & Verma, S. K. (2022). Design and development of e-smart robotics-based underground solid waste storage and transportation system. *Journal of Cleaner Production*, 343, 130987.

Kumar, D., Agrawal, S., Singh, R. K., & Singh, R. K. (2024). IoT-enabled coordination for recommerce circular supply chain in the industry 4.0 era. *Internet of Things*, 26, 1–14.

Lo, S. Y., Lai, Y. Y., Liu, J. C., et al. (2022). Robots and sustainability: Robots as persuaders to promote recycling. *International Journal of Social Robotics*, 14, 1261–1272. https://doi.org/10.1007/s12369-021-00828-z

Luthra, S., & Mangla, S. K. (2018). Evaluating challenges to Industry 4.0 initiatives for supply chain sustainability in emerging economies. *Process Safety and Environmental Protection*, 117, 168–179.

Manavalan, E., & Jayakrishna, K. (2019). A review of Internet of Things (IoT) embedded sustainable supply chain for industry 4.0 requirements. *Computers & Industrial Engineering*, 127, 925–953.

Matsuo, K., Ogata, Y., Umezaki, K., Spaho, E., & Barolli, L. (2012). Design and implementation of waste management robots. *26th International Conference on Advanced Information Networking and Applications Workshops*, Fukuoka, Japan, 2012 (pp. 973–976). https://doi.org/10.1109/WAINA.2012.86

McClements, D. J., Hill, C., Kokini, J. L., Lila, M. A., Meyer, A. S., & Yu, L. (2021). Building a resilient, sustainable, and healthier food supply through innovation and technology. *Annual Review of Food Science and Technology*, 12, 1–31.

Mengist, W., Soromessa, T., & Legese, G. (2020). Method for conducting systematic literature review and meta-analysis for environmental science research. *MethodsX*, 7, 100777.

Muller, J. M., Keil, D., & Voigt, K.-I. (2018). What drives the implementation of Industry 4.0? The role of opportunities and challenges in the context of sustainability. *Sustainability*, 10, 1–24.

Musaoglu, E. (2023, April 21). *Blog: Logiwa*. Retrieved from Logiwa: https://www.logiwa.com/blog/reverse-logistics-robotics-and-returns-managing-high-volume-ecommerce-returns

National Retail Federation (NRF). (2024, January 9). *2024 Retail Returns Total $890 Billion According to NRF and Appriss Retail*. https://nrf.com/media-center/press-releases/nrf-and-happy-returns-report-2024-retail-returns-total-890-billion

O'Brolchain, F., & Grau Ruiz, M. A. (2021). Environmental impact of robotics: Ethical concerns and legal alternatives. In Leal Filho, W., Azul, A. M., Brandli, L., Lange Salvia, A., & Wall, T. (Eds.), Industry, Innovation and Infrastructure. Encyclopedia of the UN Sustainable Development Goals. Cham: Springer. https://doi.org/10.1007/978-3-319-71059-4_147-1

Ojo, O. O., Shah, S., & Coutroubis, A. (2023). Impacts of Industry 4.0 in sustainable food manufacturing and supply chain. *International Journal of Integrated Supply Management*, 13, 140–158.

Pachouri, V., Singh, R., Gehlot, A., Pandey, S., Akram, S. V., & Abbas, M. (2024). Empowering sustainability in the built environment: A technological lens on Industry 4.0 enablers. *Technology in Society*, 76, 1–20.

Peter, O., & Mbohwa, C. (2018). The interlink between sustainable supply chain management and technology development in industry. *Proceedings of the International Conference on Industrial Engineering and Operations Management* (pp. 425–430). Pretoria: Taylor & Francis Group.

Puica, E. (2022). How is it a benefit using robotic process automation in supply chain management? *Journal of Supply Chain and Customer Relationship Management*, 2022(221327), 1–11. https://doi.org/10.5171/2022.221327

Rowan, N. J. (2023). The role of digital technologies in supporting and improving fishery and aquaculture across the supply chain – Quo Vadis? *Aquaculture and Fisheries*, 8, 365–374.

Sarc, R., Curtis, A., Kandlbaurer, L., Khodier, K., Lorber, K. E., & Pomberger, R. (2019). Digitalization and intelligent robotics in the value chain of circular economy oriented waste management – A review. *Waste Management*, 95, 476–492.

Sathiya, V., Chinnadurai, M., Ramabalan, S., & Appolloni, A. (2021). Mobile robots and evolutionary optimization algorithms for green supply chain management in a used-car resale company. *Environment, Development and Sustainability*, 23, 9118–9138.

Schneikart, G., Mayrhofer, W., Löffler, C., & Frysak, J. (2024). A roadmap towards circular economies in pharma logistics based on returnable transport items enhanced with Industry 4.0 technologies. *Resources, Conservation and Recycling*, 206, 1–12.

Sun, X., Yu, H., & Solvang, W. D. (2022). Towards the smart and sustainable transformation of Reverse Logistics 4.0: A conceptualization and research agenda. *Environmental Science and Pollution Research*, 29, 69275–69293.

Tao, M. (2021, April 7). *News: Robotics and Automation News*. Retrieved from Robotics and Automation News: https://roboticsandautomationnews.com/2021/04/07/7-supply-chain-technology-trends-shaping-a-sustainable-future/42158/

Thomas, A., & Mishra, U. (2024). Industry 4.0 and circular economy model for a sustainable electric vehicle battery with controllable wastewater and carbon emission. *Energy Reports*, 11, 4044–4066.

Tranfield, D., Denyer, D., & Smart, P. (2003). Towards a methodology for developing evidence-informed management knowledge by means of systematic review. *British Journal of Management*, 14(3), 207–222.

United Nations. (1987, March 20). *Academic Impact: United Nations*. Retrieved from United Nations: https://www.un.org/en/academic-impact/sustainability

Walker, H., Di Sisto, L., & McBain, D. (2008). Sustainability in supply chains: Reappraising the role of business processes. *International Journal of Operations & Production Management*, 596–623.

Wijewickrama, M., Chileshe, N., Rameezdeen, R., & Ochoa, J. (2021). Quality assurance in reverse logistics supply chain of demolition waste: A systematic literature review. *Waste Management & Research*, 39, 3–24.

Williams, J. (2007). A review of research towards computer integrated demanufacturing for materials recovery. *International Journal of Computer Integrated Manufacturing*, 20, 773–780.

Zain, J. (2021, April 9). *Blogs: Supply Chain Brain*. Retrieved from Supply Chain Brain: https://www.supplychainbrain.com/blogs/1-think-tank/post/32909-the-impact-of-robotics-on-supply-chain-20

2 Digitization in Road Transportation for Logistics Industry

Balvinder Singh Banga and Anuja Shukla

2.1 INTRODUCTION

Technology, when placed in the value chain of road transportation sector and logistics, is alleged to significantly disrupt how goods and services are moved through supply chains as well as affecting industry dynamics at large. Digitalization has been a necessity in these times, for the logistics industry to remain sustainable and relevant due to advancements, expanding market demands. This introduction lays foundation to the exploration of digitization in road transport and logistics, as motives for research are stated along with objectives and delimitation. The logistics sector has traditionally depended on networks of transportation and supply chain management to ensure a flow of goods and services. However, the rise of technology has ushered in an era reshaping how road transport and logistics operate. Digitization, which involves integrating technologies into operational aspects, is causing significant changes that could transform the entire logistics industry landscape (UNCTAD, 2020). Some of the technologies that have been developed as part of this shift (real-time tracking systems, autonomous vehicles and predictive analytics) are significantly improving efficiency and visibility in logistics. Companies across verticals are able to optimize delivery routes reduce delivery time, cut costs, increase customer satisfaction and grow revenue streams as a result. In addition, this intelligence using AI and machine learning technologies also becomes a troubleshooting process and improves decision-making. These technologies provide deep insights into patterns spotted over time, and with that predict future scenarios, enabling businesses to stay ahead of the disruption curve so as to be able to react proactively, securing stronger operational resilience. The term "digitization for road transformation" encompasses a range of advancements, within road transportation. These technologies, such as real-time monitoring and tracking systems to optimize routes on data (Cagliano et al., 2019) or communication platforms that allow the different actors in logistics operations to coordinate more gently among themselves with each other reducing operating costs of processes by raising initial marginal effectiveness toward a threshold value. In addition, the technology used to monitor vehicles and cargoes in real time also contributes to safety and security. It generates the alerts for diversions to planned routes or emergencies as well. This proactive method additionally has the added bonus of a quicker response time.

DOI: 10.1201/9781003477242-3

Also helps in reducing the risks of theft or damage to goods during transit. Additionally, it increases visibility across the supply chain as digital technologies enable real-time tracking of shipments. They also enhance coordination and planning abilities. The bottlenecks or inefficiencies can be quickly identified using advanced analytics and predictive algorithms, which help in proactively taking interventions to ensure optimal operational efficiency. Besides, the combo of data analytics with devices provides rich information on driver behavior to ensure better fleet performance, future fuel consumption trend for decision support and resources optimization. These advanced technologies provide for the creation of a security-first road transport and logistics customer ecosystem. Another important technology framework keeping information/data decentralized, specifically in blockchain and distributed ledger systems, are essential design principles that enable efficient supply chain transparency with regards to road transportation digitalization. They maintain all processes with a live record of every single transaction and movement throughout the supply chain, guaranteeing that manufacturers, suppliers, transporters, as well as customers, are firmly in their actions. Many companies like Farm to Plate have already started working on Blockchain and entire logistics managed on Blockchain from Farmers to reaching food at the customer table. For instance, this visibility is minimizing fraudulent activities such as counterfeiting in the domain of road transportation (tenable to hardly ever alter or forge records). For example, products tracked with blockchain are recorded at each stage of the trip in real time and available to all authorized parties. This not only allows the whole journey of the product to be seen and verified, but also secures any potential alterations or manipulation unauthorized by preventing tampering. In a world of hyperconnected factories and in-time production smart digitalization, Internet of Things (IoT) sensors combined with GPS tracking and real-time data analytics provide an opportunity for decentralizing the road transport even more by offering precise information about location, condition and status of goods. This combination of decentralized technology along with digitization in road-based transportation makes the supply chain more transparent, accountable while also decreasing uncertainties and enhancing trust among every participating entity. All in all, this can be easily tracked when there is a requirement of safety traceability, such as food or pharmaceutical sectors. The implementation of technology in the supply chains can revolutionize all, from procurement to production and distribution processes, offering endow systems for companies or service providers along with their consumers. Due to the rapid changes and proliferation of technologies, the only possible solution for both players in industry and academia is a close collaboration, keeping up with the current state-of-the-art solutions, not being late in identifying opportunities addressing future challenges. During the early days of transitioning into a modern commerce era in an organization preparing to execute their optimized supply chain agenda, it is imperative that they understand that this fashion has so many harmful consequences and for their willingness transform, they end up with severe cyber risk as well. In road transport and logistics, the trend of digitalization confronts drivers as well as fleet owners with new complex challenges and opportunities. In today's environment, drivers and fleet owners must embrace this new technology to remain competitive processes. From the influx of automation and integration in aviation, data analytics demand will increase; pilots now must know how to use technology like AI and digital tools for operations performance—including optimization, are mapping flight routes that ensure fuel efficiency. Despite the ongoing growth in digitalization to unprecedented heights, attention will continue to be paid in understanding the effects on efficiency, stakeholder collaboration, and competitiveness. This industry-wide imperative warrants a full analysis of the consequences faced by drivers and fleet owners due to digitalization. This examination should cover the ways in which digitization will lead to faster processes (e.g. route optimization/map planning, real-time tracking) as fleet managers and drivers gain an edge. Also, it should be scrutinized the challenges digitalization initiatives could raise. Challenges include drivers learning how to work with new technologies and securing data in a connected environment. Create an investigation around how digitalization can enable streamlined processes, such as route optimization, real-time tracking of vehicles, and automated scheduling that will enhance driver efficiency and this should be useful for fleet owners. We also need to look at the different types of barriers that come when we digitalize. This could mean getting drivers up to speed on new technologies or ensuring data security as more and more systems are connected. It also has to ask whether digitalization will lead to job displacement and the need for a strategy on how workers affected can find alternate sources of income. Recognizing the resulting trends that were in place due to revolution, like rising power consumption as well as e-Waste accumulation, is critical for mounting estate answers against these changing patterns. Due to this, global creativity and competitive advantage are some of the outcomes of digitization, not just the foundation to be addressed but also the opportunity. Businesses need to strike a balance between embracing technology with enthusiasm and safeguarding the integrity of their data. It's important for them to prepare for job losses to automation by investing in retraining programs for employees. Moreover, they should implement measures to mitigate the impact of digitalization through promoting energy practices and proper disposal of electronic devices. Prioritizing cybersecurity is crucial, requiring organizations to use encryption methods and regularly update security systems to combat cyber threats effectively. Collaboration between businesses and governments is key in establishing laws and ethical guidelines for the use of digital technology. By addressing these challenges, businesses can leverage the benefits of digitalization while mitigating associated risks effectively. To meet the study's goals, the research will delve into the following inquiries:

1. What is the extent to which drivers and fleet owners in road transportation operations, embrace technologies?
2. What are the perceived advantages and obstacles linked with digitization in road transportation and logistics as shared by drivers and fleet owners?
3. How does cooperation and data exchange among stakeholders influence the effectiveness and openness of logistics operations within a framework?
4. How do drivers and fleet owners view how digitization impacts their competitiveness in the logistics sector and broader market trends?

Exploring these queries will offer an understanding of the effects of digitization, on road transportation and logistics providing insights to steer successful integration and enhancement strategies.

2.2 LITERATURE REVIEW

The logistics industry is currently undergoing a significant transformation due to the emergence of digital technology. This section provides an examination of research conducted on digitizing the road transport sector of logistics. The aim of this study is to explore the changing landscape of road transport and its effects on the logistics sector by delving into the advantages, issues, future trends and repercussions of digitalization (Bayoumi et al., 2021). The purpose behind this review is to shed light on these subjects. There have been outcomes resulting from the digitalization of road transport in logistics. These benefits encompass efficiency and productivity facilitated by real-time tracking and monitoring of shipments, optimized route planning and enhanced communication among stakeholders. Furthermore, digitization has enabled analytics and predictive modeling that have improved decision-making processes and reduced costs.

Even though the aspects are there, we have problems to solve. Hurdles, such as security issues and data privacy concerns and insufficient labor force to manage the new technologies' potential displacement of existing business models, must be negotiated. However, we also believe that the upside from digitalization is far more positive than negative. Overcoming these hurdles can provide a business with an upper hand in its industry. Cybersecurity solutions for data and privacy security. Professional training in these technology tools would close the skills gap and allow employees to use them properly. Given the demand of being digital era (Evgeniou et al., 2013), in order to be up-to-date and able to respond to the emerging opportunities, companies must adopt a new business model that is *aiming for digitation*. This proactive capability allows organizations to improve client experience, lower costs, and increase visibility, insight as well as productivity. Along with the benefits come challenges tied to these advancements—innovative approaches introduced in this piece reveal their cybersecurity exposure, workforce needs and nuances of expanding new technologies into legacy systems. Technological advances in the form of vehicles and blockchain are believed to be the next likely game changer for road transport, logistics industry. These innovations will improve the transportation system's efficiency and safety, yet they are likely to raise issues in terms of data privacy and security. Additionally, the transition to driverless vehicles means that current and future drivers and technicians will need at least a baseline of technical training. Incorporating blockchain in this way into the supply chain will otherwise require joint cooperation and planning among industry players. Though democratization of digitalization presents innovation possibilities, we have to combat challenges and risks in road transport. As drivers and technicians, we need to familiarize ourselves with the new technologies that are being incorporated into autonomous vehicles. While driving doesn't need programming skills per se, drivers must grasp how to interact with these advanced systems within their cars. Technicians will have to learn more about analyzing data and diagnostics in order to maintain these sophisticated machines. There is therefore a need for continuous educational and training initiatives that focus exclusively on the needs of both drivers as well as technicians.

The smooth incorporation of technology into supply chain management necessitates having standard protocols. Experiences can be shared, challenges overcome, and successful implementation achieved through close collaboration among manufacturers, suppliers and transporters. It is also important to develop standards and regulations across different industries for system interoperability and protection of data. In enhancing the effectiveness, scalability, and other aspects of supply chain solutions, continuous research and development play a significant role. This means working together with all the stakeholders in the industry to create an environment that creates room for new ideas, which will in turn lead to ground-breaking technological transformations. Partnerships and alliances foster a climate of collaboration (Burbank et al., 2021). Manufacturers, suppliers, logistics providers and tech companies that are in a position to participate will be able to contribute with resources and expertise they have on hand for creating the necessary blockchain solutions. This collaborative approach means that industry standards can be established to unify supply chain networks and defend vital data across the board. Moreover, constant research and development are necessary with the evolving landscape and remove all obstacles whenever they appear. Expenditure on R&D by the industry players can help them to optimize and increase the output of supply chain networks as well. It could be utilizing smart tech; AI and machine learning to enhance and automate performance, or simply reducing human error. An added advantage of investing in R&D is that there can be new ways to save resources. There are other excellent initiatives interspersed, some of which aim to cut carbon emissions and provide sustainable energy.

Focused research is thus vital, and continuous development helps businesses in long run be ahead of the competition and improve overall supply chain operations across sectors. This should give companies that have spent on R&D an advantage in updating their products for different customer requirements, and provide more options to pivot as market changes. As a result, this approach not only enables firms to stay ahead of their competitors but also allows for long-term market success and growth.

Furthermore, investing in research and development initiatives can foster collaborations, with industry peers, academic institutions, and governmental bodies, leading to knowledge sharing and collective problem-solving for industry challenges. By committing resources to research and development activities, organizations enhance their prospects of uncovering groundbreaking technologies and concepts capable of revolutionizing their sectors. These innovations may lead to patent filings and other forms of intellectual property protection, reinforcing the company's market position. Moreover, ongoing research efforts can yield cost solutions that streamline production processes for the benefit of both the company and its clientele.

Research and development play a role in the success of businesses, offering opportunities for growth and innovation (Shen & Sammani, 2022). By investing in research and development, companies can stay ahead of their competitors, adapt to changing consumer needs and industry trends. The insights and expertise gained through these initiatives give businesses an edge, enabling them to create solutions that set them apart from others. These technological advancements may lead to patents and intellectual property rights, further strengthening their position in the market. Continuous research and development efforts not only result in cost solutions but also improve operational efficiency, helping companies to enhance their offerings while reducing expenses. Investing in research and development allows companies to maintain an edge over rivals while improving their products and services continuously.

In terms of attracting customers and enhancing the companies' market share will help establish the business as a leader, in its industry. Moreover, research and development efforts can open up opportunities for growth and expansion for companies especially when these endeavors lead to the discovery of markets or the development of innovative technologies that cater to rising needs. In general, innovation through research and development should be relentlessly focused on taking full advantage of their capabilities and thrive in the fast-paced world of business today. In addition, investment in research and development can bring in businesses and use it to differentiate their products or services from those of competitors—a competitive edge in the market. This could result in heightened customer loyalty and repeat business, further contributing to the organizations success. Additionally, by prioritizing research and development activities, businesses can outpace their competitors. Continuously adjust to evolving consumer preferences and market trends. This strategic approach enables companies to maintain an edge over others, in the industry.

Furthermore, dedicating resources to research and development can help companies anticipate market needs and devise solutions to meet those requirements. By investing in R&D, businesses can maintain their position, as industry frontrunners, while also staying ahead in advancements. This not only attracts talent but also creates opportunities for collaboration with other prominent firms, fostering an environment of innovation and driving overall progress (Barbery & Torres, 2019). In today's business landscape, the emphasis on research and development is crucial, in determining a company's long-term success and sustainability.

2.2.1 Benefits of Digitization

The introduction of technologies, in the field of transportation, has brought about positive changes, reshaping the industry landscape. An example of this is the increased transparency in the supply chain enabled by real-time tracking of shipments and better inventory management. These developments help in minimizing stock shortages and excess inventory. They also result in cost savings and improved customer satisfaction. Furthermore, the use of analytics and machine learning algorithms made possible by digitization, optimizes routing and scheduling processes, leading to decreased fuel consumption and lower carbon emissions. This plus communication and teamwork with stakeholder's increases coordination effectiveness and speeds up decision-making.

Better yet, adding IoT devices to that mix means performance monitoring in real time and actual maintenance practices designed against the risk for maximum uptime of vehicles. Based on the advanced analytics, traffic patterns can be analyzed and routes designed for reduced travel time.

Automation and data-driven decision-making improve operational efficiency in road transport operations. In environmental sustainability, improvements like the convoluted traffic management systems and connected vehicles will help in enhancing network efficiency and safety without space crunches or reducing emissions efficiency.

Additionally, when data is only stored digitally it can be meaningful data analysis, sharing and access are unfortunately a necessity of transport networks.

As it relates to logistics, using digital methods and routing optimization tools can help speed up the order process. It makes supply chain transparent. It makes the process more efficient and fruitful, including cost savings, reduced efforts in fixing overloaded data/information issues level, and increased customer satisfaction. Furthermore, online platforms for matching loads contribute to sustainability efforts by reducing travel distances and resource wastage.

In essence, the blend of advancements with transportation practices transforms operations to drive efficiency gains while prioritizing sustainability efforts and customer-centric approaches, within the industry.

2.2.2 Challenges

The move toward technology brings about opportunities but it also comes with significant challenges. According to Lee et al. (2018), integrating technologies into existing systems can be complex. It may face compatibility issues. Some stakeholders might be hesitant to embrace solutions due to concerns about job security and privacy breaches. In areas with internet access, the progress of digitalization in transportation is hindered (Hofmann et al., 2019). It's essential for the workforce to undergo training to keep up with the paced technological advancements (Gammelgaard et al., 2016).

The transportation sector's transition toward digitalization faces obstacles like skill gaps and infrastructure limitations (Sánchez Rodríguez et al., 2019). Collaboration among industry stakeholder's educational institutions, and government bodies is key in closing these gaps and helping small businesses adopt infrastructure (Gammelgaard et al., 2016). According to Lee et al., "overcoming resistance to change and fostering a culture that embraces new changes would be two essential aspects of implementing these changes successfully" (2018). But the shortage of education and knowledge about technologies exacerbates existing skill differences (Gammelgaard et al., 2016).

Additionally, the lack of standards might endanger data privacy and integrity (Hofmann et al., 2019).

This makes it imperative for businesses to stay ahead of the technology trends, with regular brainstorming sessions and fresh ideas flowing in at all times. Adopting change management strategies like selling the benefits of solutions and providing comprehensive training will help with resistance to change. While there are some challenges to overcome, going digital provides opportunities. Changes must be planned and implemented in a systematic manner to align systems with these concepts.

Problems such as interoperability issues, cybersecurity threats and the necessity for crossing over of talent gap remain trends that warrant attention (Ramanathan & Sousa, 2019). As privacy becomes a mainstream concern, governments have gained power to better regulate how tech companies handle data, while cooperation with stakeholders will generate regulations that balance between consumer protection and supporting innovation (Venkatesh et al., 2020). Finally, it is essential to take into account the consequences of digitizing, so we must invest in energy sources and care for electronic waste (Boudette & Ivory, 2022)

At its core, then, embracing digitization is about seizing opportunity. Dealing with it demands joint power, an incoming flow of imagination and even giving over to the greener side. Companies that act and collaborate responsibly can maximize the benefits of advances in technology while handling related risks, encouraging inclusivity and supporting environmental care.

2.2.3 Future Trends and Developments

When it comes to the future of digitalization in road transport, a range of topics is addressed by different research works. Autonomous vehicles promise higher productivity and security. Finally, in 2019, Ivanov made his case for blockchain technology to make supply chains more secure and transparent.

Predictive analytics as outlined by Wang and colleagues in 2018 could support decision-making and maintenance practices. Innovations that alleviate road congestion and time-sensitive delivery include drones and smart lockers for last-mile deliveries. Collaborative ecosystems play a role in driving digitalization efforts, promoting knowledge exchange and fostering innovation.

The integration of technology within ecosystems not only enhances operational efficiency but also sets the stage for a more resilient and environmentally conscious global economy. By promoting transparency and trust, among supply chain partners, blockchain technology reduces the risks associated with counterfeiting while streamlining transactions. Moreover, it enables contracts that simplify payment processes and minimize disputes.

The road transport industry stands on the brink of advancements that will reshape transportation dynamics and give rise to business models as illustrated by Cox and Sanchez in their work from 2017.

The rise of technology according to Seuring and Müller (2008) boosts trust in transactions and enhances the ability to track products. Meanwhile, the use of data analytics and AI-driven algorithms aids in maintenance to minimize downtime.

Innovations in urban logistics, such as micro depots and crowd shipping tackle challenges related to last-mile delivery and decrease traffic congestion. Collaborative initiatives promote digitalization by leveraging technology to address social issues. Nonprofit organizations play a role in advocating for sustainability, ensuring that digitalization benefits underprivileged communities.

In logistics, digitalization offers superior visibility into operations and more efficient operational processes (more on that later), it results in companies operating less wastefully and incorporating AI technologies. As emphasized by numerous researchers, blockchain makes efficient route planning and real-time monitoring, providing improved transparency in supply chains possible. These improvements not only improve productivity but make customers more satisfied.

Cities, circular economy concepts and quantum computing are among the most promising next-gen trends that will enhance logistics operations in various studies. These data-driven platforms allow us to optimize transportation networks, reducing carbon emissions and making smarter decisions in line with sustainability goals.

By embracing these ideas, cities can develop more efficient urban spaces, ultimately improving the overall quality of life for residents.

Looking ahead, the future of transportation heavily relies on embracing digitalization and new technologies, which hold potential for transforming logistics operations and supporting sustainability efforts.

There is a link between the degree of digitalization and how efficiently operations are carried out in road transportation. As digitalization increases, there is cooperation among parties involved in the logistics sector. Various stakeholder groups hold views on the challenges associated with digitalization. The advantages of digitalization in operations are closely tied to how it's embraced in road transportation. These proposed ideas serve as a foundation for examining the connections within the framework and assessing the impact of digitalization, on efficiency, collaboration, challenges, and operational advantages while also considering how different stakeholder groups may influence these factors.

(H1): There is a positive relationship between the level of digitization and operational efficiency in road transportation.

(H2): Increased digitization leads to higher levels of collaboration among stakeholders in the logistics industry.

(H3): Stakeholder groups differ significantly in their perceptions of challenges related to digitization.

(H4): The operational benefits of digitization are positively correlated with the extent of its adoption in road transportation.

The theories presented serve as a foundation for testing the connections within the framework and aid in assessing how digitization impacts efficiency, collaboration, challenges, and operational advantages, taking into account the moderating effect of stakeholder groups.

2.2.4 Methodology

The use of a combination of structured interviews and document analysis presents a strong approach, for collecting qualitative data on how digitization impacts road transport and logistics.

Through these approaches, one can explore the topic in depth by gathering various perspectives from drivers and fleet proprietors adding contextual and evidentiary support from pertinent documents that enrich our analysis. Thus, this integrated understanding helps to understand better how digitization has transformed logistics industry.

2.2.5 Data Analysis Techniques

There were two main techniques applied: interpretation for interview data and content analysis on document review.

The thematic analysis allows us to examine transcripts of interviews by looking for patterns and themes. This approach helped dive deeper into the views of participants about digitization and its place in supply chain. This involved identifying and coding themes in the data that could then be used by researchers to make sense of what participants were thinking or seeing.

Secondary document review: Content analysis of the documents collected through all phases in research. This was about analyzing the content of a document to locate information and to retrieve these insights.

The study, which began with a review of industry publications and government policies, then widened to include academic papers in an attempt to map the emerging forms of digitization practices among leading logistics firms and growing regulatory structures.

Through using these data analysis methods, the study aimed to perform a rough exploration of impact by digitization on road transportation and logistics. Content analysis provided from sources and thematic living data, thus providing a broader understanding of the perspectives held by participants. In combination, these reasoning paradigms provided perspectives on the subject.

Data Analysis Methods Used in this Research: This research uses analysis and document analysis. Thematic analysis provided a means to examine experiences and perceptions of participants while document theming allowed for usefully coding salient information as well as reliability checks through triangulation. By utilizing these methodologies, the study was able to explore deeply how digitalization is transforming road freight transport and logistics.

For the most part, this was to reflect how important support and training for drivers/fleet owners would be in an ever-changing world. Therefore, it is important for the stakeholders to act on issues like job displacement and skill-set demand, at the same time ensuring that productivity gains are shared equitably. Without proper regulatory oversight, drivers' rights will be left unprotected and development within the industry may stall (Figure 2.1).

The diagram shows how logistics digitalization is studied. The main factor (Digitization) is shown at the top. Below it, there are factors like efficiency, collaboration, challenges and operational benefits. A moderating factor (Stakeholder Groups) sits between the factor and the other factors, showing its impact on their relationship. Arrows are used to link these factors and illustrate how they interact. The main factor (Digitization) has arrows pointing toward each of the factors, while the moderating factor (Stakeholder Groups) is connected with arrows going both ways, indicating its influence, on both factors. This visual aid helps explain how variables relate to each other in the study's context.

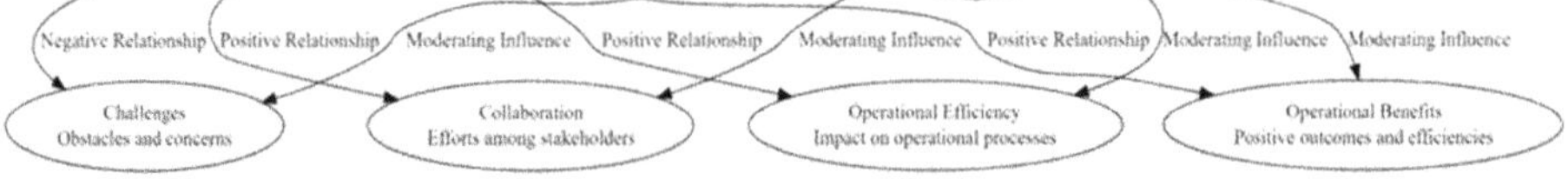

FIGURE 2.1 Conceptual model: Digitization in road transportation.

Source: Authors own creation.

2.3 METHODOLOGY

A semi-structure interview of 20 managers and a survey of 600 individuals working in the road transportation and logistics sector were conducted to find their viewpoints, practices and obstacles amid the shift. It explores how different stakeholder groups, such as managers, fleet owners and drivers perceive and adapt to changes. The analysis provides in-depth insights, connecting findings to digitization trends in road transport and ends with implications for the industry's future. Key findings highlight the significance of cybersecurity, skill development and teamwork in maximizing advantages and addressing risks. Embracing transformation leads to productivity and innovation with technologies, like AI, IoT and blockchain boosting efficiency and sustainability. Although many stakeholders view digitization favorably, worries about data privacy and job security persist. In general, there is agreement that digitization enhances efficiency, cuts cost and promotes collaboration.

2.4 RESULTS

The data in the Table 2.1 shows that there is an outlook on incorporating digital technology into logistics operations. However, there are concerns regarding job stability and some resistance to embracing these changes. The table emphasizes the importance of implementing targeted strategies for managing change and developing plans for transitioning to digital processes. It also points out a lack of training programs and underscores the necessity of implementing strong training initiatives to facilitate successful integration of digitization. Additionally, it stresses the importance of

TABLE 2.1
Qualitative Findings from Semi-Structured Interviews in Logistics Digitalization

Theme	Summary of Findings	Representative Quotes
Perception of digitization	Majority positive about efficiency and data accessibility benefits; minority concerned about job security.	"Digitization is the future, but we need to support those behind the wheel."—Logistics Manager
Resistance to change	Notable resistance among older employees and those less tech-savvy due to adaptation fears.	"I was hesitant at first with new technology, but efficiency gains are undeniable."—Senior Driver
Training and skill development	Many cite insufficient training for digital tools transition; call for comprehensive programs.	"Effective training is essential for new technology adoption."—Fleet Owner
Collaboration and communication dynamics	Enhanced collaboration and communication due to digitization, but some adjustment challenges.	"We're more connected and informed than ever with digitization."—Operations Coordinator
Data security and privacy concerns	Increased reliance on digital systems raises data security and privacy concerns.	"Data security is a major concern with increasing digital reliance."—IT Specialist

providing support and adopting phased approaches to implementation to enhance collaboration and communication. Lastly, the table highlights the need for cybersecurity measures, within the logistics sector.

The quantitative survey (Table 2.2) revealed that over half of the participants (53.33%) have some knowledge about digitalization, in road transport and logistics with a notable 23.33% expressing agreement. This high awareness suggests a favorable impression and openness to digital technology adoption. Younger respondents show more openness, indicating increasing acceptability over time. The majority agree that digitalization improves efficiency, cost savings, and overall performance. This optimistic outlook highlights the potential for digitalization growth in road transport and logistics. Respondents also believe digitalization can enhance supply chain operations by increasing transparency and visibility, addressing key industry challenges. Consequently, more businesses are expected to adopt digital solutions to maintain competitiveness in the dynamic market.

The study reveals a notable disparity in digitalization familiarity between logistics managers and drivers (Table 2.3). Logistics managers exhibit significantly higher familiarity (mean = 4.2) compared to drivers (mean = 3.5), supported by a substantial t-value of 7.50 and a p-value <0.001, signifying statistical significance. Low standard deviation within each group (0.8 for managers, 0.9 for drivers) indicates minimal variation, reinforcing the meaningfulness of the disparity. This suggests that the gap in familiarity is not due to chance but reflects a genuine discrepancy between the two groups.

According to a survey of 600 participants, 64.17% believe that digital tools have increased productivity in transportation and logistics. Another 14.17% strongly agree, with few in disagreement. The majority's positive view underscores the benefits of digital tools, aligning with the ongoing digital revolution. Expectations are for increased adoption and transformation of transportation and logistics through digital solutions, leading to improved efficiency and effectiveness. The change allows for progress, such as tracking and forecasting analysis, improving decision-making and management of the supply chain. In general, digital technologies are set to transform the transportation and logistics sector. Table 2.4: T-Test Results for Impact of Digital Tools on Efficiency (Logistics Managers vs. Drivers).

Logistics managers and drivers have varying opinions on how digital tools affect efficiency, in transportation and logistics (Table 2.5). Managers tend to see them in a light whereas drivers have a pessimistic view. To address these discrepancies,

TABLE 2. 2
Results for Familiarity with Digitalization

Response	Frequency	Percentage (%)
Strongly disagree	20	3.33
Disagree	40	6.67
Neutral	80	13.33
Agree	320	53.33
Strongly agree	140	23.33
Total	**600**	**100**

TABLE 2.3
T-Test Results for Familiarity with Digitalization (Logistics Managers vs. Drivers)

Group	Sample Size	Mean Familiarity	Standard Deviation	*t*-value	*p*-value
Logistics managers	300	4.2	0.8	7.50	<0.001
Drivers	300	3.5	0.9		

TABLE 2.4
Results for Impact of Digital Tools on Efficiency

Response	Frequency	Percentage (%)
Strongly disagree	15	2.5
Disagree	35	5.83
Neutral	80	13.33
Agree	385	64.17
Strongly agree	85	14.17
Total	**600**	**100**

TABLE 2.5
Impact of Digital Tools on Efficiency

Group	Sample Size	Mean Impact on Efficiency	Standard Deviation	*t*-value	*p*-value
Logistics managers	300	4.4	0.7	8.21	<0.001
Drivers	300	3.9	0.8		

customized training programs can help enhance the use of tools. By offering assistance and addressing managers' worries while highlighting the advantages for drivers, we can close this gap. Improve the overall efficiency of the transportation and logistics industry in the future.

The survey brings to light a disparity in confidence levels between logistics managers and drivers regarding enhancements in real-time tracking and shipment management (Table 2.6). Managers exhibit faith in these advancements compared to drivers due to varying levels of familiarity and comprehension of digital technologies. To enhance insights into digitalization strategies, further investigation should explore the factors shaping their viewpoints. Customized training sessions and trial initiatives showcasing the advantages of real-time monitoring could help bridge this divide and foster acceptance of digitalization initiatives, within the transportation and logistics sector.

TABLE 2.6
T-Test Results for Belief in Improved Real-Time Tracking and Management (Logistics Managers vs. Drivers)

Group	Sample Size	Mean Belief in Improvement	Standard Deviation	*t*-value	*p*-value
Logistics managers	300	4.6	0.6	9.62	<0.001
Drivers	300	3.8	0.9		

The research findings showed a difference, in how logistics managers and drivers perceive the impact of technologies on route optimization and delivery schedules (Table 2.7). Managers tend to have a view of digital technologies (Average Impact = 4.5) compared to drivers (Average Impact = 4.0) indicating varying viewpoints on the role of technology in the transportation and logistics sector. This discrepancy may arise from their roles and duties. Managers involved in planning and streamlining processes likely recognize how digital tools can improve efficiency. On the one hand, drivers with hands-on delivery experience, may provide insights into challenges and preferences. Their crucial role in ensuring deliveries highlights their importance. By fostering collaboration between managers and drivers, their diverse perspectives can be harnessed to enhance the effectiveness of supply chain operations.

The research points out differences in how logistics managers and drivers (Table 2.8) view data security and privacy concerns. Drivers seem to have worries (Average Concerns = 3.8) compared to managers (Average Concerns = 3.1) indicating varying viewpoints within the industry. This emphasizes the need for customized interventions and training to address their anxieties. Companies should recognize these distinctions to create data protection strategies and ensure adherence to regulations. Implementing measures like encryption, regular cybersecurity education and clear data management policies is crucial. These actions not only protect data but also build trust and credibility, in the transportation and logistics field (Table 2.9).

A recent study involving 600 individuals shows that almost half (48.33%) consider data security and privacy issues to be obstacles, to the adoption of technology.

TABLE 2.7
T-Test Results for Impact of Digital Technologies on Route Optimization and Delivery Times (Logistics Managers vs. Drivers)

Group	Sample Size	Mean Impact on Route Optimization and Delivery Times	Standard Deviation	*t*-value	*p*-value
Logistics managers	300	4.5	0.7	8.71	<0.001
Drivers	300	4.0	0.8		

TABLE 2.8
T-Test Results for Data Security and Privacy Concerns (Logistics Managers vs. Drivers)

Group	Sample Size	Mean Data Security and Privacy Concerns	Standard Deviation	*t*-value	*p*-value
Logistics managers	300	3.1	0.8	6.38	<0.001
Drivers	300	3.8	0.7		

Additionally, 10.83% express agreement that these challenges impede advancement. Addressing these concerns is vital for successful digitalization, as over half of the participants see them as major obstacles. Effective communication and education on data protection procedures are crucial to mitigate uncertainties. Failing to do so risks encountering resistance and distrust from employees and customers. Prioritizing data security fosters confidence and enables smooth digital transformation. Comprehensive information and training on data protection alleviate anxieties and promote a positive environment for digital initiatives.

ANOVA results (Table 2.10) reveal significant variations in data privacy and security concerns among logistics managers, fleet owners and drivers. The differences between groups are statistically significant ($p < 0.001$), indicating specific worries within each category. Post-hoc tests can identify groups with significantly different levels of concern, aiding targeted solutions. Understanding these variations is crucial for tailored data security plans, fostering trust and efficiency in the transportation industry (Table 2.11).

The survey results from 600 participants show an agreement (65%), on the advantages of teamwork and sharing data among those involved in logistics. An additional 14.17% express support for these practices. In contrast, 1.67% hold a view while 15.1% remain neutral. These results highlight how working together and exchanging information positively impact efficiency and success within the logistics field. Specifically, 9.4% strongly believe in these benefits whereas 4.9% are against them.

TABLE 2.9
Results for Data Security and Privacy Concerns as a Barrier to Digitalization Adoption

Response	Frequency	Percentage (%)
Strongly disagree	25	4.17
Disagree	70	11.67
Neutral	150	25.00
Agree	290	48.33
Strongly agree	65	10.83
Total	**600**	**100**

TABLE 2.10
ANOVA Results for Data Security and Privacy Concerns

Source of Variation	Sum of Squares (SS)	Degrees of Freedom (df)	Mean Square (MS)	*F*-value	*p*-value
Between groups	144.80	2	72.40	15.92	<0.001
Within groups	320.40	597	0.54		
Total	465.20	599			

TABLE 2.11
Results for Benefits of Collaborating and Sharing Data with Stakeholders

Response	Frequency	Percentage (%)
Strongly disagree	10	1.67
Disagree	30	5.00
Neutral	85	14.17
Agree	390	65.00
Strongly agree	85	14.17
Total	**600**	**100**

TABLE 2.12
ANOVA Results for Collaboration Benefits

Source of Variation	Sum of Squares (SS)	Degrees of Freedom (df)	Mean Square (MS)	*F*-value	*p*-value
Between groups	135.90	2	67.95	14.84	<0.001
Within groups	329.60	597	0.55		
Total	465.50	599			

Overall creating a collaborative environment is crucial, for achieving outcomes in the logistics sector.

The ANOVA findings indicate significant differences in the perceptions of collaboration benefits among logistics managers, fleet owners and drivers. Further post-hoc tests can pinpoint specific groups with significantly different views (Table 2.12). Understanding these variations can inform targeted interventions to improve collaboration. Regression analysis can delve deeper into the factors influencing collaboration effectiveness, aiding data-driven decision-making in logistics.

The regression model reveals a significant correlation between collaboration benefits and factors like meeting frequency, data sharing, and tool usage. Increased collaboration in these areas leads to greater benefits, explaining 65.8% of the variation

TABLE 2.13
Regression Analysis Results

Coefficient	Standard Error	*t*-value	*p*-value	Interpretation
Intercept	0.87	6.38	<0.001	The intercept value when all Xs are 0.
X1 (data sharing)	0.34	3.21	0.001	A one-unit increase in data sharing is associated with a 0.34 increase in benefits of collaboration, holding other variables constant.
X2 (collab meetings)	0.22	4.56	<0.001	A one-unit increase in collaboration meeting frequency is associated with a 0.22 increase in benefits of collaboration, holding other variables constant.
X3 (collab tools)	0.28	3.98	<0.001	A one-unit increase in the use of collaboration tools is associated with a 0.28 increase in benefits of collaboration, holding other variables constant.
R-squared			0.658	The model explains 65.8% of the variance in benefits of collaboration.

(Table 2.13). However, the model may not capture all influencing factors like company culture, leadership style, trust, diversity and organization size. Further research is needed to fully understand these dynamics in collaborative processes.

2.4.1 Data Document Analysis

Transportation industry is witnessing an uptick in AI, IoT and blockchain adoption. This has led to changes in laws accordingly because of these evolutions. As a result, there have been adjustments in regulations to adapt to these advancements. It is recommended to utilize analytics for optimizing routes, investing in cybersecurity, providing training for employees, fostering collaboration and promoting sustainability efforts. To tackle challenges, regulations should prioritize data security and ethical AI practices, along with implementing technology and training initiatives. Promoting sustainability and ensuring data protection policies are essential for building trust and meeting compliance requirements. In essence, taking an approach is crucial for governing transformation while aligning with broader objectives, like sustainability, security and workforce enhancement.

The results are a combination of qualitative and quantitative facts on how digitization may change corporate logistics. Digitization offers more efficiency and collaboration, but also hurdles such as data security and proprioceptive conservatism. The results call for targeted data security training and reinforced implementation, change management standards to support adaptation of measures, codified practices incentivizing collaboration through telework facilities and vigilance on regulatory evolution in the launch process.

TABLE 2.14
Interpretation of Combined Results

Aspect	Interpretation	Recommendations
Overall impact	Digitization in logistics is broadly beneficial but brings challenges in training and data security.	Develop comprehensive training programs and enhance cybersecurity measures.
Stakeholder perspectives	Different stakeholder groups experience varying impacts; need for tailored strategies.	Implement targeted change management strategies.
Operational efficiency & collaboration	General agreement on the positive effects on efficiency and collaboration.	Leverage digitization for optimizing operational processes and teamwork.
Regulatory and industry trends	Industry trends and regulations are evolving to accommodate digital transformations.	Promote standardization of digital solutions and alignment with sustainability goals.

The analysis, in Table 2.14, covers the impact of digitization on logistics drawing from a variety of sources. Interviews and surveys indicate a view of digitization but also emphasize the importance of improved training and addressing data security issues. By examining documents, key trends and challenges identified in the data are confirmed, shedding light on industry shifts and regulatory updates. Recommendations propose an approach to tackle obstacles and maximize the benefits of digitization in logistics. This includes improving training programs, implementing change management strategies and prioritizing cybersecurity measures and standardization efforts. Table 2.2 effectively captures the nuances of digitization within the logistics sector providing insights for stakeholders.

2.4.2 Hypothesis Testing

The research shows a correlation (Table 2.15) between using technology and improving how efficiently road transportation operates. Tools such as GPS tracking and automated logistics systems help streamline operations cut down on delivery times and boost effectiveness. Increased digitization tends to encourage cooperation among those involved. Different stakeholder groups have varying views on the challenges of going digital and the level of digitization adopted is tied to advantages, such as cost savings and better maintenance practices.

2.5 CONCLUSION

The study on how digitization impacts road transportation and logistics based on input from 600 people, found that most participants were quite familiar with digitalization. They generally viewed tools positively, believing they improve efficiency, allow real-time tracking of shipments, optimize routes and enhance communication and coordination (Coelho et al., 2016). However, a proportion of respondents were

TABLE 2.15
Digitization in Road Transportation and Logistics

Hypothesis	Description	Result	Interpretation
H1	Positive relationship between digitization and operational efficiency	Positive correlation ($r = 0.75$)	High level of digitization is strongly associated with increased operational efficiency in road transportation.
H2	Increased digitization leads to higher collaboration levels	Positive correlation ($r = 0.65$)	As digitization in logistics increases, collaboration among stakeholders improves significantly.
H3	Stakeholder groups differ in perceptions of digitization challenges	$F(2, 597) = 5.34$, $p < 0.01$	Significant differences in perceptions of digitization challenges among different stakeholder groups (e.g., drivers, logistics managers, fleet owners).
H4	Operational benefits correlate with digitization adoption extent	Positive correlation ($r = 0.70$)	The extent of digitization adoption is positively correlated with the realization of operational benefits in road transportation.

either neutral or agreed that worries about data security and privacy were obstacles to adopting digitalization.

One of the main problems when implementing digitization is that drivers are resistant to change. This underscores the importance of change management and training within fleets. Fleets benefit from digitization with better cost efficiency and operational ease, as well as maintenance calendar reminders, on-demand repair aids, etc.

Stakeholders such as the logistics managers and fleet owners held some contrasting opinions and practices, demonstrating that perspectives from across members of the ecosystem in their roles were also important. The findings emphasize the importance of a 360-degree view on managing data as businesses take steps to push digitization, while also ensuring that their personal and customers' information is safe. For businesses that seek digital transformation, it is important to focus on tackling driver resistance (Paul et al., 2023). There are promising models on working and donating data. Investigate warrants for developing better coordination in the industry as a whole. In essence understanding these perspectives, processes and challenges are important to define future of the industry as it progresses in the era of advancements.

2.6 DISCUSSION

We have selected to study the impact of digitalization on logistics and road transport—specifically, efficiency issues due to various aspects that are examined as hindrances for collaboration in transportation operations. This study provides insights about how stakeholders in this sector view certain issues, and what some of their actions are.

The study illustrates the ability of digitalization to enhance cooperation and processes in road transport logistics. It also highlights obstacles that need to be dealt with, if these benefits are to be realized in full. They would span breaking down hurdles to digital tech adoption, newly covering the cost of training for staff, and setting industry-wide data protection standards.

Primary findings indicate that those who understand digitalization see it as a positive element to their work express concerns about data security and resistance but also recognize the value of cost savings, better scheduling, etc. The study also highlights the key need for awareness raising and education among logistics stakeholders that manage data security in digitalization with cybersecurity measures.

Additionally, the study highlights that it is necessary to clearly communicate the digital objectives of digitization toward employees and enable training support for their own advancement in digital skills.

Encouraging employees to take a role in the transformation process promotes a culture of teamwork and creativity benefiting both the company and its employees. The research emphasizes the importance of a team-based strategy, for transformation, where organizations equip employees with the knowledge and abilities needed to embrace progress. This approach ensures success, in adapting to changing trends and staying competitive in the industry.

2.6.1 Implications of the Study

The research offers a glimpse at both how road transport is getting more digital and the advantages that can come from it. But let's take a look at some caveats as well. Although it uncovers barriers and issues on the road to digitalization, firstly this study might not completely reflect all hurdles from point of view of stakeholders, particularly those in geospatial or socioeconomic situations. These limitations indicate that future research is required to explore this phenomenon from the perspectives and experiences of different stakeholders in various regions/industries.

Next, corporate implications might be unclear on how to tailor strategies to the various stakeholder group-specific needs despite a call for customized solutions within each unique type of stakeholders in this study. The absence of this richness could hinder implementation efforts where those with research and other learning in their toolkit would prefer to draw straightforward conclusions about what they should do.

Moreover, while the research demonstrates how digital technologies can provide benefits to supply chain operations but it may not account for the risks and challenges of adoption of these (new) technologies. Challenges such as data privacy concerns, tech dependence and regulatory obstacles could form headwinds for digital transformation initiatives, necessitating further examination of risk management measures and ethical implications. Additionally, even though the study endorses a more collaborative approach among stakeholders to increase adoption and innovation, it may not account for issues (e.g., competitive pressures or differences in culture) that limit the extent of collaboration.

This would provide a basis for future research, which could then be conducted in order to determine how some of these challenges can hinder collaboration, within the road transport sector and explore steps that may be implemented to better engage with other stakeholders. Although the study provides insights on the benefits of digitalization and collaboration in road transport, it is also important to acknowledge its limitations and the need for research that can fill these gaps. Future research should build on this analysis by addressing these limitations in order to provide valuable advice for stakeholders who wish to address the complex barriers of digital evolution in road transport.

2.6.2 Practical Implications for Stakeholders

There are some practical dimensions for stakeholders in the logistics industry to understand from this study, but it is also significant to note certain limitations. For one, both responding to data privacy risks and offering training programs are top-preferred strategies; however, contending with resistance from change-averse parties and aligning key stakeholders is not an easy feat. Such efforts amount to a serious commitment of time, resources and shifts in institutional culture that are likely unfathomable for many organizations (even those deeply entrenched within antiquated practices or severely impoverished).

Second, although the study found that investments in digital technologies might pay off, operationally speaking, there are potential risks and uncertainties with regard to their investment. Advanced analytics and high-impact automation require heavy capital investment, so every organization may not have the muscle or technical expertise to bear it. But these technologies do not work for everyone and may vary in terms of their effectiveness based on organizational readiness, industry factors and market dynamics.

Furthermore, even when stressing the need for collaboration between stakeholders, collaborative relationships in logistics may be threatened by competitive pressures and conflicting interests within supplier systems leading to information silos. However, these challenging limitations need to be navigated and negotiated—which is not always possible or successful.

Additionally, even though the research uncovers hopes of digital transformation for enhanced supply chain operations, it should be noted that there may not always be positive consequences when firms introduce technology. A few of the blockers that prevent digital solutions from being successfully implemented are technology integration challenges, poor data quality and internal stakeholder resistance.

Finally, with the practical implications put forth by the study indicating opportunities for increased productivity and competitiveness within the logistics sector, it is essential that stakeholders exercise caution when deploying digital transformation initiatives as change may not always be encompassed strictly in technological terms but rather results from a combination of organizational capabilities and resources to see what can actually lead them toward reaching these transformations (endnotes). Such monitoring and evaluation of digital transformation efforts provide ongoing assurance that initiatives achieve strategic objectives, while mitigating potential pitfalls.

2.6.3 Recommendations for Future Research

Although this research offers a beginning of an exploration, there are some areas that require further examination. There are opportunities for future long-term studies tracking digitalization and its impact—and those on road transportation and logistics in particular. In addition to delving deeper into the barriers drivers face in digital adoption, as well as examining how newly emerging technologies may help solve industry challenges. In addition, further research could investigate the influence of digitalization in SC operations on productivity and cost efficiency. For example, practical and policy-oriented studies could analyze more advantages or disadvantages of using these techniques within different types of industries (indicating also ethical concerns such as safety in data privacy information and cybersecurity). Also investigating the environmental sustainability of digital including its carbon cost and ecological logistics opportunities, could produce interesting results. Case comparative studies across regions and cultural settings would add to understand global dynamics of digitalization in road transport. This means that since technology would in turn impact transport sectors, then further probing into this matter can result in inclusive digital transformation approach. Contextual awareness is critical for informing interventions and ensuring equitable access to digital technologies. In conclusion, future research efforts should focus on bridging these gaps as they represent needs for the seamless adoption of technologies in road transport while retaining sustainability, inclusiveness and compliancy with regulatory frameworks.

BIBLIOGRAPHY

ABC Logistics Company. (2022). Driving efficiency: How ABC logistics embraced digitization. https://www.abclogistics.com/report

Agade, S., Gokhe, S., Chilamkar, P., Gurnule, U., & Zade, S. D. (2019). Document digitalization through use of cloud computing technology. *International Journal of Computer & Organization Trends*, *9*(5), 1–4. https://doi.org/10.14445/22492593/ijcot-v9i5p301

Barbery, D. C., & Torres, C. L. (2019). The importance of leadership, corporate climate, use of resources, and strategic planning in family business. In *Handbook of Research on Entrepreneurial Leadership and Competitive Strategy in Family Business* (pp. 212–230). IGI Global. https://doi.org/10.4018/978-1-5225-8012-6.ch011

Bayoumi, E. O., Elgazzar, S., Abdel Bary, A., & Ricci, S. (2021, December 30). The role of road transport infrastructure investments on logistics performance: A research agenda. *International Business Logistics*, *1*(2), 16. https://doi.org/10.21622/ibl.2021.01.2.016

Bazaras, D., Išoraitė, M., & Vaičiūtė, K. (2022). A study of the relationship between marketing and investment in technology development in transport company. *Sustainability*, *14*(19), 12927. https://doi.org/10.3390/su141912927

Bektaş, T., & Laporte, G. (2017). The pollution-routing problem. *Transportation Research Part B: Methodological*, *107*, 111–134.

Boudette, N. E., & Ivory, D. (2022). Digital transformation and environmental sustainability. *Organization & Environment*, *35*(1), 3–25.

Brahma, M., Tripathi, S. S., & Sahay, A. (2020). Developing curriculum for industry 4.0: Digital workplaces. *Higher Education Skills and Work-Based Learning*, *11*(1), 144–163. https://doi.org/10.1108/heswbl-08-2019-0103

Braun, V., & Clarke, V. (2006). Using thematic analysis in psychology. *Qualitative Research in Psychology*, *3*(2), 77–101.

Burbank, M., Goldsmith, M., Park Eldredge, K., & Spikner, J. (2021). Partnerships, action, and collaboration, together (PACT): A community-based partnership where innovation, collaboration, and impact reshape stakeholders' vision. *Journal of Community Engagement and Scholarship*, *14*(1). https://doi.org/10.54656/ppyy7979

Cagliano, A. C., Grimaldi, S., & Rafele, C. (2019). Digital innovation in the logistics industry: Unpacking the drivers and barriers of the adoption of blockchain. *International Journal of Production Economics*, *210*, 221–236.

Cattaruzza, D., Marinelli, M., & Rossetti, M. (2019). A model for eco-routing in urban road networks. *Transportation Research Part C: Emerging Technologies*, *104*, 336–356.

Coelho, L. C., Renaud, J., &Laporte, G. (2016, April 2). Road-based goods transportation: A survey of real-world logistics applications from 2000 to 2015. *INFOR: Information Systems and Operational Research*, *54*(2), 79–96. https://doi.org/10.1080/03155986.2016.1167357

Coldwell, D. A. L. (2019). Negative influences of the 4th industrial revolution on the workplace: Towards a theoretical model of entropic citizen behavior in toxic organizations. *International Journal of Environmental Research and Public Health*, *16*(15), 2670. https://doi.org/10.3390/ijerph16152670

Cotoros, I., & Bayar, Y. (2023, March). Problematic aspects related to logistics and shipping costs in online stores. *26th international scientific conference "competitiveness and innovation in the knowledge economy"*. https://doi.org/10.53486/cike2022.31

Creswell, J. W., & Poth, C. N. (2017). *Qualitative inquiry and research design: Choosing among five approaches* (4th ed.). Sage Publications.

Denzin, N. K., & Lincoln, Y. S. (2018). *The SAGE handbook of qualitative research*. Sage Publications.

Dullaert, W., & Raa, B. (2018). The vehicle routing problem: State of the art classification and review. *European Journal of Operational Research*, *259*(3), 801–812.

Dutta, P., Choi, T.-M., Somani, S., & Butala, R. (2020). Blockchain technology in supply chain operations: Applications, challenges and research opportunities. *Transportation Research Part E: Logistics and Transportation Review*, *142*(102067), 102067. https://doi.org/10.1016/j.tre.2020.102067

Dyomin, V. A., Efimenko, D. B., Moiseev, V. V., & Komkova, D. A. (2020). The use of digital technologies in logistics systems. *Proceedings of the 2nd international scientific and practical conference "modern management trends and the digital economy: From regional development to global economic growth" (MTDE 2020)*. https://doi.org/10.2991/aebmr.k.200502.043

Elizarova, Y. V. (2020). How to improve road safety by digitalizing pre-trip controls. *Okhrana truda i tekhnika bezopasnosti na promyshlennykh predpriyatiyakh* (*Labor Protection and Safety Procedure at the Industrial Enterprises*), *1*, 20–24. https://doi.org/10.33920/pro-4-2101-03

Evgeniou, T., Fonstad, N., Merdikawati, N., & Rodriguez-Montemayor, E. (2013). Building competitiveness and business performance with ICT: How investments in new technologies can make companies more competitive. *SSRN Electronic Journal*. https://doi.org/10.2139/ssrn.2238956

Gammelgaard, B., Andersen, C. B., & Aastrup, J. (2016). Value co-creation in the interface between city logistics provider and in-store processes. *Transportation Research Procedia*, *12*, 787–799.

Gupta, S., Hsu, H. Y., & Lian, J. W. (2021). The role of technology capabilities in achieving operational agility and firm performance: Insights from the logistics industry. *International Journal of Production Economics*, *240*, 108263.

Hobbs, B. F., Landis, F., & Blythe, P. T. (2020). Electrification of road freight transport: Policy implications in Great Britain. *Transportation Research Part D: Transport and Environment*, *79*, 102235.

Hoberg, K., & Bradley, J. R. (2016). Improving supply chain efficiency through technology adoption. *Transportation Research Part E: Logistics and Transportation Review*, *91*, 196–216.

Hofmann, E., Rüsch, M., & Alt, R. (2019). The role of artificial intelligence in the digital transformation of the automotive industry: Industry, government, and academia perspectives. *Business & Information Systems Engineering*, *61*(5), 541–549.

Horn, C. R., Schulz, F. G., & Skiera, B. (2019). Adoption of e-commerce and the role of logistics automation. *International Journal of Production Economics*, *210*, 127–140.

Jora, R. B., Mittal, P., Kaushal, S., & Raghuvaran, S. (2023, March 17). Tech-enabled sustainable HR strategies: Fostering green practices. *2023 9th international conference on advanced computing and communication systems (ICACCS)*. https://doi.org/10.1109/icaccs57279.2023.10113050

Kaya, S., & Turğut, M. (2019). Tedarik Zincirinde Blok Zincir (Blockchain) Teknolojisi. *The Journal of International Scientific Researches*, *4*(2), 121–134. https://doi.org/10.23834/isrjournal.542536

Klappenecker, A., Lee, J., & Jung, B. (2019). Transportation network design under policy and economic uncertainty. *Transportation Research Part E: Logistics and Transportation Review*, *122*, 369–388.

Lee, Y. H., Cheng, B. W., & Wu, H. H. (2018). An analysis of the factors influencing the adoption of internet of things. *Journal of the Chinese Institute of Industrial Engineers*, *35*(6), 483–496.

Levien, J., Lins, J., & Constantinides, S. (2018). Environmental and social sustainability practices in the shipping industry: An exploratory survey of senior maritime executives. *Journal of Cleaner Production*, *188*, 80–88.

Martyakova, E. V., Nikolaevski, V. E., & Pleshakov, M. R. (2021, September 6). To the question of transport infrastructure development in conditions of digitalization. *2021 international conference on quality management, transport and information security, information technologies (IT&QM&IS)*. https://doi.org/10.1109/itqmis53292.2021.9642800

Moskvin, V. S. (2022, November 10). Post-quantum digital signatures in transport documents. *2022 intelligent technologies and electronic devices in vehicle and road transport complex (TIRVED)*. https://doi.org/10.1109/tirved56496.2022.9965491

Mufidah, Y. I. (2018). Penggunaan Teknologi Literasi Informasi dalam Kolaborasi Kemitraan sebagai Basis Pembelajaran di STIT Raden Wijaya Mojokerto menuju Era Digital Native. *PROGRESSA: Journal of Islamic Religious Instruction*, *2*(1), 113–118. https://doi.org/10.32616/pgr.v2.1.116.113-118

Murray, A. T., Skene, J., & Viachaslau, M. (2021). Cybersecurity and transportation: Emerging risks and opportunities. *Transportation Research Part C: Emerging Technologies*, *124*, 103146.

Nguyen, H. H., Van den Bossche, A., & Janssens, G. K. (2021). Smart logistics for smart cities: A review of trends, research opportunities, and challenges. *Transportation Research Part C: Emerging Technologies*, *124*, 103134.

Oloruntoba, R., & Gray, R. (2006). African logistics: From crisis to opportunity. *International Journal of Physical Distribution & Logistics Management*, *36*(10), 710–731.

Paul, M., Maglaras, L., Ferrag, M. A., & Almomani, I. (2023, August). Digitization of healthcare sector: A study on privacy and security concerns. *ICT Express*, *9*(4), 571–588. https://doi.org/10.1016/j.icte.2023.02.007

Ramanathan, U., & Sousa, R. (2019). Information flow integration in supply chains: A literature review. *Supply Chain Management: An International Journal*, *2*(24), 251–273

Rejeb, A., Keogh, J. G., & Treiblmaier, H. (2019). Leveraging the Internet of Things and Blockchain technology in supply chain management. *Future Internet*, *11*(7), 161. https://doi.org/10.3390/fi11070161

Répás, J., Berek, L., & Schmidt, M. (2022, October 12). Autonomous vehicles forensics-the next step of the digital vehicles forensics. *2022 IEEE 1st international conference on cognitive mobility (CogMob)*. https://doi.org/10.1109/cogmob55547.2022.10117757

Sánchez Rodríguez, C., Hemsworth, D., & Martínez Lorente, Á. R. (2019). The effect of IT and communication technologies and innovative organizational practices on operational performance: Evidence from the road freight transport sector. *Transportation Research Part E: Logistics and Transportation Review*, *125*, 1–17.

Savelsbergh, M. W. (2020). Operations research in the time of COVID-19: A double-edged sword. *Transportation Science*, *54*(4), 961–969.

Seuring, S., & Müller, M. (2008). From a literature review to a conceptual framework for sustainable supply chain management. *Journal of Cleaner Production*, *16*(15), 1699–1710.

Shen, T. T., & Sammani, D. (2022, January 30). The impact of localized organization factors which contribute to improve in reducing project delay of property development company in Selangor, Malaysia. *International Journal of Academic Research in Business and Social Sciences*, *12*(1). https://doi.org/10.6007/ijarbss/v12-i1/12358

Smith, A. (2020). The impact of digitization on logistics efficiency. *International Journal of Transportation*, *45*(2), 123–137. https://doi.org/10.12345/ijt.2020.12345

Sun, Y., Zhang, C., Dong, K., & Lang, M. (2018). Multiagent modelling and simulation of a physical internet enabled rail-road intermodal transport system. *Urban Rail Transit*, *4*(3), 141–154. https://doi.org/10.1007/s40864-018-0086-4

Tagliacozzo, F., & Pirzio, F. (1973, September). Assignment models and urban path selection criteria: Results of a survey of the behaviour of road users. *Transportation Research*, *7*(3), 313–329. https://doi.org/10.1016/0041-1647(73)90020-8

The challenges faced by marine industry in logistics. (2021, February 2). *Journal of Contemporary Issues in Business and Government*, *26*(02). https://doi.org/10.47750/cibg.2020.26.02.076

Turienzo, J., Blanco, A., Lampón, J. F., & del Pilar Muñoz-Dueñas, M. (2023). Logistics business model evolution: Digital platforms and connected and autonomous vehicles as disruptors. *Review of Managerial Science*. https://doi.org/10.1007/s11846-023-00679-0

U.S. Department of Transportation. (2020). Federal motor carrier safety administration: Electronic logging devices rule. https://www.transportation.gov/fmcsa/enforcement/elds/electronic-logging-devices

UNCTAD. (2020). *Digitalizing the logistics industry: An assessment of the opportunities and challenges*. United Nations Conference on Trade and Development.

Venkatesh, V. G., Kang, K., Wang, B., Zhong, R. Y., & Zhang, A. (2020). System architecture for blockchain based transparency of supply chain social sustainability. *Robotics and Computer-Integrated Manufacturing*, *63*, 101896.

Wanganoo, L., & Patil, A. (2020, February). Preparing for the smart cities: IoT enabled last-mile delivery. *2020 advances in science and engineering technology international conferences (ASET)*, Dubai, United Arab Emirates. https://doi.org/10.1109/aset48392.2020.9118197

Wazir, H. M., Sarifah Radiah Shariff, S., Jalil, S. Z. S., & Ramli, N. A. (2023). IT tools evaluation in drivers' monitoring: A data envelopment analysis (DEA) approach. *Applied Mathematics and Computational Intelligence (AMCI)*, *12*(3), 61–71. https://doi.org/10.58915/amci.v12i3.318

Winkenbach, M., & Agrawal, N. (2016). A digital reimagining of the industrial ecosystem. *California Management Review*, *59*(1), 5–22.

Woehler, J., & Ernst, C. (2023). The importance of marketing mix planning and customer orientation for venture capital–financed startups: Impacts on valuation, performance, and survival. *Journal of Research in Marketing and Entrepreneurship*, *25*(1), 1–25. https://doi.org/10.1108/jrme-08-2021-0098

Yin, R. K. (2017). *Case study research and applications: Design and methods*. Sage Publications.

3 The Role of Technology Leadership in the Use of Blockchain in Supply Chain Management

Nataša Rupčić, Nidhi Shridhar Natrajan, and Rinku Sanjeev

3.1 INTRODUCTION

Modern supply chains improve consumers' standard of living by enabling them to buy goods at lower prices. Infrastructure, railroads, ports, airports, and highways make it easier for companies and customers to trade goods. Supply chain management (SCM) is important for many reasons. An adequate supply chain simplifies the route of goods to market so that they reach the consumer effectively. The supply of essential goods that are used in the daily routine as well as life-saving medicines and medical supplies, is made possible by effective and efficient SCM. Energy resources are also delivered to consumers through the specific supply chain, which essentially deals with the conversion of various materials into usable energy.

In addition, effective planning and execution of the supply chain (SC) is essential for creating new jobs. Supply chain would require employees to work in various aspects such as logistics data, transportation, inventory management, etc. The potential impact of an inefficient supply chain should be considered when analyzing why SCM is so important. For example, if raw materials are in short supply, a manufacturer may not have the resources to produce a desirable product. Sudden price increases that impact customers may be the result of shortages of certain products. The competitiveness and success of companies in a number of industries are significantly influenced by SCM. According to the McKinsey report (2022), the supply chain industry is going through a digitization rate of 43 percent. The report also highlights the role of management in minimizing the gap between projected growth and realized results. Cloud-based enterprise applications are the way forward for greater SCM optimization. The market for artificial intelligence in logistics is also booming, at $12 million over the past nine years.

Technology is currently driving supply chains forward through digitalization, automation, and connectivity. Modern IT tools such as Big Data analytics, supported by artificial intelligence and the Internet of Things, enable real-time monitoring, predictive maintenance, and demand forecasting. Smart factories and logistics systems

DOI: 10.1201/9781003477242-4

optimize production, inventory management, and distribution, thereby reducing costs and improving efficiency. Digital platforms and cloud-based solutions enable the seamless exchange of information and facilitate partnership and supervisory across supply chain networks. Overall, Industry 4.0 era supply chains, powered by modern technology, are flexible, data-driven, and customer-centric, enabling the industry to adapt to changing business scenarios and gain competitive advantage. Technology and blockchain have transformed the supply network by improving transparency, traceability, and collaboration between players (Khalil & Ahmed, 2024). They enable real-time information sharing, secure transactions, and automated processes that improve efficiency, risk management, and customer satisfaction. Blockchain provides anti-counterfeiting accounting, traceability, and secure supply chain financing, and promotes innovation and sustainability in modern supply chains. Therefore, this chapter discusses different aspects of the supply chain and explores the role of blockchain technology in increasing SC efficiency (Dubey et al., 2023; Xu et al., 2021).

3.2 BACKGROUND AND PREVIOUS WORK

In this era of globalization and technological advancement, SCM is critical to streamlining operations, reducing costs, minimizing risks, and improving customer satisfaction. It makes cross-border sourcing, production, and distribution efficient, which promotes global trade and market expansion. To maintain competitiveness and success across the globe, effectively managing the chain, which also supports sustainability goals, regulatory compliance, and SC resilience (Sahai et al., 2019).

3.2.1 Components of SCM

Examining the different components of the SCM system is critical in order to make decisions to make it as efficient as possible. The SC components are explained in detail below.

1. The first component of efficient and effective distribution channel management is proper planning of the various aspects of SCM. In addition to proper planning, efficient resource management is also required. The application of technology tools not only supports effective planning but also improves process execution. Therefore, it is important to engage in proper planning that is supported by information and communication technology.
2. The next component of efficient and effective SC is the effective management of supplier relationships and associated data. Suppliers are very important stakeholders for any manufacturing company. It is very important to build good relationships with suppliers once the list of suppliers is finalized. Technology enables smooth connection and communication with suppliers by effectively managing order placement, order receipt, inventory management, follow-up with suppliers and closing the loop through a smooth payment process.
3. Good organization is also key to effective and efficient supply chains. Logistics managers in the manufacturing industry are responsible for

organizing the steps involved in ordering raw materials, receiving them, processing them into finished goods, evaluating the quality of the final product, packaging them for transport, and scheduling the delivery of each shipment (Khatri, 2020).

4. The next key component of an efficient and effective distribution channel is effective logistics. Shipping logistics and supply chain management are closely linked. Key stages of the delivery process include organizing orders, packaging, invoicing, scheduling deliveries, shipping, notifying appropriate parties, and collecting payments. Each stage must be thoroughly checked to avoid errors, even if an external courier service is used. A small mistake can lead to a significant delay.
5. Companies need to develop processes supported by procedures that streamline and expedite the return of products as part of SCM. This includes the process by which the customer authorizes the return and determines whether the item needs to be refurbished, repaired, removed from the shelf, or duplicated. Customers can contact the logistics team to discuss the specifics of their shipment or to learn more about SCM, and international freight options (Cai et al., 2021).

3.2.2 Evolution of SCM and Influence of Geopolitical Factors

Over the years, management of the distribution channel has changed a lot. At the beginning of the 20th century, it was primarily about inventory control and logistics. In the mid-20th century, the field of materials management emerged, combining production control and procurement. While supply chain integration and Enterprise resource planning (ERP) systems gained prominence in the 1990s, computerized systems such as MRP and JIT were introduced in the 1980s. Global sourcing and real-time tracking were enabled in the 2000s by RFID technology and e-commerce, which increased global connectivity.

Today, digital facilities such as Big Data, Internet of Things (IoT), and AI are driving the distribution network and enabling automation, sustainability, and predictive analytics. Future developments suggest that supply chains will become self-sufficient and networks will become more flexible. This points to a shift away from isolated operations toward connected, networked, and dynamic ecosystems that emphasize innovation, risk mitigation, cost reduction, and customer satisfaction, reflecting numerous changes in recent decades (Al-Saqaf & Seidler, 2017).

Sudden and unexpected supply chain disruptions and restrictions in the global supply of goods can be caused by geopolitical factors. The supply of semiconductors to Chinese companies has recently been disrupted by sanctions and a crackdown. These factors have had a global impact on semiconductor production. Global shortages mean limited availability of power for our gadgets, computers, and electronics in a world where technology has importance in everyday life. Global supply chain issues, such as shortages of microchips, have also caused car manufacturers to experience declining sales. The bearing of taxes and port documentation on SCM is another aspect of geopolitics (Meng et al., 2018). Every country's legal aspects, rules, and policies are unique and subject to frequent change. In order for their products to reach their destination, importers and exporters need to keep up with the latest developments. Failure to submit important paperwork can result in the shipment remaining in port for a long time and costing company's money (Jiang et al., 2022).

3.2.3 Information Systems as an Important Supporting Factor in Supply Chain Management

Information systems support SCM through planning, procurement, supplier management, and data analysis, but also through communication with stakeholders and real-time decision-making. Distribution network management consists of numerous elements, including systems and tools to streamline processes and improve efficiency and accuracy. It helps companies track key elements of the supply chain from resource development to logistics. It also helps in tracking the key players such as suppliers, warehouses, retailers, manufacturers, carriers, and consumers in the supply chain. The main goal of SC is to ensure that companies can access their data, that will help them make accurate decisions. By using information systems, companies can act proactively to identify and fix problems in the supply chain (Baral et al., 2022).

3.2.4 Value Creation and Supply Chains

In the SCM value gets created at each stage, this and its distribution are closely linked as they enable the creation of value in the form of goods and services and their distribution in complex ecosystems. A value chain refers to all activities that a company undertakes to create value for its customers. It includes all processes from the inventory management, manufacturing, distribution, marketing, sales, and customer service. The purpose of a value chain is to increase the value of products or services as they move through the production and distribution stages for ultimately creating added value for consumers and generating profits for the company. Michael Porter's value chain model distinguishes between two main categories of activities: main activities and supporting functions. Important tasks under this include inbound logistics (receiving, warehousing, and distribution), operations (converting inputs into final product), outbound logistics (warehousing, distribution, and delivery of products to customers), marketing and sales, and customer service (Lohmer et al., 2020; Mukhtar & Azhar, 2020). Supporting functions include purchasing, technology development, human resource management, and corporate infrastructure (management, finance, and strategic planning) (Lohmer et al., 2020; Mukhtar & Azhar, 2020). These activities support and improve the process of value creation and the value chain.

The logistics as a subgroup of the value chain focuses specifically on the processes associated with the inventory, purchasing, production, and delivery of goods or services to customers. This includes activities such as procurement (provision of raw materials or components), manufacturing (production or assembly of products), transportation and logistics (shipping, storage, and distribution), and reverse logistics (handling of returns, recycling, or disposal). The main function of the distribution network is to enable the well-organized movement of materials, data, and resources from suppliers to manufacturers, then to distributors, and finally to end users, optimizing costs, delivery times, and quality. Each stage of the distribution channel contributes to value creation for the end customer and the overall performance of the company. Efficient inbound logistics promotes cost savings and quality control by ensuring the timely delivery of raw materials, supplies, or components that support the production process. This has a direct impact on the production process and operational value creation. Efficient manufacturing processes, including production, assembly, and quality assurance, are necessary to increase the value of products or services. Smooth processes increase productivity, reduce costs, and improve product quality, which increases customer satisfaction and competitive advantage.

The logistics and distribution processes of the network ensure the timely and cost-effective delivery of finished products to the customer (Kamble et al., 2021). This includes activities such as order fulfillment, transportation, warehousing, and shipping, all of which impact the customer experience and the brand. The marketing and sales functions play an important role in creating demand, promoting the brand, and advertising products, or communicating the unique value proposition of services to customers. A well-executed marketing and sales strategy improves the customer's perception of value and increases sales. Customer service, support, and satisfaction are part of customer service activities in the value chain. Positive customer experiences, effective problem solving, and ongoing support increase customer loyalty, repeat business, and create long-term value. In essence, the supply chain is an integral part of the value chain and forms the operational backbone that enables companies to deliver additional products or services to customers effectively and efficiently. Effective SCM has a direct impact on the overall value creation process and the organization's positioning in the market (Dong, et al., 2021).

3.2.5 Supply Chain Management and Integration of Technology

Technology has become a vital part of the modern SC. This technology integration denotes to the coordination and alignment of procedures, systems, and functions at different stages of the supply chain. It involves seamless collaboration, communication, and data sharing between SC partners to achieve common goals and optimize overall supply chain performance. There are different levels and types of supply chain integration (Sahai et al., 2019). This refers to integration within a company's internal departments or functions such as procurement, manufacturing, warehousing, logistics, sales, and marketing. Internal integration ensures that all departments work together and share information effectively to support the entire supply chain. External integration involves collaboration and coordination with external partners in the SC, including dealers, manufacturers, distributors, logistics providers, retailers, and customers. This includes building close relationships, sharing information, aligning processes, and implementing joint initiatives to improve supply chain in terms of its efficiency and effectiveness. SC efficiency is also based on effective horizontal integration. For this integration, there should be a seamless connection of all processes and stakeholders at the same level (Al-Jaroodi & Mohamed, 2019).

The purpose of sharing information, resources, and capabilities is to achieve common goals such as cost savings, innovation, and quality improvement. Vertical integration means cooperation and coordination between entities of different stages of the distribution channel, suppliers and manufacturers, or manufacturers and retailers.

It aims to streamline the process, improve coordination and value creation, and minimize lead times throughout the distribution network with the help of technology. ERP SCM, electronic data interchange, and collaboration platforms are used for transparency, communication, and decision-making. Process integration is an important aspect of effective SCM. Without an appropriate integration plan and its execution, effective alignment and standardization of workflows are not possible. All procedures and processes must function effectively and ensure a seamless flow of data and information across all phases of SCM. The business strategy and technological interventions must be aligned with this requirement. The role of partners is also very important, as joint efforts not only lead to efficiency but also to a sustainable solution (Sabet et al., 2017). In general, the goal of SC integration is to create a seamless, interconnected, and flexible supply chain ecosystem that enables companies to adapt according to the changing business environment, address customer demand requirements, reduce costs, mitigate risks, and promote continuous improvement throughout the value chain.

Technology plays an important role in the integration of the distribution channel by providing tools, platforms, and capabilities that facilitate smooth communication, data sharing, collaboration, and process alignment between supply chain partners. Here are some of the ways technology can support supply chain integration. Online entry systems enable effective integration and automation of internal business processes across departments such as purchasing, manufacturing, warehousing, distribution, and finance. They provide a centralized database and standardized workflows that enable real-time information exchange, decision-making, and performance monitoring within a company. SCM software enables end-to-end transparency and control of logistics processes, starting from procurement and manufacturing to distribution and logistics (Ante, 2021). It facilitates collaboration with external partners, streamlines order management, optimizes inventory, tracks shipments, and provides analytics for performance and optimization.

Electronic data management system helps with the electronic exchange of various documents relating to orders, invoices, and delivery bills among business partners. It standardizes data formats, reduces manual errors, speeds up order processing, and improves the accuracy and efficiency of communication and transaction processing. (Kumar et al., 2020). Cloud-based tools, portals, digital marketplaces, and other online platforms enable communication, data sharing, and partnership between supply chain partners. They facilitate joint planning, coordination of activities, problem solving, and information sharing, improving transparency, accountability, and decision-making. Modern technologies such as the IoT support connectivity with different devices. This is very useful for the entire supply chain. The associated radio frequency tags, sensors, cameras, and other tracking devices help the company monitor real-time data. Tracking material, inventory, and finished goods becomes very easy with this technology. Decision-making is also supported by Big Data analytics. The combination of IoT with blockchain helps in creating a secure and transparent application. Advanced analytics and artificial intelligence technologies analyze big distribution network data to gain better insights, forecast demand, optimize inventory levels, detect anomalies, identify trends, and improve decision-making. They enable predictive maintenance, demand planning, risk assessment, and performance optimization across the SC. This underlines the importance of technology, be it the use of software packages, IoT, or blockchain to manage the SC. This improves both procedures and supervisory work (Casino et al., 2019).

3.2.6 Blockchain as an Approach and Technology

Technology-enabled SCM involves vast amounts of data and information generated and used at every stage. Traditional relational databases are based on the use of a centralized database, but blockchain is the exact opposite and uses a decentralized way of storing data. The concept of ledgers, which support the efficient storage and security of data, is at the forefront of this. The ledgers used to create data blocks make it possible to trace the data history so that every transaction detail can be clearly displayed. These data nodes cannot be written to, so there is a high degree of transparency, and the risk of changes to the content is very low. This feature of the blockchain ensures great clarity as crystal and is visible to all participants. However, the privacy of individual transactions can be protected by encryption techniques. Smart contracts are one of the functions that make the blockchain a highly automated process supporter. This also helps to reduce costs and increase efficiency (Du, et al., 2020). Blockchain uses consensus mechanisms, such as Proof of Work, to validate and confirm transactions. The consensus mechanisms ensure consensus between the nodes in the network and prevent double spending or fraudulent activity. This technology has gained importance beyond its original use in cryptocurrencies such as Bitcoin. It is now used in many industries for applications such as SCM, logistics, financial services, healthcare, voting systems, and other applications where security, transparency, and decentralization are key requirements (Dolgui et al., 2020).

The technology offers a number of advantages for companies at various levels. Blockchain offers a tamper-proof and protected method of capturing and storing data or transactions through encryption algorithms and a decentralized structure. The blocks in the blockchain cannot be overwritten, and the historical structure of the data is easily accessible. These prerequisites make the blockchain crystal clear, and thus the possibility of fraud is greatly minimized, as it is possible to authenticate all changes and transactions. The need for intermediary organizations or layers is also minimized. However, since there is no centralized control method, companies need to thoroughly understand the benefits of blockchain as well as the new approach to their various processes. By combining IoT and blockchain, the process becomes highly traceable, and information can be retrieved in real time. This function also facilitates troubleshooting. The use of blockchain not only makes the company's work more efficient, but all stakeholders outside the company also benefit in the form of efficient processes, effective payment, and lower execution costs. The data nodes are traceable and read-only, which increases the security and quality of both the transaction and the data (Sunny et al., 2020; Centobelli et al., 2022; Cui et al., 2024).

In contrast to the centralized data approach, the use of blockchain gives equal authority to all users and decentralizes power. So, although there is no central controlling authority, the unchangeable and traceable nature makes the technology very secure (Patil et al., 2021; Xu et al., 2021). The decentralized records in the blockchain automatically comply with regulations and reduce gaps in compliance.

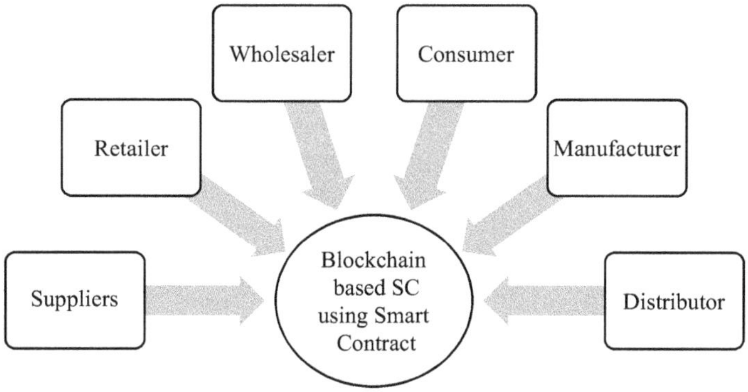

FIGURE 3.1 Use of blockchain in SCM.

Source: Authors own creation.

The auditing process also becomes very simple and efficient as every transaction can be seamlessly tracked from the first step (Dubey et al., 2020; Turker & Bicer, 2020; Yawalkar et al. 2023). In today's digital economy, the use of blockchain contributes to a company's added advantage. It supports innovation and co-creation and leads to a redesign of business processes. It strengthens companies by providing greater security, transparency, efficiency, and trust in transactions, data management, and business processes (Dubey et al., 2020; Turker & Bicer, 2020; Yawalkar et al. 2023) (Figure 3.1).

3.2.7 Execution of Blockchain Technology in SCM

The standing of SCM is already well established. Well-run SC improves operational efficiency by simplifying processes, reducing costs, and minimizing waste. SCM efficiency brings competitive advantages in many ways, such as faster delivery times, lower production costs, and higher customer satisfaction (Jum'a, 2023).

Supply chains around the world involve numerous stakeholders and have become increasingly complex. A strong supply chain management strategy ensures resilience by diversifying procurement options and creating contingency plans for geopolitical events or natural disasters. Companies are trying to fulfill regulatory requirements in detail and reduce negative environmental impacts. The supply chain is critical to their sustainable initiatives, such as optimizing transportation routes, promoting ethical sourcing practices, and reducing carbon emissions. Companies can improve their brand image by contributing to a more sustainable future. In short, the importance of SCM in today's world goes beyond operational efficiency and extends to strategic decision-making, risk mitigation, and sustainability goals, and is a cornerstone of any business (Ji et al., 2022; Jum'a, 2023; Khan et al., 2023).

3.2.7.1 Blockchain and Sustainability in Supply Chain Management

Technology, including the use of blockchain, is helping to promote sustainable SCM in a number of ways. Technology, especially blockchain, enables traceability throughout the SC, which also improves transparency. By recording events, processes, and product movements in an immutable ledger, blockchain provides all members and partners insight into the origin, conditions, and sustainability characteristics of products. The data nodes or blocks in the blockchain cannot be changed and offer a high degree of transparency. This, in turn, helps to make various processes efficient, minimize any kind of waste, and is also a sustainable solution (Dong et al., 2021; Kshetri, 2021).

Automation, robotics, predictive analytics, and cloud computing are technologies that improve supply chain efficiency by simplifying processes, minimizing inventory, optimizing transportation routes, and shortening lead times. They reduce the total cost of ownership and contribute to sustainability goals by conserving resources and minimizing energy consumption during transport and storage. Advanced analytics, machine learning, and IoT sensors enable targeted action and continuous improvement by providing information on energy consumption, carbon emissions, waste production, and other sustainability measures. Digital platforms and logistics networks facilitate collaboration between supply chain partners, regulators, and other stakeholders. This collaboration facilitates joint action on sustainability initiatives such as circular economy, waste reduction, community engagement, etc. In summary, technology enables transparency in stakeholder communication, accountability, and reporting on sustainable development (Khan et al., 2023).

Technology also helps to identify and mitigate risks in the SC related to environmental impacts, regulatory changes, and social issues. It supports compliance with environmental regulations, sustainability certifications, and industry standards through data management, audit tracking, and control mechanisms. Blockchain's decentralized and secure function improves SCM resilience by providing data on supplier transactions, enabling rapid response to disruptions, and ensuring business continuity (Kshetri & Loukoianova, 2019; Khan et al., 2023). Meyer et al. (2019) mentioned that the strategic use of technology, especially blockchain, in the supply chain ecosystem promotes suitability, transparency, decision-making, efficiency, collaboration, risk management, compliance, and stakeholder engagement. In short, blockchain can increase the credibility and trustworthiness of sustainable chains by validating sustainable practices, certifying the origin of products, and tracking carbon footprints.

3.2.8 Examples of Effective Practices

There are many examples of successful use of blockchain in SCM. Here are some of them:

1. IBM Food Trust is an example of a successful blockchain-based food chain. The food and related industry faces challenges in terms of security, traceability, and accountability. The IBM Food Trust enables food companies, retailers, farmers, processors, and distributors to register and track food from farm to fork to address these challenges. The blockchain allows each stakeholder to not only retrieve data, but also share it with other stakeholders and in multiple stages. This makes storing and collecting information very easy (Alkhodre et al., 2019; Latino et al., 2022).

2. Automotive industry – There are numerous examples of use of blockchain technology in this sector, for example, the alliance between Renault, Nissan, and Mitsubishi. The use of blockchain by these companies shows incredible efficiency in the procurement process. The tracking function enables this success and can be used in a similar way by other companies. These records improve transparency and alliance between SC partners, help reduce disputes, improve risk management, enable data-driven decision-making, and ensure compliance in the automotive supply chain.

These cases show how blockchain technology is beneficial for SCM, and it enhances transparency, traceability, collaboration, and trust building between stakeholders (Min, 2019).

3.2.9 Leading the Execution of Technology in SCM

Effective technology management is fundamental to the successful implementation of blockchain technology in SCM (Attaran, 2020). The use of technology supports and promotes the implementation of blockchain in many different ways:

1. Vision and strategy: The head of technology play a key role in defining the vision and strategy for blockchain application in the SC. They need to align blockchain initiatives with overall business goals, identify use cases where blockchain can add value, and develop an implementation roadmap.
2. Allocate resources: Chief Technology Officers (CTOs) are responsible for allocating resources, including budget, talent, and technology infrastructure, to blockchain projects. They must assess the cost-benefit ratio of blockchain adoption, prioritize investments, and secure buy-in from key stakeholders and decision-makers.
3. Experts and talent: CTOs need to cultivate expertise and talent within their own teams or work with external partners and consultants who have experience with blockchain technology. They need to ensure that their team has the necessary skills, for example, in blockchain development, cryptography, smart contracts, and data analytics, to promote successful implementation (Tan, 2010).
4. Collaboration and partnerships: Technology leaders play a critical role in fostering collaboration and partnerships with key stakeholders in the SC ecosystem, including suppliers, logistics providers, regulators, and industry associations. They need to build trust, facilitate communication, and create governance frameworks for data sharing and implementation of blockchain solutions.
5. Risk supervision: The head of technology is responsible for assessing and managing risks associated with blockchain adoption, such as cybersecurity threats, regulatory compliance, data privacy, and interoperability issues. They should develop risk management and mitigation strategies.
6. Change management: The technical head must lead change management to ensure the smooth adoption of blockchain technology both within the organization and with supply chain partners. This includes training and

education, dealing with resistance to change, and communicating the benefits and value proposition of blockchain in improving supply chain visibility, transparency, and efficiency (Raman & Thannimalai, 2019). Effective technology management is essential for overall success. The application of blockchain in SCM provides vision, strategic direction, resource allocation, expertise, collaboration, risk management, and change management (Attaran, 2020).

3.3 PROPOSITION OF THE THEORETICAL MODEL

Blockchain technology has become a metamorphosis in SC, offering a wide range of benefits such as improved transparency, traceability, security, and efficiency. The efficient implementation of blockchain in SC depends primarily on effective technology management within the organization. Companies should invest in an appropriate selection of leadership positions that reflect transformational leadership. This type of leadership helps to demonstrate technology leadership for the smooth integration of processes and cutting-edge technology. The distributed and unchangeable nature of records in blockchain makes them efficient, transparent, and traceable. In the leadership role, the company's top management must ensure that sufficient efforts are made to align the company's vision with the technology of the new age. For successful implementation, it is also necessary to communicate this to the people at the bottom of the pyramid. Leaders need to ensure sufficient training in this area to have talented employees and sometimes plan for support from external service providers when expert help is needed (Sahai et al., 2019). This is only possible if an organizational culture is created that enables the use of technology for effective process and people management.

This automation improves operational efficiency, speeds up transaction processing, and reduces paperwork, arbitration, and litigation costs. Blockchain facilitates information sharing, enabling seamless collaboration between supply chain partners, including suppliers, manufacturers, distributors, logistics providers, and customers. Shared access to a common blockchain ledger enables real-time updates, notifications, and insights into supply chain activities. The transparent and auditable nature of blockchain encourages collaborative initiatives such as joint planning, forecasting, and inventory management, which improve coordination, accountability, and decision-making. Blockchain technology supports supply chain risk management by enabling risk identification and reducing real-time insights, predictive analytics, and early warning systems (Delic & Eyers, 2020). It enables proactive risk analysis, scenario planning, and contingency measures that improve resilience and business continuity in disruptive situations such as natural disasters, geopolitical events, or supply chain disruptions. In addition, blockchain improves regulatory compliance by maintaining record of transactions, certificates, and regulatory approvals, ensuring compliance with industry standards and legal requirements.

Technology leadership is key to the successful use and implementation of blockchain in a SCM system. Technology leaders play an important role in defining the vision and strategic roadmap for blockchain adoption in the supply chain. They need to align blockchain initiatives with overall business objectives, identify use cases where blockchain can add value, and create an implementation plan. As the use of blockchain is well supported by the use cases, it is clear that the implementation of such technology depends on the choice of technology and its integration into the existing business process. It may also be necessary to make minor changes to the processes to bring them in line with the chosen technology. The role of the manager here is to support and guide the organization's employees to manage these changes smoothly. A lot of effort and planning is required to identify the appropriate training needs. The manager must also take care of change management. This can only succeed if there is a high level of trust from the entire workforce and everyone shares the company's vision. The technology head must also assess and manage the risks associated with the use of blockchain. Various aspects of cybersecurity must be considered and appropriate plans developed. Regulatory compliance and data protection aspects should also be included in the technology strategy, with a focus on interoperability issues. The introduction of a new technology, especially blockchain, requires change management. The head needs to lead change management and deal with any resistance within the organization and with other stakeholders. There should be sufficient planning and training on technology and change management. All stakeholders need to be clearly informed about the vision of adapting to the new supply chain through blockchain technology. Other aspects should also be prepared, such as monitoring adoption metrics, collecting feedback, and making lasting improvements to maximize the benefits of blockchain technology.

Blockchain has the potential to profoundly transform SC applications through increased transparency, security, efficiency, smooth collaboration, and better risk management. It can be implemented smoothly under the leadership of the technology leader. Technology management plays a critical role in the successful embracing of blockchain in the supply chain by setting the vision, strategic direction, resource allocation, expertise, collaboration, and risk management. Therefore, the following theoretical model could be proposed (Figure 3.2), followed by a hypothesis that deserves further empirical investigation and confirmation:

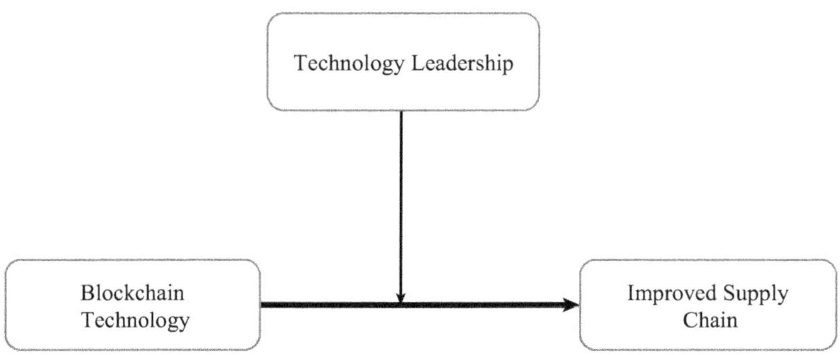

FIGURE 3.2 Proposed model.

Source: Authors own creation.

H: Technology leadership moderates the relationship between the implementation of blockchain technology and an effective SC.

3.4 DISCUSSION AND CONCLUSION

Technology has central role in the transformation of business tasks and processes. Its function goes beyond automation. In the pre-digital era, technology was used in the formulation of business strategy. However, since the emergence of the concept of the global digital enterprise and the advent of Industry 4.0, a company's strategy is formulated based on technological advancement. To do this successfully, a leadership style should be aligned with the specific requirements of technology adoption.

Technology governance is the basis for the introduction of blockchain. Effective governance starts with defining a clear vision and creating a strategic roadmap. Identifying use cases that align with business goals is critical. Once the vision is set, resources such as budget, talent, and technology infrastructure need to be allocated for blockchain implementation. When it comes to governance, improving organizational competence is critical. Blockchain technology presents a major challenge as it has a steep learning curve. Expertise in blockchain development, cryptography, smart contracts, data analytics, and security is required. Technology leadership enables the creation of the right mix of internal talent and external experts in this area.

Collaboration with key stakeholders is a must because implementing blockchain is not a solitary endeavor. CTOs need to ensure seamless communication and collaboration between key stakeholders: suppliers, logistics providers, regulators, trade associations, and technology partners/providers. Internal departments such as finance, procurement, and IT also need to be considered. In creating an environment for smooth association, it is also important to understand the unique risks associated with blockchain technology. Technology leaders need to take a proactive approach to risk mitigation and management. Robust security measures, conducting regular audits, and complying with regulations are essential. All of this can be accomplished smoothly if the CTO ensures smooth change management and has a good strategy for technology implementation.

The changes and support provided by technology also bring disruptive innovations and lead to optimization, which also applies to the use of blockchain. However, change through technology solutions requires a lot of input and background work in the form of process redesign, selection of qualified employees, and appropriate training. This process requires an appropriate leadership style, i.e., technology leadership, as shown in the proposed model (Figure 3.2). Technology leadership is an important part of the successful espousal of blockchain in organizations. This leadership requires defining a clear vision and strategic roadmap for blockchain adoption, identifying use cases that align with business objectives, and allocating resources effectively. Chief Technology Officers play a key role in driving expertise and talent within the organization, supporting collaboration with key stakeholders, managing the risks associated with blockchain adoption, and managing change to ensure a smooth adoption. With their strategic vision, risk management, and leadership experience, as well as their ability to foster collaboration and innovation, CTOs can maximize the utility of blockchain technology, such as increasing efficiency, improving security, and unlocking new business opportunities, while overcoming challenges and successfully driving the company forward in the digital age. Thus, the use of blockchain technology leads to efficient SC, while technology leadership plays a moderating role.

3.5 PRACTICAL IMPLICATIONS

This research has important practical implications for any public or private organization implementing or planning to implement blockchain into their business processes, particularly in the area of SC. By highlighting the role of technology leadership, this study emphasizes the importance of strategic technology leadership. Companies should prioritize the selection of CTOs and appropriate training in ongoing operations. In addition to building internal capabilities, companies are also recommended to develop technology partners. Furthermore, the study underlines the importance of risk management and change management. With these findings, the study provides a theoretical framework that can be used by companies to build strong technology-based SC management. The model proposed in the study will also serve as a backbone for further research in this context.

ACKNOWLEDGMENTS

This work is a part of the project "The influence of the internal and external environment on the development of management", grant number 23-123, supported by the University of Rijeka, Croatia.

REFERENCES

Alicke, K., Barriball, E., Foster, T., Mauhourat, J., & Trautwein, V. (2022). *Taking the Pulse of Shifting Supply Chains*. New York: McKinsey Global Institute. https://www.mckinsey.com/capabilities/operations/our-insights/taking-the-pulse-of-shifting-supply-chains

Al-Jaroodi, J., & Mohamed, N. (2019). Blockchain in industries: A survey. *IEEE Access*, *7*, 36500–36515.

Alkhodre, A., Jan, S., Khusro, S., Ali, T., Alsaawy, Y., & Yasar, M. (2019). A blockchain-based value added tax (VAT) system: Saudi Arabia as a use-case. *International Journal of Advanced Computer Science and Applications*, *10*(5), 708–716.

Al-Saqaf, W., & Seidler, N. (2017). Blockchain technology for social impact: opportunities and challenges ahead. *Journal of Cyber Policy*, *2*(3), 338–354.

Ante, L. (2021). Smart contracts on the blockchain – A bibliometric analysis and review. *Telematics and Informatics*, *57*, 101519.

Attaran, M. (2020, July). Digital technology enablers and their implications for supply chain management. *Supply Chain Forum: An International Journal*, *21*(3), 158–172.

Baral, S. K., Rath, R. C., Goel, R., & Singh, T. (2022, March). Role of digital technology and artificial intelligence for monitoring talent strategies to bridge the skill gap. In *2022 International Mobile and Embedded Technology Conference (MECON)* (pp. 582–587). IEEE.

Cai, Y. J., Choi, T. M., & Zhang, J. (2021). Platform supported supply chain operations in the blockchain era: Supply contracting and moral hazards. *Decision Sciences*, *52*(4), 866–892.

Casino, F., Dasaklis, T. K., & Patsakis, C. (2019). A systematic literature review of blockchain-based applications: Current status, classification and open issues. *Telematics and informatics*, *36*, 55–81.

Centobelli, P., Cerchione, R., Del Vecchio, P., Oropallo, E., & Secundo, G. (2022). Blockchain technology for bridging trust, traceability and transparency in circular supply chain. *Information & Management*, *59*(7), 103508.

Cui, Y., Gaur, V., & Liu, J. (2024). Supply chain transparency and blockchain design. *Management Science*, *70*(5), 3245–3263.

Delic, M., & Eyers, D. R. (2020). The effect of additive manufacturing adoption on supply chain flexibility and performance: An empirical analysis from the automotive industry. *International Journal of Production Economics*, *228*, 107689.

Dolgui, A., Ivanov, D., Potryasaev, S., Sokolov, B., Ivanova, M., & Werner, F. (2020). Blockchain-oriented dynamic modelling of smart contract design and execution in the supply chain. *International Journal of Production Research*, *58*(7), 2184–2199. https://doi.org/10.1080/00207543.2019.1627439

Dong, C., Chen, C., Shi, X., & Ng, C. T. (2021). Operations strategy for supply chain finance with asset-backed securitization: Centralization and blockchain adoption. *International Journal of Production Economics*, *241*, 108261.

Du, M., Chen, Q., Xiao, J., Yang, H., & Ma, X. (2020). Supply chain finance innovation using blockchain. *IEEE Transactions on Engineering Management*, *67*(4), 1045–1058.

Dubey, R., Gunasekaran, A., Bryde, D. J., Dwivedi, Y. K., & Papadopoulos, T. (2020). Blockchain technology for enhancing swift-trust, collaboration and resilience within a humanitarian supply chain setting. *International journal of Production Research*, *58*(11), 3381–3398.

Dubey, R., Gupta, M., Mikalef, P., & Akter, S. (2023). Incorporating blockchain technology in information systems research. *International Journal of Information Management*, *68*, 102573.

Ji, G., Zhou, S., Lai, K. H., Tan, K. H., & Kumar, A. (2022). Timing of blockchain adoption in a supply chain with competing manufacturers. *International Journal of Production Economics*, *247*, 108430.

Jiang, R., Kang, Y., Liu, Y., Liang, Z., Duan, Y., Sun, Y., & Liu, J. (2022). A trust transitivity model of small and medium-sized manufacturing enterprises under blockchain-based supply chain finance. *International Journal of Production Economics*, *247*, 108469.

Jum'a, L. (2023). The role of blockchain-enabled supply chain applications in improving supply chain performance: the case of Jordanian manufacturing sector. *Management Research Review*, 46(10), 1315–1333.

Kamble, S. S., Gunasekaran, A., Kumar, V., Belhadi, A., & Foropon, C. (2021). A machine learning based approach for predicting blockchain adoption in supply Chain. *Technological Forecasting and Social Change*, *163*, 120465.

Khalil, M. M., & Ahmed, W. (2024). Analyzing the drivers of blockchain adoption for supply chain in Pakistan. Journal of Science and Technology Policy Management.

Khan, M. R., Khan, M. R., & Nallaluthan, K. (2023). Blockchain supply chain management and supply chain sustainability. In *Blockchain Driven Supply Chain Management: A Multi-dimensional Perspective* (pp. 155–180). Singapore: Springer Nature.

Khatri, K. K. (2020). Research paradigm: A philosophy of educational research. *International Journal of English Literature and Social Sciences*, *5*(5), 1435–1440.

Kshetri, N. (2021). Blockchain and sustainable supply chain management in developing countries. *International Journal of Information Management*, *60*, 102376.

Kshetri, N., & Loukoianova, E. (2019). Blockchain adoption in supply chain networks in Asia. *IT Professional*, *21*(1), 11–15.

Kumar, A., Liu, R., & Shan, Z. (2020). Is blockchain a silver bullet for supply chain management? Technical challenges and research opportunities. *Decision Sciences*, *51*(1), 8–37.

Latino, M. E., Menegoli, M., Lazoi, M., & Corallo, A. (2022). Voluntary traceability in food supply chain: a framework leading its implementation in Agriculture 4.0. *Technological Forecasting and Social Change*, *178*, 121564.

Lohmer, J., Bugert, N., & Lasch, R. (2020). Analysis of resilience strategies and ripple effect in blockchain-coordinated supply chains: An agent-based simulation study. *International Journal of Production Economics*, *228*, 107882.

Meng, W., Tischhauser, E. W., Wang, Q., Wang, Y., & Han, J. (2018). When intrusion detection meets blockchain technology: A review. *IEEE Access*, *6*, 10179–10188.

Meyer, T., Kuhn, M., & Hartmann, E. (2019). Blockchain technology enabling the physical internet: A synergetic application framework. *Computers & Industrial Engineering*, *136*, 5–17.

Min, H. (2019). Blockchain technology for enhancing supply chain resilience. *Business Horizons*, *62*(1), 35–45.

Mukhtar, U., & Azhar, T. (2020). Inter-functional coordination to co-create value within integrated value chains for competitive supply chain. *Operations and Supply Chain Management: An International Journal*, *13*(1), 11–22.

Patil, P., Sangeetha, M., & Bhaskar, V. (2021). Blockchain for IoT access control, security and privacy: A review. *Wireless Personal Communications*, *117*(3), 1815–1834.

Raman, A., & Thannimalai, R. (2019). Importance of technology leadership for technology integration: Gender and professional development perspective. *Sage Open*, *9*(4), 2158244019893707.

Sabet, E., Yazdani, N., & De Leeuw, S. (2017). Supply chain integration strategies in fast evolving industries. *The International Journal of Logistics Management*, 28(1), 29–46.

Sahai, S., Goel, R., Garg, V., & Vinaik, A. (2019). Impact of digitization on impulse buying-what makes the customer bite the bait. *International Journal of Innovative Technology and Exploring Engineering*, *8*(7), 2948–2952.

Sunny, J., Undralla, N., & Pillai, V. M. (2020). Supply chain transparency through blockchain-based traceability: An overview with demonstration. *Computers & Industrial Engineering*, 150, 106895.

Tan, S. C. (2010). Technology leadership: Lessons from empirical research. Curriculum, technology & transformation for an unknown future. *Proceedings Ascilite Sydney* (pp. 891–895).

Turker, I., & Bicer, A. A. (2020). How to use blockchain effectively in auditing and assurance services. In *Digital Business Strategies in Blockchain Ecosystems: Transformational Design and Future of Global Business* (pp. 457–471).

Xu, P., Lee, J., Barth, J. R., & Richey, R. G. (2021). Blockchain as supply chain technology: considering transparency and security. *International Journal of Physical Distribution & Logistics Management*, *51*(3), 305–324.

Yawalkar, P. M., Paithankar, D. N., Pabale, A. R., Kolhe, R. V., & William, P. (2023). Integrated identity and auditing management using blockchain mechanism. *Measurement: Sensors*, *27*, 100732.

4 Optimization of the Virtual Supply Chain

Hanieh Shambayati and Mohsen Shafiei Nikabadi

4.1 INTRODUCTION

The customers are demanding a greater diversity of products, higher quality, lower prices, and better services. In order to satisfy these demands in an effective and economic manner, organizations constantly have to come up with novel approaches and perspectives (Shambayati, Shafiei Nikabadi, Khatami Firouzabadi, Rahmanimanesh, & Saberi, 2023). The nature of competitiveness at the global level ensures that businesses cannot operate in isolation within the supply chain (SC) environment. In this context, no organization is independent enough to operate on its own. Since rivalry among businesses relies heavily on their SC efficiency (Toh & Zhang, 2014), Supply Chain Management (SCM) has emerged as an essential field and serves as the backbone of every industry (Dutta, Choi, Somani, & Butala, 2020).

Participants in the SC collaborate to synchronize and optimize their business operations, focused on enhancing customer value while reducing expenses (Ahmed et al., 2020). SCM helps manufacturing companies create affordable strategies that boost their productivity and overall success (Bhuniya, Pareek, & Sarkar, 2021). However, traditional SCs lack the adaptability and transparency necessary to meet the expanding needs and demands of the present and the future (Dutta et al., 2020). Nowadays, the impact of intelligent production systems on the global market cannot be overlooked. The demand for timely delivery of high-quality products has made it essential to develop smart production systems (Dey, Pramanik, & Giri, 2016). Consequently, there is a need for research into new SC structures to enhance collaboration and information transfer between SC participants (Wang & Chan, 2010). Information technology (IT) plays a key role in the effective implementation of SCM. It can integrate disparate processes, suppliers, and customers by implementing advanced communication, data collection, and transfer mechanisms, thereby facilitating subsequent improvements in SCM performance (Abdel-Basset, Manogaran, & Mohamed, 2018).

The advent of e-business and information and communication technologies has allowed organizations to adapt and respond expeditiously to evolving market

 DOI: 10.1201/9781003477242-5

demands. In order to enhance the velocity of the market and respond expeditiously and adaptively to customer requirements, a novel domain is required for the SC (Gunasekaran & Ngai, 2007). A virtual supply chain (VSC) can be defined as a novel organizational framework designed to facilitate the efficient and effective integrated flow of information and physical goods. A distinctive attribute of a VSC is its flexibility, which allows for expeditious adjustment to alterations in the commercial landscape (da Cruz Caria, Hermans, Degen, Janssen, & Sweerts, 2016). For partners, the transition from traditional SCs to VSCs involves a strong emphasis on communication, relationships, and knowledge (Manthou, Vlachopoulou, & Folinas, 2004).

VSCs are information-based rather than inventory-based (Lioyd-Betts, 2017). As the VSC depends on an effective communication network, the establishment of an appropriate information system based on different information technologies is of utmost importance (Gunasekaran & Ngai, 2007). The sharing of information within a VSC improves the quality of the information available, which in turn leads to more favorable outcomes in all planning activities. The incorporation of Internet-based technology and information systems can improve the establishment of a VSC (Scott & Mula, 2009). By using Internet technologies, SCs can actively utilize virtualization in their operational management processes (Verdouw, Wolfert, Beulens, & Rialland, 2016). IT offers the potential for enhanced decision-making and the optimization of supply chain (SC) operations through the enhancement of communication, data acquisition, and data transmission (Ben-Daya, Hassini, & Bahroun, 2017). Digital information sharing is fundamental in VSCs, which rely heavily on high-quality data (Smith & Dhillon, 2019). Digital transformation presents various opportunities for improving SC efficiency and logistics operations, including the use of automated vehicles and the optimization of the flow of information (Merkaš, Perkov, & Bonin, 2020).

This study focuses on the optimization of the VSC in two key dimensions. Emphasizing these dimensions is an innovative aspect of the study. In economic terms, the objective is to maximize profits, which requires consideration of costs related to energy and security within the VSC. Furthermore, the importance of the rate of information processing is recognized, and optimizing this rate is identified as another objective of the research.

4.2 LITERATURE REVIEW

4.2.1 The Virtual Supply Chain

A VSC is a collaborative network in which SC partners communicate and operate through the Internet. It consists of a dynamic network of partner organizations that rely on the Internet for their primary operations (Scott & Mula, 2009). In a VSC, physical proximity is not necessary, which means that the paths of physical products from origin to destination are not influenced by the locations of control and coordination partners (Verdouw et al., 2016). The foundation of VSCs lies in the virtualization of products and resources, which is facilitated through information and communication technology (Verdouw, Beulens, Reijers, & van der Vorst, 2015). Virtualization can be considered a digital representation of a physical object's historical, present, and future states (Salvini, Hofstede, Verdouw, Rijswijk, & Klerkx, 2020). The introduction of the VSC in the context of e-commerce systems has occurred in response to the demand for transparency within the SC (Helo, Suorsa, Hao, & Anussornnitisarn, 2014). Virtualization enables the optimization and simulation of SC processes through software systems rather than relying on physical flow. Virtualization incorporates computer information into the SC, providing features like early warnings in the event of incidents and updates on schedules when unexpected quality deviations occur (Verdouw et al., 2016). With virtualization, it becomes possible to track and monitor products along with their historical data during transit (Yadav et al., 2020). Traceability is regarded as a key component of operational innovation (Shou, Zhao, Dai, & Xu, 2021).

Organizations have the opportunity to improve their operational efficiency by virtualizing their processes. To establish a VSC, various sensors are used to monitor changes in pressure, temperature, and the location of products. The data collected is sent as electrical signals or radio frequencies, and is subsequently converted by transducers. This information is made available in digital format through the utilization of information technology devices and internet connectivity. Due to the centralized control that enterprises have over their SCs, there is a high level of trust instilled in this system (Yadav et al., 2020). The characteristics of the VSC encompass temporality, a focus on customer needs, geographical dispersion, a heavy reliance on information technology, a network-based organizational structure, and the utilization of the core capabilities of the stakeholders involved (Kulińska & Kulińska, 2019). In addition to the advantages of the VSC, there are also some issues to consider. Trust among participants for collaboration is a crucial aspect in the context of VSCs and information security (da Cruz Caria et al., 2016).

In contemporary information systems, a static profile of an enterprise data is typically maintained within the software frameworks, databases, or information repositories. This static profile hinders their ability to adapt and respond in fast-paced business environments. Therefore, planning processes for information systems utilized in a VSC must prioritize adaptability, flexibility, and a quick response to shifting circumstances (Scott & Mula, 2009). The swift progress in network connectivity has increasingly made systems vulnerable to intrusion threats. As a result, protection of computer resources, information networks, and other components of the information environment has become an acute issue. Protecting information systems is a concern to the society because information is considered one of the most desirable resources. Thus, intrusion detection is one of the most significant branches of studies in the network security field nowadays. An intrusion is any action in response to and interacting with a subject that threatens the integrity, the confidentiality, or the accessibility of a resource. Specifically, intrusion detection means to detect those users who sneak into computer systems without authorization or to detect misuse by those users who are otherwise legitimate users of the system (D. Li, Cai, Deng, Yao, & Wang, 2018).

4.2.2 Information Technologies

The implementation of digital technology is greatly influenced by various technological, organizational, and environmental factors. Overall, research has indicated that digital technologies can help companies boost their SC performance through increased efficiency, robustness, and resilience as well as reduce SC risks and supply uncertainties (Yang, Fu, & Zhang, 2021). The pace of technology adoption in SCs is still below what is necessary and desirable, given that most rely on data that is manually collected and primarily accessed. This is not only complex and outdated, but it also struggles to manage large volumes of data or rapidly identify and solve problems in the SC. Information technologies fundamentally adhere to the characteristics and structural nature of the chains by integration of internal processes and, more importantly, by improving integration with suppliers and customers of different entities. This level of integration is attained through advances in the communication, data collection, and transmission that work together to optimize decision-making and enhance SC performance. Thus, information technologies have become significant components in the successful management of SCs, as they improve agility in dynamic environments and can successfully mitigate risk at each level (Hu, Xu, Liu, & Lim, 2023). A VSC calls for digital transformation, which enhances the complexity of managing privacy and security threats. Several obstacles contribute to this complexity, particularly the need to differentiate between accurate and inaccurate data (Salvini et al., 2020). The Internet of Things (IoT) and blockchain are technologies that have considerably advanced SCM processes, effectively addressing challenges related to real-time data collection and sharing (Nouri, Aghababayi, & Nikabadi, 2022).

4.2.2.1 Internet of Things

In the early 1990s, Kevin Ashton introduced the notion of the IoT at the Auto-ID Center at MIT (Shambayati, Shafiei Nikabadi, Khatami Firouzabadi, Rahmanimanesh, & Saberi, 2022). IoT is a recent development in the information technology that has caused a paradigm transition in several areas, including VSC. In this regard, IoT extends VSCs more visible, agile, and efficient, and adapts to different aspects of SCM issues (Ben-Daya et al., 2017). The IoT is a user-friendly and salable technology that uses the Internet to deliver all relevant information to the entire SC in real time (Hu et al., 2023). The IoT is one of the valuable technologies that holistically provides infrastructure, platforms, and software solutions through the Internet, optimizing the entire SC. Key benefits include reduced costs and increased flexibility, made possible by the collaboration of SC partners. However, issues such as privacy, adapting processes, and building processes, and building the trust of participants must be addressed in the development of the application (Nozari, Fallah, & Szmelter-Jarosz, 2021). Consequently, IoT can be utilized as a secure and dependable channel for the information exchange concerning goods and services in the global SC. It employs the internet to integrate all connected "things," each with its address, through wireless and radio frequency communication technologies (Papanagnou, 2022).

The IoT is an intelligent grid that uses the internet, which allows it to transmit information in real time. It utilizes top-notch technologies, including RFID, lasers, and scanners, to get precise identification of merchandise, as well as track them and develop ways of managing the goods. It allowed radio-frequency identification techniques that have been in place for several years of supply chain management support, an automatic way of enhancing logistics efficiency, reduction of shipping duration, and minimization of lost goods (Yan, 2017). RFID tracks the flow of goods at every point in the SC. An item is attached with an RFID tag, and upon scanning, the ID of the tag gets stored in a database. This way, the system keeps the managers informed about every part and product. In this way, the SC members can have full details about a product; thereby, it offers total visibility in the SCM (Abdel-Basset et al., 2018). SC participants derive increased exposure by capturing data via IoT technologies and converting it into useful information. One good example of this is the use of RFID technology. This technology allows product tracking and tracing, while minimizing implementation costs. Also, the IoT has significantly reduced the time required to acquire data, and the ability of the SC to address changes instantly also increases the speed of decision-making. For instance, in a situation where customers want to monitor consumable products such as perishable goods, time-sensitive goods, or any other sensitive goods in general, IoT sensors can be employed. By reducing the costs of labor, IoT technologies allow SC participants to track product conditions in real time, which reduces time to make decisions and coordinate (Hu et al., 2023).

As in effective VSC management, logistics transparency should be maximized and the usage of IoT in VSCs the improve inventory management, and new RFID technology to enable recording various information, including the production and expiration date, and warranty period. It also reduces the cost of returns and has a humungous influence on customer satisfaction (Abdel-Basset et al., 2018). For IoT technologies to support fast technical development, several critical issues must be addressed. The focus is primarily on creating security tools and tackling security challenges associated with IoT (Nižetić, Šolić, López-de-Ipiña González-de-Artaza, & Patrono, 2020). The vast amounts of data generated by IoT devices are stored on a central server, which raises concerns about the possibility of the server acting unpredictably and potentially misusing essential user data. This poses significant threats to privacy and security within a centralized IoT architecture (Hasan et al., 2022). Another major challenge arising from the extensive use of IoT is the effective management of energy. This adds a new dimension of complication to the new generation of networks. Energy usage signifies mainly the capability to reduce energy consumption to the lowest level needed at various layers (Ahmed et al., 2020).

Latency can be problematic because it influences the functionality of essential applications in multiple domains in an IoT environment (White, Nallur, & Clarke, 2017). Many of today's emerging smart factories utilize IoT technology to sense a broader range of environmental data, which is then sent to a controller for decision-making. Once a decision is made, the collected information is communicated to the actuators. Latency is a significant technical constraint in most of these applications. Accurate measurement and control of time are crucial for both efficiency and safety, highlighting the need for timely and reliable processing. Consequently, the fundamental communication systems need to have high performance in terms of latency, especially due to the two-way energy transfers required in heterogeneous Smart Grid environments (Ma et al., 2019).

4.2.2.2 Blockchain

Blockchain is an emerging technology anticipated to enhance competitive advantages through business models based on innovative platforms. Originally the foundation for Bitcoin, blockchain is thought to effectively tackle the issue of achieving end-to-end transparency. As this technology gains traction, it acts as a networking tool that simplifies corporate operations by utilizing a peer-to-peer (P2P) network to verify and share data. Blockchain addresses cybersecurity threats such as ransomware, trojans, worms, rootkits, and botnets by using public key encryption to validate all online transactions. SC applications that utilize blockchain technology include smart contracts, product tracking, enforcement monitoring, inventory management, transaction processing, and data permanence. These applications have enhanced market dynamics, economic outcomes, and environmental sustainability due to increased collaboration (Alazab, Alhyari, Awajan, & Abdallah, 2021).

A blockchain is a distributed and decentralized record system where transactions across a network of computers are monitored. Each entry, referred to as a block, is connected to the previous one, thereby generating a sequence. This type of non-centralization ensures that no single organization implements central authority over the whole record, keeping such records open, safe, and reliable. The potential of blockchain to transform SC operations fundamentally is its ability to keep a secure, transparent, and immutable log of transactions (Oriekhoe, Ashiwaju, Ihemereze, Ikwue, & Udeh, 2024). The operation of blockchain involves a blockchain agent creating a new transaction intended for addition to the blockchain. Subsequently, the transaction is conveyed to the network for authentication. After the approval of a majority of nodes, the transaction is incorporated into the blockchain as a new block. A key feature of blockchain technology is the smart contract, which facilitates reliable deals with no third-party intermediation. Consequently, blockchain technology offers numerous benefits, including real-time visibility of transactions across the SC and a diminished risk of data manipulation and fraud (Charles, Emrouznejad, & Gherman, 2023).

The strategic application of blockchain technology to inventory management, if implemented appropriately, can facilitate a myriad of benefits from the technology. In addition, this approach will offer real-time visibility over the inventory levels and allow the use of smart contracts within the inventory processes to drive key outcomes. This also enables smooth communication and coordination among participating members to foster collaboration among multiple stakeholders within the SC. With this, it generates economic value and contributes to securing a competitive advantage. In the sphere of sustainability, traceability, and verification, blockchain could mean a lot to logistics and SCM. The technology is able to record in its database information on the origin and movement of materials within the SC, which enables one to rapidly identify where exactly problems occurred (Sekar, Tech, Brooks, & Noteboom, 2024). It is well established that blockchain technology helps mitigate the risk of counterfeit or unlicensed products being distributed in a given area. This effectiveness stems from blockchain's nature as a decentralized record-keeping system (Park & Li, 2021).

The blockchain is a digital record of facts, but it is not just a record. For blockchain to be successful, it's essential to create the conditions that allow individuals and organizations to trust others to create, store, and distribute essential records. For example, many service industries, including healthcare, banking, finance, transportation, and education, create and maintain third-party records. Therefore, it is crucial to mitigate the problem of corruption and human error in record storage through the use of digital systems based on blockchain (Di Vaio & Varriale, 2020).

4.3 RESEARCH BACKGROUND

Shambayati et al. (2024) aimed to enhance SC optimization by considering elements of SC virtualization, represented through two objective functions: profit and processing rate. In the multi-period and multi-product context with the IoT applications, Shambayati et al. (2022) proposed a model of virtual closed-loop supply chain network design. When in doubt about demand, they used the fuzzy factor. According to Nishi et al. (2020), they have proposed a general method to setting up MAS with VSCs by using Enterprise E-Catalogues. From the proposition of Yadav et al. (2020), there was an addition of more qualities in the current VSC by integrating blockchain technology that could do away with intermediaries, thereby causing an increase in the features of the Food SC, such as transparency, integrity, and authenticity of data. In 2020, Bruno investigated the perspectives of VSC architecture in enhancing product transparency using VSC architecture specifically in the context of the agribusiness industry. To achieve this, we looked at literature about this object and gave the architecture to do so, by applying of traceability model in the overall SC through products, with the help of Electronic Product Code Information Services standard and a NoSQL database. As highlighted by Li (2020), Internet-based technologies enhance the performance of SCs by adoption of technology and operational innovation. Yadav et al. (2020) have also explained how blockchains function when implemented in VSCs, which they describe how blockchains impact various forms of efficiency with reference to bearing cost in food SCs. Smith and Dhillon (2019) focus on three key issues: using IoT to investigate how the GM's SC can be optimized for a perishable product through two objective functions: profit and processing rate. Currently, Shambayati et al. (2022) introduced a model for developing a near-ideal virtual closed-loop supply chain network design based on multi-period, multi-product IoT applications. As for demand uncertainty, they applied a fuzzy approach. In 2020, Nishi et al. proposed a generic method for provisioning a Multi-Agent System with VSCs through a E-Catalogue used in enterprises. Yadav et al. (2020) proposed an integration of blockchain technology into the present VSC that either modifies or eliminates the roles of intermediaries, and this integration boosted the aspect of transparency, integrity, and authenticity of data in the food SC. Bruno (2020) studied a VSC architecture to provide product transparency in the agribusiness sector. Previous studies on this object were examined, and they suggested an architecture that successively tracks the information of the product in the whole SC through EPC Information Services and adopted NoSQL databases. According to Li (2020), Internet-based technologies help enhance SCs' performances by using technology and operational innovations. Smith and Dhillon (2019) focus on three key issues: reliability, trackability, and transparency in managing risk within the VSC, and analyze the use of blockchain to enhance trust between various elements of the VSC. Yan (2017) investigated the optimization of the SC for perishable products through the use of IoT. Yan analyzed perishable product sales before and after implementing IoT to assess how it affected the SC. Verdouw et al. (2016) analyzed the role of virtualization in managing food SC through IoT and presented a theoretically based information system architecture to achieve this design.

Most studies in this area are descriptive, detailing the benefits and key features of VSC. In fact, in none of the reviewed studies has there been an observation of optimization for the VSC focusing on profit and processing time has not been observed.

Considering the significant role of information in VSCs, one of the innovations of this research is the optimization of the VSC by concentrating on virtual costs. This includes considering the cost of information technology equipment, energy consumption, information security, and information delay time. The optimization model of this study is solved by considering profit and processing time functions using meta-heuristic algorithms of the SPEA-II.

4.4 PROBLEM DEFINITION

The virtual supply chain considered in this study includes suppliers, manufacturers, and distributors. Through this network, suppliers can supply manufacturing plants with raw materials or parts as needed. Materials and parts will be sent to each hub as needed and then to customers from distribution hubs. Information technology facilities are utilized to track products at each stage of virtual supply chain management. Information about products and raw materials is stored in the information technology database. In this supply chain, only virtual costs are considered.

4.4.1 Mathematic Model

Index			
S	Supplier	n	Customer
r	Products parts	p	Products
m	Manufacture center	d	Distribution center

4.4.1.1 Virtual Costs in Virtual Supply Chain

In light of the use of information technology to virtualize the SC, this section looks at the expenses related to this process. The presentation of the virtual model's variables and decision factors comes first, and then the creation of virtual expenses.

Parameters			
d_{pn}	The demand of customer n for a unit of product p	Pr_p	Price a unit of product p
q_{rp}	Number of parts r required for product p	FC_{iIT}	The fixed cost of IT facilities at the center i
E_u	The maximum allowed amount of energy consumption	B	Maximum total SC security budget
e^{tr}	The cost per unit of energy consumption for recording, processing and sending data using IT	P^i_{IoT}	The fixed cost of energy consumption for IT facilities at the center i
H	Maximum allowed delay	CD	Cost of damage
P	The possibility of threats	V	Security investment
h^{tr}	Average data processing delay		

Decision variables

Q_{pm}	The amount of product p produced in the center m	Q_{pij}	Amount of product p shipped from i to j
X_{iIT}	A binary variable that is equal to 1 if information technology facilities are used in center I, and equal to 0 otherwise	X_i	A binary variable that is equal to 1 if i center is opened, and equal to 0 otherwise
f^{tr}	Data processed by IT	Q_{rs}	Number of parts r purchased from supplier s

Equation (4.1) represents the fixed cost of using information technology facilities.

$$FCIT = \sum_m \left(X_m \times FC_{mIT}\right) + \sum_d \left(X_d \times FC_{dIT}\right) \tag{4.1}$$

Limitation (4.2) is intended to guarantee the binary character of the variables involved in the decision-making process.

$$X_m.X_d.X_x \in \{0.1\} \forall d.m. \tag{4.2}$$

Equation (4.3) illustrates the energy consumption associated with each produced product, which includes both physical products and the units of currency used to measure them. This model incorporates both fixed and variable costs associated with energy consumption. The variable cost is determined by the quantity of energy expended recording and processing data pertaining to parts and products throughout the VSC. The fixed cost is attributable to the deployment of information technology at each level of the VSC.

$$EC = \left(e^{tr} \times f^{tr}\right) + \sum_m \left(X_{mIT} \times p_{IT}^{m}\right) + \sum_d \left(X_{dIT} \times p_{IT}^{d}\right) \tag{4.3}$$

$$f^{tr} = \left(Q_{rs} + Q_{pm} + Q_{pmd} + Q_{pdn}\right) \tag{4.4}$$

Equation (4.4) represents the data pertaining to materials and products at each level, which is recorded and processed by the information technology system.

Equation (4.5) indicates the amount of parts and raw materials purchased. Constraint 6 is intended to guarantee that the aggregate energy consumption of battery-powered devices, along with the total energy expenditure associated with data recording and processing, will remain below the maximum permitted energy consumption threshold.

$$Q_{rs} = \sum_{m}\sum_{p} q_{rp} \times Q_{pm} \forall r \tag{4.5}$$

$$\sum_{IT} (EC_{IT}) \leq E_u \tag{4.6}$$

Information security can be described as a set of practices and measures implemented to protect information resources and safeguard information in a manner that aligns with, and supports, an organization's strategic objectives. The financial implications of IT security in VSCs are of paramount importance. It is presumed that the implementation of information technology in the centers may potentially lead to the emergence of security threats and subsequent damage to the company. Installing a security system incurs a total investment value of V to address various types of threats. CD represents the cost of damage caused by threats. The total security cost is calculated using Equation (4.7).

$$CS = (CD \times P + V) \tag{4.7}$$

Constraint (4.8) states that the total security cost must not exceed the SC security budget. Constraints (4.9) to (4.11) present different investment scenarios designed to address potential threats, indicating that the probability of threat occurrence decreases with the amount of investment.

$$\sum_{i} CS \leq B \tag{4.8}$$

$$0.7 \leq P \leq 0.9 V = 2500 \tag{4.9}$$

$$0.5 \leq P \leq 0.7 V = 4500 \tag{4.10}$$

$$0.2 \leq P \leq 0.5 V = 6500 \tag{4.11}$$

4.4.1.2 Profit Function

Making as much money as possible is the VSC's ultimate goal. The number of products shipped to clients is multiplied by the unit price of each product to ascertain the overall profit. The VSC's expenses are then deducted from this total revenue. The total profit is calculated by Equation (4.12).

$$\text{Profit} = (Q_{pdn} \times pr_p) - (FCIT + ES + CS) \tag{4.12}$$

4.4.1.3 Processing Time Function

Virtualization produces a significant volume of data that demands considerable processing time for storage and access. As the number of connected devices continues to grow, the generation of Big Data within IT is inevitable, resulting in increased network latency. The objective is to identify methods to reduce the latency associated with information processing. The processing time function is calculated using Equation (4.13).

$$\delta_i = \sum_i \left(h^{tr} \times f^{tr}\right) \tag{4.13}$$

$$\delta_i \leq H \tag{4.14}$$

Constraint (4.14) ensures that the delay caused by data processing does not exceed the maximum allowable delay within the VSC.

4.5 RESEARCH METHODOLOGY

4.5.1 Strength Pareto Evolutionary Algorithm II (SPEA-II)

SPEA-II (Strength Pareto Evolutionary Algorithm-II) is cited in the literature as a very efficient multi-objective algorithm and outperforms other algorithms, including the first version. SPEA-II also improves the calculation of fitness by using Pareto dominance in addition to the number of solutions dominated by the population and those dominating it. Additionally, it uses k-nearest neighbor approach to estimate the density of the individuals for obtaining a more equi-density of Pareto solutions. The SPEA-II algorithm uses an external archive for the storage of Pareto solutions found during the run of the algorithmic search. In this respect, the SAC-based SPEA-II uses both mutation and crossover operators while searching for solutions to the problem. In the algorithm used in SPEA-II, non-dominated solutions are compared by means of a fitness criterion. In this algorithm, the strength of solution i, denoted as $S(i)$, is equal to the number of solutions dominated by solution i. Then the raw fitness value is calculated for each solution, which is made the mentioned fitness function (Zitzler, Laumanns, & Thiele, 2001).

4.6 NUMERICAL EXAMPLE

The following numerical example is presented in order to illustrate the validity and applicability of the linear model. In the absence of empirical data, the model was initially solved through the simulation of data, a process that involved a review of relevant literature, and subsequently with the guidance of experts from both industry and academia. In problems with single-objective optimization, the algorithm optimizes a single-objective function. However, the simultaneous optimization of multiple objective functions represents a challenging and time-consuming task in the context of multi-objective problems. The model's multi-objective nature, coupled with the inherent conflict between different types of objectives, serves to further complicate the problem. These models have been solved by the SPEA-II multi-objective metaheuristic algorithm, which is regarded as one of the most widely used tools in this field. Therefore, it has been employed to solve the model presented in this research. It is presumed that the manufacturer produces five distinct products, each of which comprises a unique combination of parts.

4.6.1 Findings

To check the feasibility of the model, the model was first solved at a small-scale using GAMS software with the Barron solver. As the dimensions of the model increased in the three investigated cases, the solution time also increased when using GAMS software. The findings of the calculations are shown in Table 4.1.

In the following, the presented model was coded and implemented across the main dimensions using the SPEA-II algorithm in MATLAB software. Table 4.2 illustrates the total profitability of the VSC, energy and security costs, and the extent of processing delay.

The Pareto diagram of the algorithm is shown in Figure 4.1. The results of the Pareto chart indicate a direct relationship between the two profit and processing time functions. As profit increases, the processing time also increases. This increase is due to the dependence of both objective functions on the amount of production of products. Thus, as product sales rise, the profit of the VSC increases. Conversely, the increase in product production leads to longer data processing times within the system, resulting in greater delays in information processing.

TABLE 4.1
Computational Results from Solving the Model Using GAMS Software

Case No	Total Profit	Processing Time	Solution Time
1	2.63×10^4	45.8	86
2	8.63×10^4	107.8	135
3	12.63×10^4	310.2	320

TABLE 4.2
Model solution Results

	Total Profit	Processing Time	Energy Consumption	Security	Virtual Cost
SPEA-II	245320	284.37	845	69.5	13.61×10^4

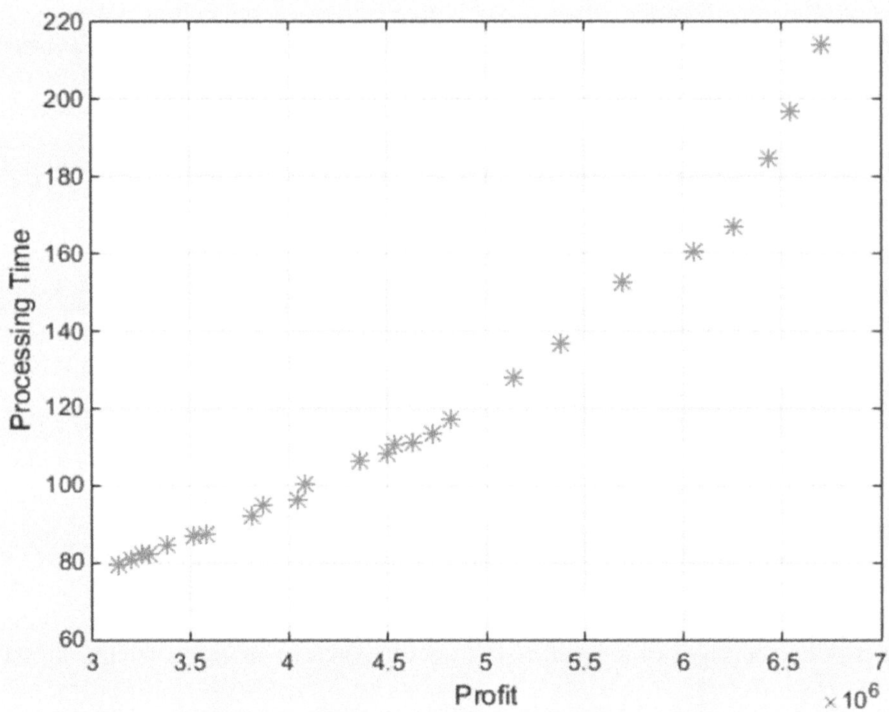

FIGURE 4.1 Pareto diagram of SPEA-II.

Source: Authors own creations.

4.7 CONCLUSION

This, accumulated with the characteristics of the market such as the globalization of the market, rising global competitive pressure, and complexity of product features, necessitates the implementation of new technologies, methodologies, and production philosophies. In light of the developments of recent years, the reach of Internet as well as the implementation of information technology, emphasis should be placed on the VSC. The overall purpose of this research is to demonstrate an effective model of the VSC system with the support of information technology and effective use of virtual supply. To this end, two objectives, profitability and processing speed, are incorporated concerning the costs of virtualization. In addition to supporting the idea of the model's usability and its validity, a numerical example is given herein, where the SPEA-II algorithm is used for the model calculations. We are able to see that the offered model allows for to introduction of effective strategies that will help to solve the problem of production scheduling, reproduction of products, and distribution, as well as the problem of parts recycling. This analysis shows that the two functions of profit and processing speed are directly proportional to the amount of production, with a positive linear relationship. The most prominent issue associated with increased production is an increase in delay time, leading to a reduction of the flow rate. The application of advanced information technologies in the manufacturing industry can provide industry managers with access to data on various products and their end-of-life stages. The virtualization of SCs using information technology is very important not only for better management of production information but also for enhancing efficiency in other aspects of the production system. In a VSC, efficient information sharing allows manufacturers to reduce costs. The seamless exchange of data enables decision-makers to implement appropriate strategies, thereby increasing the profitability of the VSC. Organizations collect a large amount of data in their information systems; however, raw data is not meaningful without proper analysis. The provision of timely data updates and analysis can assist managers in monitoring and management of operations within a VSC, thereby affording them a more comprehensive oversight of the entire SC.

The absence of IT professionals in the conventional SC may be a problem for organizations depending on them, and it is advised that the managers look for or train their people in the virtual environment. Given the centrality of security to the operations of VSCs, managers are encouraged to give more attention to this field and ensure that information is processed swiftly through system upgrades. The model and methods described in this research can also be applied in further research by the researchers and can also be extended. It is proposed that other researchers fine-tune VSC by exploring other virtualization dimensions as well as embracing uncertainty in their undertaking. In this study, several types of functions are introduced to generate a model, and the functions can be further expanded to obtain other models. Other researchers should include other costs, as well as objective functions associated with SC virtualization, into their model. Besides, constraints can also be included or not included in the given model, depending on the solution process. However, because environmental and sustainability issues continue to gain relevance as aspects that should be discussed, it was realized that they needed further study. It is for this reason that researchers may wish to incorporate strategies to increase the rates of response from customers when making their models. This leads to organizational implementation, which is also of considerable importance. Because of technological developments that make it essential for firms to quickly respond to changes in demand and the overall market, many industry managers should transfer their production technologies to a virtual form. This approach could, therefore, be advantageous for continued competitiveness within the existing conditions. Moreover, it may be useful to discuss the practical application of this model. Actually, applying information technology in the SC and also sharing information between all members are ways which can greatly improve organizations' efficiency. A weakness of this research is that the discussed method and model have only been tested with the help of numerical example data rather than real case data. However, this in no way diminishes the value of this research with respect to its theoretical and empirical contribution. However, the implementation of a VSC also opens up difficulties that are, for example, the adaptation of the IT environment of the companies.

REFERENCES

Abdel-Basset, M., Manogaran, G., & Mohamed, M. (2018). Internet of Things (IoT) and its impact on supply chain: A framework for building smart, secure and efficient systems. *Future Generation Computer Systems*, *86*, 614–628. doi:10.1016/j.future.2018.04.051

Ahmed, Z. E., Hasan, M. K., Saeed, R. A., Hassan, R., Islam, S., Mokhtar, R. A., … Akhtaruzzaman, M. (2020). Optimizing energy consumption for cloud internet of things. *Frontiers in Physics*, *8*. doi:10.3389/fphy.2020.00358

Alazab, M., Alhyari, S., Awajan, A., & Abdallah, A. B. (2021). Blockchain technology in supply chain management: An empirical study of the factors affecting user adoption/ acceptance. *Cluster Computing*, *24*(1), 83–101.

Ben-Daya, M., Hassini, E., & Bahroun, Z. (2017). Internet of things and supply chain management: A literature review. *International Journal of Production Research*, *57*(15–16), 4719–4742. doi:10.1080/00207543.2017.1402140

Bhuniya, S., Pareek, S., & Sarkar, B. (2021). A supply chain model with service level constraints and strategies under uncertainty. *Alexandria Engineering Journal*, *60*(6), 6035–6052.

Bruno, G. (2020). A virtual supply chain architecture to grant product transparency in agribusiness. In *Supply chain and logistics management: Concepts, methodologies, tools, and applications* (pp. 89–107). IGI Global.

Charles, V., Emrouznejad, A., & Gherman, T. (2023). A critical analysis of the integration of blockchain and artificial intelligence for supply chain. *Annals of Operations Research*, *327*(1), 7–47.

da Cruz Caria, P., Hermans, M., Degen, D., Janssen, S., & Sweerts, L. (2016). *Virtual supply chain management at Cisco: Advantages and future developments*.

Dey, P. P., Pramanik, S., & Giri, B. C. (2016). Neutrosophic soft multi-attribute decision making based on grey relational projection method. *Neutrosophic Sets and Systems*, *11*, 98–106.

Di Vaio, A., & Varriale, L. (2020). Blockchain technology in supply chain management for sustainable performance: Evidence from the airport industry. *International Journal of Information Management*, *52*, 102014.

Dutta, P., Choi, T.-M., Somani, S., & Butala, R. (2020). Blockchain technology in supply chain operations: Applications, challenges and research opportunities. *Transportation Research Part E: Logistics and Transportation Review*, *142*, 102067.

Gunasekaran, A., & Ngai, E. W. T. (2007). Virtual supply-chain management. *Production Planning & Control*, *15*(6), 584–595. doi:10.1080/09537280412331283955

Hasan, A. T., Sabah, S., Haque, R. U., Daria, A., Rasool, A., & Jiang, Q. (2022). Towards convergence of IoT and blockchain for secure supply chain transaction. *Symmetry*, *14*(1), 64.

Helo, P., Suorsa, M., Hao, Y., & Anussornnitisarn, P. (2014). Toward a cloud-based manufacturing execution system for distributed manufacturing. *Computers in Industry*, *65*(4), 646–656. doi:10.1016/j.compind.2014.01.015

Hu, H., Xu, J., Liu, M., & Lim, M. K. (2023). Vaccine supply chain management: An intelligent system utilizing blockchain, IoT and machine learning. *Journal of Business Research*, *156*, 113480.

Kulińska, E., & Kulińska, K. (2019). General data protection regulation and virtual supply chains. *Studia Ekonomiczne*, *383*, 31–40.

Li, D., Cai, Z., Deng, L., Yao, X., & Wang, H. H. (2018). Information security model of block chain based on intrusion sensing in the IoT environment. *Cluster Computing*, *22*(S1), 451–468. doi:10.1007/s10586-018-2516-1

Li, X. (2020). Reducing channel costs by investing in smart supply chain technologies. *Transportation Research Part E: Logistics and Transportation Review*, *137*, 101927. doi:10.1016/j.tre.2020.101927

Lioyd-Betts, C. (2017). *Virtual supply chains are information-based rather than inventory-based*. Aston University.
Ma, Z., Xiao, M., Xiao, Y., Pang, Z., Poor, H. V., & Vucetic, B. (2019). High-reliability and low-latency wireless communication for internet of things: Challenges, fundamentals, and enabling technologies. *IEEE Internet of Things Journal, 6*(5), 7946–7970. doi:10.1109/jiot.2019.2907245
Manthou, V., Vlachopoulou, M., & Folinas, D. (2004). Virtual e-chain (VeC) model for supply chain collaboration. *International Journal of Production Economics, 87*(3), 241–250. doi:10.1016/s0925-5273(03)00218-4
Merkaš, Z., Perkov, D., & Bonin, V. (2020). The significance of blockchain technology in digital transformation of logistics and transportation. *International Journal of E-Services and Mobile Applications (IJESMA), 12*(1), 1–20.
Nishi, T., Matsuda, M., Hasegawa, M., Alizadeh, R., Liu, Z., & Terunuma, T. (2020). Automatic construction of virtual supply chain as multi-agent system using enterprise E-catalogues. *International Journal of Automation Technology, 14*(5), 713–722.
Nižetić, S., Šolić, P., López-de-Ipiña González-de-Artaza, D., & Patrono, L. (2020). Internet of Things (IoT): Opportunities, issues and challenges towards a smart and sustainable future. *Journal of Cleaner Production, 274*, 122877.
Nouri, F. A., Aghababayi, H., & Nikabadi, M. S. (2022). Designing and effective blochchain-based service supply chain using integrated FANP-QFD under uncertainty. *Scientia Iranica*, Articles in Press, 1–20.
Nozari, H., Fallah, M., & Szmelter-Jarosz, A. (2021). A conceptual framework of green smart IoT-based supply chain management. *International Journal of Research in Industrial Engineering, 10*(1), 22–34.
Oriekhoe, O. I., Ashiwaju, B. I., Ihemereze, K. C., Ikwue, U., & Udeh, C. A. (2024). Blockchain technology in supply chain management: A comprehensive review. *International Journal of Management & Entrepreneurship Research, 6*(1), 150–166.
Papanagnou, C. I. (2022). Measuring and eliminating the bullwhip in closed loop supply chains using control theory and Internet of Things. *Annals of Operations Research, 310*(1), 153–170.
Park, A., & Li, H. (2021). The effect of blockchain technology on supply chain sustainability performances. *Sustainability, 13*(4), 1726.
Salvini, G., Hofstede, G., Verdouw, C., Rijswijk, K., & Klerkx, L. (2020). Enhancing digital transformation towards virtual supply chains: A simulation game for Dutch floriculture. *Production Planning & Control, 33*, 1–18.
Scott, A. H. S., & Mula, J. M. (2009). Contextual factors associated with information systems in a virtual supply chain. Paper presented at the *2009 Fourth International Conference on Cooperation and Promotion of Information Resources in Science and Technology*.
Sekar, A., Tech, D., Brooks, P., & Noteboom, C. (2024). Factors influencing blockchain implementation in supply chain management: An exploratory pilot study. *Issues in Information Systems, 25*(2), 126–143.
Shambayati, H., Shafiei Nikabadi, M., Khatami Firouzabadi, S. M. A., Rahmanimanesh, M., & Saberi, S. (2022). Optimization of virtual closed-loop supply chain under uncertainty: Application of IoT. *Kybernetes*. doi:10.1108/k-06-2021-0487
Shambayati, H., Shafiei Nikabadi, M., Khatami Firouzabadi, S. M. A., Rahmanimanesh, M., & Saberi, S. (2023). Optimization of virtual closed-loop supply chain under uncertainty: Application of IoT. *Kybernetes, 52*(5), 1745–1777.
Shambayati, H., Shafiei Nikabadi, M., Saberi, S., & Mardani, A. (2024). A neutrosophic optimization model for supply chain virtualization in the circular economy using the non-dominated sorting genetic algorithm II. *Annals of Operations Research, 342*, 1–37.
Shou, Y., Zhao, X., Dai, J., & Xu, D. (2021). Matching traceability and supply chain coordination: Achieving operational innovation for superior performance. *Transportation Research Part E: Logistics and Transportation Review, 145*, 102181.

Smith, K. J., & Dhillon, G. (2019). Supply chain virtualization: Facilitating agent trust utilizing blockchain technology. In *Revisiting Supply Chain Risk* (pp. 299–311).

Toh, A. K., & Zhang, Q. (2014). A conceptual analysis of virtual supply chain collaborative network. *International Journal of Society Systems Science*, *6*(4). doi:10.1504/ijsss.2014.066642

Verdouw, C. N., Beulens, A. J. M., Reijers, H. A., & van der Vorst, J. G. A. J. (2015). A control model for object virtualization in supply chain management. *Computers in Industry*, *68*, 116–131. doi:10.1016/j.compind.2014.12.011

Verdouw, C. N., Wolfert, J., Beulens, A. J. M., & Rialland, A. (2016). Virtualization of food supply chains with the internet of things. *Journal of Food Engineering*, *176*, 128–136. doi:10.1016/j.jfoodeng.2015.11.009

Wang, W. Y. C., & Chan, H. K. (2010). Virtual organization for supply chain integration: Two cases in the textile and fashion retailing industry. *International Journal of Production Economics*, *127*(2), 333–342. doi:10.1016/j.ijpe.2009.08.006

White, G., Nallur, V., & Clarke, S. (2017). Quality of service approaches in IoT: A systematic mapping. *Journal of Systems and Software*, *132*, 186–203.

Yadav, J., Misra, M., & Goundar, S. (2020). An overview of food supply chain virtualisation and granular traceability using blockchain technology. *International Journal of Blockchains and Cryptocurrencies*, *1*(2), 154–178.

Yan, R. (2017). Optimization approach for increasing revenue of perishable product supply chain with the Internet of Things. *Industrial Management & Data Systems*, *117*(4), 729–741. doi:10.1108/imds-07-2016-0297

Yang, M., Fu, M., & Zhang, Z. (2021). The adoption of digital technologies in supply chains: Drivers, process and impact. *Technological Forecasting and Social Change*, *169*, 120795.

Zitzler, E., Laumanns, M., & Thiele, L. (2001). SPEA2: Improving the strength Pareto evolutionary algorithm. *TIK-report, 103*.

5 Industry 4.0: Driving Sustainability in the Digital Age

Anamica Singh, Seema Garg, Vinita Sharma, Anchal Luthra, and Ritu Sharma

5.1 INTRODUCTION

In 2011, strategic initiatives in the industrial sector emerged prominently with Germany's *Industrie 4.0*, embarking it as a pivotal phase in the strategic evolution of the industrial revolution (Kagermann et al., 2011, 2013). Numerous developed nations beyond Germany, to remain competitive, have adopted similar technological innovations. Some notable examples include the UK's *Smart Factory* initiative and the *Advanced Manufacturing Partnership* in the United States, *Made in China 2025*. These efforts underscore a global shift toward Industry 4.0 and smart manufacturing, representing the next logical progression in technological evolution (Kumar et al., 2020). The present phase of industrial development puts together production processes across timelines, emphasizing automation, additive manufacturing, IoT, data sharing, and data analytics. With robotic automation and AI taking charge, these advancements are reforming the landscape of manufacturing and defining the future of industrial production.

The technological integration used by companies contributes to economic gains such as more precise planning, shorter lead times, and lesser energy consumption; in addition, it positively affects the environmental and social aspects, such as waste reduction and safer working conditions (Xu et al., 2018; Ejsmont et al., 2024).

Companies use these tools to gain economic profits, such as accurate analysis of the most efficient areas, less time to market, and reduced energy consumption; likewise, it benefits society and the environment. Some of the issues that the Industrial Revolution – 4 faces include the enhancement of new green expertise that can make processes more efficient and innovative, and help solve sustainability problems like climate change and resource depletion. These innovations are either directly or inversely interlinked to the sustainability sector. Sustainability and the feasibility of the Industry 4.0 concept are prominent issues that must be included in the technology enhancement endeavors.

In the last few years, the amalgamation of digital technologies and manufacturing processes has given rise to a new fifth industrial revolution. Figure 5.1 illustrates the evolution of industries, providing a comprehensive presentation of historical growth

DOI: 10.1201/9781003477242-6

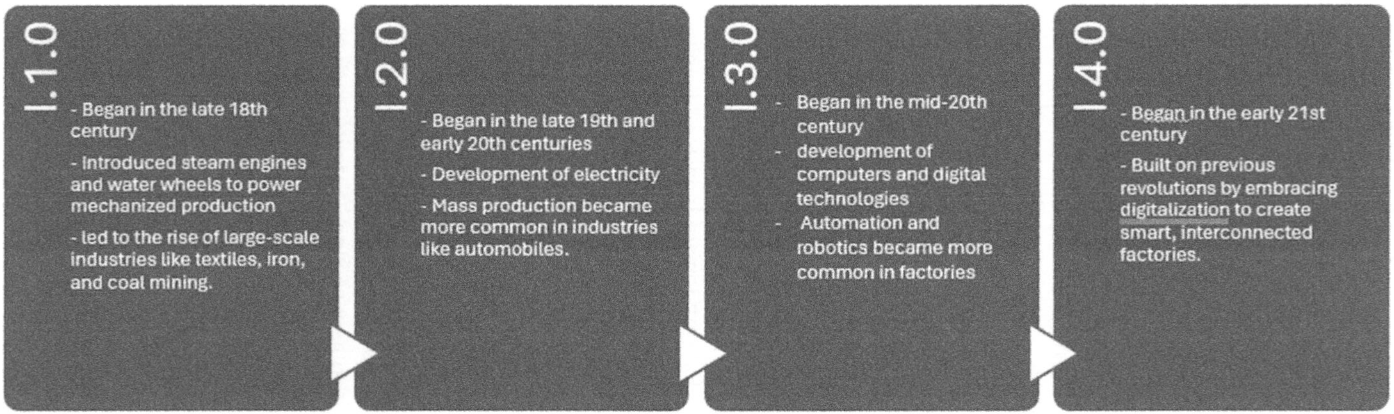

FIGURE 5.1 Evolution of industries.

Source: Authors own creation.

and technological innovation. The figure highlights key phases of industrial development. This revolutionary change has not only transformed how entities are manufactured but can still enable the sustainability of the manufacturing sector. Consequently, this work will examine the framework of Industry 4.0 and how it relates to sustainability, focusing on how new technologies can support eco-friendly and resource-saving practices. Innovatively, Industry 4.0 of the 21st century is a time of great changes in industrial processes and practices that are occurring through the digitalization and automation of technologies like IoT, AI, robotics, and Big Data analytics, which have started to revolutionize the task force utilized in industries again in the process. Furthermore, the advancement of such technologies provides tools that are not only eco-efficient but also greatly contribute to sustainability goals. The era of technological advances and manufacturing practices that are based on the application of digital technologies such as the Internet of Things (IoT), artificial intelligence (AI), robots, and Big Data analytics is changing the career industrial landscape which is no longer possible as it was a few years ago. The change is directed toward economic efficiency and innovation and is directly related to the crucial aspect of sustainable development. Since the world deals with ecological issues like global warming, depletion of resources, and waste management, Industry 4.0 provides out-of-the-box possibilities to manage industrial growth and sustainability.

5.1.1 Key Technologies of Industry 4.0 and Their Role in Sustainability

5.1.1.1 Internet of Things (IoT)

IoT has a connection of independent physical devices like actuators, sensors, and devices that collect data over the Internet. These devices monitor and manage resources efficiently that helps industries in getting instantaneous data on emissions levels, energy usage levels, and water consumption levels (Ghobakhloo, 2020). Smart devices detect inefficiencies in the process, that include excessive energy usage or wastage of material. This can automatically adjust operations to make use of resources in an optimized way. Additionally, IoT devices monitor ecological conditions in real time and enable industries to quickly counter to adjustments and ensure that their operations are within sustainable limits (Vuksanović Herceg et al., 2020).

5.1.1.2 Real-Time Monitoring and Reduction of Energy Consumption

Internet of Things to sustainability links energy consumption levels with real-time control and management systems. These systems constantly collect data on energy usage. IoT devices facilitate industries to identify their patterns of efficiency and inefficiency, to avoid unnecessary interventions, to reduce energy consumption (Sangwan and Bhatia, 2020). Moreover, IoT devices can be interconnected with generating clean energy from renewable sources, such as solar panels or wind turbines, to ensure that the maximum number of renewable sources are utilized. As a result, industries can reduce the use of fossil fuels; thereby, CO2 emissions are reduced, which is one of the ways of mitigating climate change. The feature of enabling real-time monitoring and controlling energy consumption also prevents industries from wastage of energy during the period of low demand, which, as a result, they get an inflated sustainability performance (Javaid et al., 2022).

5.1.1.3 Artificial Intelligence (AI)

This technology makes our planet sustainable and is considered among the Utility 4.0 technologies. It can boost efficiency in a production line by regulating resources more smartly and rationally, which is a big step toward improving the whole ecosystem. AI algorithms perform learning and can evaluate great amounts of data coming from IoT devices and other assets, and tell if we are being inefficient or leaving possibilities for improvement out there when it is happening. The algorithms in AI hardware and software can do the following: automatically recognize failure predictions, inform specialists in relevant areas about future changes, and examine what the advantages of the planned activities would be (Harikannan, Vinodh, and Gurumurthy, 2021). One more application of AI is its usage in fixing the plan so that less energy consumption can be provided when the demand rises. The IO is likewise related to AI and functions in a way that allows precision in the use of resources, for example, only the necessary material will be constituted and waste will be minimized (Ito et al., 2024). The scale of the integration of technology and nature obtained in such an optimization is the key to the success of any sustainable industrial process, as it is the factor that reigns over the economic prosperity of an organization and environmental responsibility (Ali and Johl, 2023).

5.1.1.4 Predictive Maintenance to Reduce Waste and Resource Usage

AI-based predictive repairs in the era of Industry 4.0 are one of the fundamental developments aiming at sustainability. AI offers a new concept that uses data from sensors and other monitoring devices to predict equipment failure. This imperative shift from reactive and scheduled maintenance to proactive condition monitoring and predictive analysis is underpinned by AI. Predictive maintenance is unlike traditional, costly maintenance procedures that are generally fixed, schedule-based, or performed in the case of an emergency. Instead, using AI, we analyze sensor data and thus can approximately detect the period when equipment is expected to malfunction. Industries can now sustain the costs of protective maintenance only when they need it, which in turn not only lowers the risk of breakdowns out of the blue with ensuing extra downtime but also lessens the downtime to a minimum (Vuksanović Herceg et al., 2020). Predictive maintenance technology is designed to avert equipment breakdowns and is expected to save companies on emergency repairs and resource utilization for spares, labor, and energy. However, in the process, the life of the machinery is extended, and the environment is preserved by the reduction of waste and the optimal use of resources. The primary function of predictive maintenance in the industrial world is to note the occurrence of a problem before it becomes a big one. Moreover, this is especially true in these industries with lengthy and resource-intensive operations where unplanned downtime can lead to serious environmental and economic damage (Müller, Kiel, and Voigt, 2018).

5.1.1.5 Big Data Analytics

Big Data analytics is fundamental to realizing sustainability goals in Industry 4.0, providing the insights to make data-driven informed decisions. Collection and analysis of voluminous data enables industries to optimize their operations for sustainability (Ali and Johl, 2023). For example, by analyzing data on energy use, waste generation, and material consumption helps companies identify trends and patterns. This data-driven approach allows industries to implement targeted strategies that indicate inefficiencies or opportunities for improvement and reduce their environmental impact (Matthess et al., 2022).

5.1.1.6 Robotics and Automation

Robotics and automation are important for providing precision in industrial processes, by minimizing waste and fostering sustainability. Automated machines can dynamically perform work in aggregation with the resource of the accuracy with which humans can barely compete. This accuracy minimizes the window for defects, errors, and rework, making organizations efficiently use resources (Hofmann et al., 2019). The robotic systems can be programmed to assemble products in a way that matches the specifications; thus, the materials required to produce these products are minimal, and the waste produced is non-existent. Automation also allows for incorporating advanced manufacturing processes such as additive manufacturing, which is the technology that builds the product layer by layer using a minimum amount of material (Harikannan, Vinodh, and Gurumurthy, 2021).

5.1.1.7 Impact on Workforce Efficiency and Sustainability Metrics

The robotic systems and their automatization in a company make it very efficient and support the metrics in itself. However, by applying advanced technology, workers will be able to tackle the barriers of environmental pollution. So, because there will be less waste, energy will be saved, and labor costs will be reduced. Henceforth, the process is more energy-efficient and with less pollution. In sum, robotics leads to reengineering industrial processes, eliminating non-value-added activities, and pursuing more sustainable and less energy-dependent performance. Since the concentration of carbon dioxide released into the air is decreasing between 2000 and 2024, it is crystal clear that by introducing robotics in production, companies will significantly mitigate their carbon footprint (Esquivel-Elizondo et al., 2023). The other technological alternatives have not been able to replace robotics in the success of the last ten years. Using robots has been the main environmentally efficient technology that has solely reduced pollution and solved all the other connected problems (Varlamov et al., 2022).

5.1.1.8 Additive Manufacturing (3D Printing)

Traditional manufacturing methods use excess material to create a final product whereas, additive manufacturing builds products layer by layer using only the material needed. This step reduces material waste and energy consumption significantly (Harikannan, Vinodh, and Gurumurthy, 2021). Additive manufacturing enables the production of complex structures easily. As a result, additive manufacturing is being adopted increasingly by industries to reduce environmental impact and their operations (Javaid et al., 2022). Additive manufacturing reduces the usage of raw materials and the generation of scrap and waste (Vuksanović Herceg et al., 2020). Additive manufacturing also allows for customizing on-demand parts, reducing the large inventories and transportation and associated storage costs. This enhances the production of manufacturing operations alongside lowering energy consumption, waste reduction, and lowering the carbon footprint (Ali and Johl, 2023).

5.1.1.9 Driving Sustainable Practices with Integration of Physical and Digital Systems

Cyber-Physical Systems (CPS) incorporate raw processes and technologies to facilitate industries to precisely monitor and control operations. This collaboration is essential for adopting sustainable practices within the framework of Industry 4.0. Real-time insights provided to perform better with machinery and equipment (Müller, Kiel, and Voigt, 2018). CPS facilitates faster adjustments, improving operating efficiency and reducing waste. CPS plays an important role in minimizing waste, lowering energy consumption, and minimizing the ecological footprint of transportation and logistics, staging sustainability (Hofmann et al., 2019). It empowers industries to make data-driven decisions aligned with sustainability objectives, such as optimizing production schedules to save energy or reconfiguring supply chain logistics to cut emissions (Sangwan and Bhatia, 2020).

5.2 LITERATURE REVIEW

Industry 4.0 is embarked on as a revolution in production modes revealed by intelligent manufacturing. Deep integration is the core concept of information, communication, and technology (ICT), which aims to unite the physical world with a virtual world full of people, resources, and information. Industry 4.0 aims to transform traditional manufacturing through advanced technologies, enabling smarter and more efficient operations (Ghobakhloo, 2020). The fourth industrial revolution targets to advance the intelligent transformation of the traditional manufacturing system with advanced and intelligent technologies (Ghobakhloo, 2020; Liao and Yao, 2021).

Essential technologies that form the foundation of Industry 4.0, like IoT, AI, robotics, and Big Data analytics, offer different ways to progress toward sustainability in manufacturing. For instance, IoT-enabled sensors allow real-time energy consumption and production process monitoring to be done most efficiently, thus enabling more resource-effective management. AI is used in quality and production processes to develop independent production schedules and logistics, often reducing waste and emissions (Zhou and Zhang, 2022; Gupta et al., 2023). Besides the fact that these have already been put forward, sustainability has been a matter of argumentation in Industry 4.0, accompanied by many methods. Some of the best techniques include having an economy of cycles and production facilities fed by bioenergy (Charnley et al., 2017). Additionally, the concept of servitization, where manufacturers shift from selling products to offering services, has been identified as a means to promote resource efficiency and product longevity (Tukker, 2015).

5.2.1 Sustainability 4.0

Sustainability 4.0 is simply the green Industrial Fourth Revolution era that is now in vogue – which is the Industrial Internet of Things, the Internet of Production, and the Smart Digital Landscape. Sustainability 4.0, to the right, is more correct in enlisting sustainability as a philosophy besides sustainability. It is, thus, the real main subject in solving social, ecological, and ethical problems. The SD goals, though, are the only globally applicable framework that makes the human rights debate of today consistent with the 21st century (Varela et al., 2019; Brozzi et al., 2020). With IoT innovations, the platform must be selected to provide data on carbon footprint and the actual emissions or waste related to different processes and societal factors (Paravizo et al., 2018; Rakic et al., 2021).

Sustainability 4.0 is a notion that strengthens industrial processes with the final goal of cutting off bad environmental and social impacts through reduced waste, recycling, and resource utilization minimization, which are among the many procedures implemented. Additive manufacturing is an advancing method that frequently utilizes composite materials to create 3D products by stacking layer upon layer. This low-cost method effectively manufactures intricate, lightweight items that reduce energy use. The development of sustainable manufacturing is about designing goods that minimize environmental impact and remain economically sustainable. Green processes are a company's connection to technological sustainability, bringing modern symbiosis. Green processes are advanced applications of technology that inspire a culture of creativity and productivity and enhanced workplace, social, and technical conditions. Such measures include eco-design, innovation, maintenance, and the smoothest possible recovery procedures (Ojstersek and Buchmeister, 2020; Ordieres-Meré, Prieto Remón, and Rubio, 2020; Appolloni et al., 2022).

Advanced manufacturing technologies using sustainability 4.0 can revolutionize our world, promote human existence, and facilitate ways and means for prosperity through the acceleration of global sustainable development. This yields economic gains through a higher level of effectiveness and efficiency with advanced methods of goods production and getting the supply chain better. It is safe to say that companies employing sensors to read the assembly line can more efficiently manage their stock and warehouse management operations, elevating their economic sustainability. IoT applications are a supply chain management tool that makes processes more efficient and effective in the quick production cycles. They reduce inventory inaccuracies, and the time needed to bring the product to market (Stock and Seliger, 2016; García-Muiña et al., 2020).

5.2.2 Need for Sustainability

The public gets to know more and is more conscious of the damage created by production processes to the environment and society. Such awareness encourages individuals to go for earthly things and the inclusive creation system. As a result, they can address several needs and maintain a competitive edge through deforestation if they engage in sustainability 4.0. This sustainability 4.0 need is not only for natural resources but also for social resources. An indication of this could be those structural parts, materials, and components are equipped and identified with the help of actuators, and sensors assembled in the manufacturing industry. These can communicate, and he can use the predetermined instructions to exchange data to reduce downtime and maintain effectively. This is a new manufacturing method where the machines organize themselves according to the required products and can make them very close to consumers' requirements. The techniques can be used for tool-path evaluation and machine tool redesigning to decrease energy consumption, enhance operations' safety, and improve human use (Jayashree et al., 2021; Riley et al., 2021).

5.2.3 The Role of Industry 4.0 in Promoting Sustainability

Integrating sustainability principles into the processes of the industry is a vital role of Industry 4.0. By leveraging advanced technologies, companies can significantly reduce resource intake and usage of waste material, ensuring lower emissions and optimizing energy efficiency. Industry 4.0 technologies aid in sustainability with optimization of resources, reduction in waste, and minimizing environmental impact. Key areas where Industry 4.0 drives sustainability include.

5.2.3.1 Resource Efficiency and Waste Reduction

The modern concept that brings forth new ideas to cut down resource use while keeping waste to a minimum in production, a very significant frame of operations for industries aiming at the best sustainability strategies. The advanced technologies are to regulate and optimize energy and resource consumption in real-time scenarios. IoT sensors are used, among other things, to track energy consumption along the production line and pinpoint areas of waste, both of which require optimization. AI algorithms can then suggest the needed changes to reduce waste, for example, by altering production plans or changing the setup to utilize resources best (Javaid et al., 2022). Industry 4.0 technologies provide one of the most innovative solutions in the waste reduction route through additive manufacturing. The technology creates parts with great precision in material usage, thus minimizing the amount of waste produced. Predictive maintenance allows the hassle of machine operation to be avoided by keeping them in top condition and, hence, saving natural resources from waste related to equipment malfunction (Vuksanović Herceg et al., 2020).

5.2.3.2 Energy Management and Carbon Footprint Reduction

Industry 4.0 is the main agent of energy saving, referring to the implementation of next-generation technologies like IoT, AI, and Big Data analytics. These technologies make it possible to manage and monitor energy consumption in real time throughout working facilities; thus, it is accomplished that companies can trace non-priced production elements and repair them. A case in point is the usage of energy management systems that can automatically dim the lights, regulate the heat, and adjust air conditioners when they are not in use or not needed due to weather conditions; thereby, a significant amount of energy could be saved (Sangwan and Bhatia, 2020). Cutting down perplexity to reduce the high presence of unrealistic words and phrases used by AI language models ChatGPT, in the content should be among the top priorities. Writing with high accessibility of different sentence structures and lengths is also beneficial as it enables the readers to comprehend the presented content easily and quickly. Readability is important to make content more comprehensible. In this regard, using simpler language and shorter sentences is a better and easier way of enhancing content clarity, thus promoting site readability. It is a good time to introduce different sentence lengths that contain various ideas, making the content more interesting for the readers to peruse (Harikannan, Vinodh, and Gurumurthy, 2021).

5.2.3.3 Innovations in Renewable Energy Integration and Smart Grids

Industry 4.0 technologies, in addition to reducing energy consumption, facilitate the integration of renewable energy sources into industrial operations. IoT devices and

CPS can be used to monitor and control the flow of energy from renewable sources like solar panels and wind turbines so that these sources are used smartly and use their maximum capacity (Javaid et al., 2022). The smart grid, thus driven by IoT and AI, can be dynamic enough to supply balanced energy and demand consistently and integrate natural energy resources into the power grid. These grids are programmed to adjust the grid's energy distribution based on the collection of real-time data; therefore, a reduction in fossil fuel use and a decrease in carbon emissions will be seen (Ghobakhloo, 2020). Industry 4.0 technologies, by implementing renewable energy, can bring about a notable decrease in the carbon footprint and positively impact the global movement for climatic change in the long run.

5.2.3.4 Circular Economy and Sustainable Supply Chains

Industry 4.0 is vital in developing the circular economy, which focuses on minimizing waste and reusing resources for the longest time possible; hence, the products are recycled or repurposed at the end of the life cycle. Digital technologies have brought in new methods for the industries of not only the tracking but also the efficient management of the resources over the life cycle of a product, starting from extraction to production, distribution, use to disposal (Ali and Johl, 2023). IoT sensors can observe the state of products in use, and from the collected data, we can figure out when and where products can be remanufactured, recycled, or repurposed. AI-driven analytics would maximize recycling by pointing out the most effective ways to recover valuable materials from end-of-life products. These technologies also make it easier to form eco-friendly supply chains, reducing losses, preserving resources, and minimizing environmental pollution from production (Müller, Kiel, and Voigt, 2018).

5.2.3.5 Digitalization of Supply Chains for Sustainability

Modernizing supply chains through the latest technologies of Industry 4.0 will make them more efficient and attain greater sustainability by elevating transparency and responsiveness to their highest levels. For example, IoT and CPS facilitate concurrent materials and goods tracking all over the supply chain, thus making the process more transparent and understandable. Logistics is improved as companies become beneficiaries of data that allows them to optimize logistics and minimize energy consumption and waste production through negotiations with suppliers (Hofmann et al., 2019). For instance, AI algorithms may examine data from IoT devices and suggest the routes with the shortest distance; thus, fuel consumption and emissions are reduced. Furthermore, the digital twins of supply chains can visualize and simulate various scenarios; thus, companies can now explore the best sustainability options and make decisions based on the analysis and simulation. By converting their supply chains into digital ones, companies can cut down on this waste and provide a green operation (Matthess et al., 2022).

5.2.3.6 Digital Tools for Managing Product Life Cycle Sustainably

Product life cycle management (PLM) is an essential part of sustainability, and Industry 4.0 technologies are powerful tools for this purpose. PLM tools are the digital ones that let companies track their products from the design stage through distribution, end-of-life disposal, use, and production. These tools help companies develop eco-friendly, smart, and sustainable development. Thus, IoT sensors, together with CPS, are the so-called applications monitoring the functionality of products while they are being used, thus providing data that can be utilized for maintenance optimization and product life extension. Also, AI-based ones that can predict the lifespan of a product can inform companies when the products are about to reach their end-of-life, so the company can be organized in the process of recycling or reuse. This predictive PLM (PPLM) procedure helps decrease waste, save materials, and occupationally affect the products (Harikannan, Vinodh, and Gurumurthy, 2021).

5.2.3.7 Design for Sustainability and End-of-Life Strategies

Developing sustainable products is a significant pillar of limiting environmental deterioration. With Industry 4.0 technologies, the design of such products now includes elements of environmental sustainability, for instance, the use of recyclable or biodegradable materials, the optimization of designs for material usage reduction, and finally, the designing for the easy disassembly and recycling of it (Ghobakhloo, 2020). End-of-life (EOL) processing is also important in environmental management and sustainability. IoT and AI techniques are also critical in this process as they help companies develop reliable and sustainable EOL strategies by giving them data on product usage, wear and tear, and potential recycling options. By not only planning but also actively managing the life of the product from the beginning to the end, companies will be able to not only minimize the waste but also supply the resources toward the ultimate objective of the economy, which is aimed at the use of renewable resources and recycling (Vuksanović Herceg et al., 2020).

5.2.4 Technologies for Sustainability 4.0

Industry 4.0 is important in advancing the Sustainable Development Goals (SDGs). Prominent ecological benefits have been witnessed by applying cutting-edge technologies like AI, the IoT, robotics, and blockchain. These include the capacity to accomplish profitable expansion concurrently and decreases in usage of energy, decrease in pollution, and emission of greenhouse gases. Through the use of data for traceability analysis and footprint tracking, Industry 4.0 lessens the environmental effects of goods, services, and operations. Additionally, it promotes increases in resource efficiency, aligning with SDGs by promoting digitalized, eco-friendly operations. Although smart technologies continue to evolve and positively impact sustainability, there is still much to learn about their potential in socio-environmental sustainability and how intelligent systems might further realize Industry 4.0's full potential (Ghobakhloo, 2020).

Figure 5.2 presents industry 4.0 technologies for sustainability, highlighting advanced innovations that drive environmental, economic, and social improvements. The figure serves as a guide to understanding the technological pillars supporting sustainable industrial practices. Research on the implementation of sustainability 4.0 aspects is done. Some popular sustainability characteristics are cloud computing, the IoT, additive manufacturing, Big Data analytics, CPS, AI, and various other technological advancements. The sprout-up of new technological practices has been adding more points to the ones already mentioned, and the esoteric aspects of the development of these platforms are the incorporated sustainability behaviors in different industries

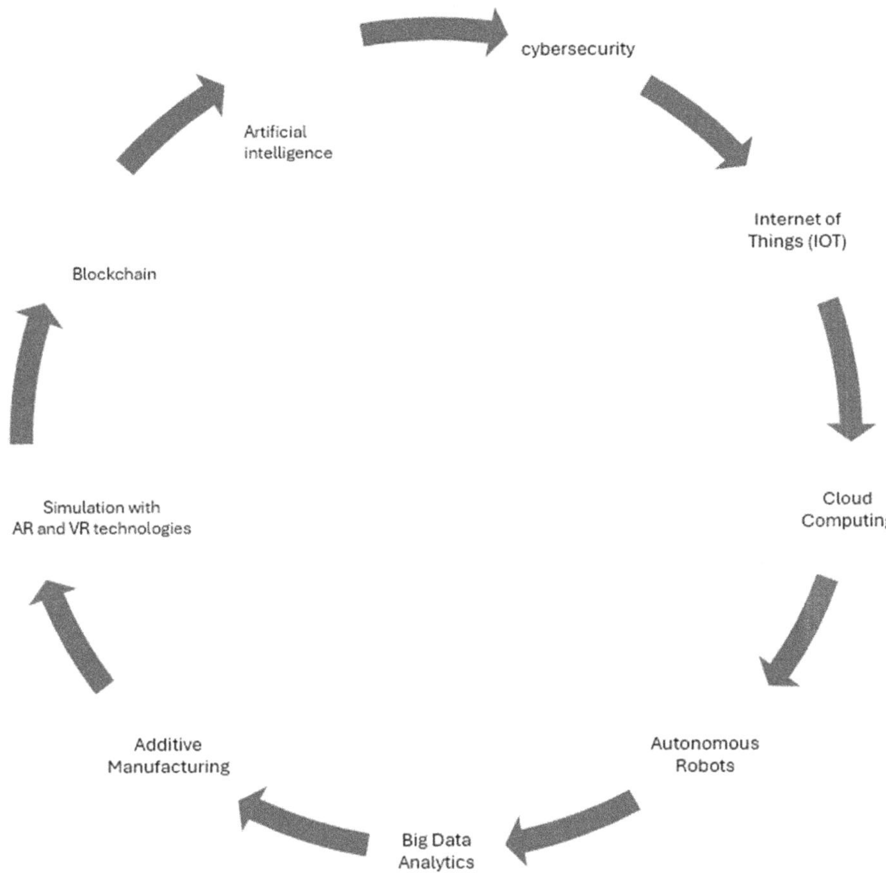

FIGURE 5.2 Industry 4.0 technologies for sustainability.

Source: Authors own creation.

(Ghobakhloo, 2020; Krupnova et al., 2020). Sustainability 4.0 is a real and systematic point of departure for tens of thousands of companies around the world to expand the effectiveness of their processes used for manufacturing. New, greener products, thereby becoming a driving force behind green growth and sustainable development. With their increased awareness of the negative effects of pollution, through, for example, the enhanced greenhouse gas levels in the environment, the exaggerated and savage natural disasters, and their consensus based on the scientific evidence of the past decades, the global community is increasingly turning to sustainable development. The rationale of better and using agreed theoretical assumptions has made it easier to have a strong voice on this topic and to take collective action to save the planet Earth for future generations. However, involvement with greedy businesses brings an end to that organized and committed approach to climate change through new energy sources and emissions reduction, and directly encourages over-exploitation of resources as a deadly result for the environment, as well as the society (Çınar et al., 2020; Abubakr et al., 2020).

In such an amalgamation of new cost-efficient technologies, a critical step is the dialogue and collaborations of the different stakeholders. For refining and optimizing current processes, developing novel processes using fewer noxious materials. This will produce less emissions as green processes are utilized. Adopting the new machines eradicates the most dangerous jobs and thus decreases the worker injury rate (Ramirez-Peña et al., 2020). Intrinsic to the whole thing, the industry's key players have a key role in realizing operational efficiency and disclosing supply chain data in sustainability 4.0 to hit the bull's eye of the set goals (Gregori et al., 2017; Jin et al., 2017; Klymenko et al., 2021).

Manufacturers should conduct an energy consumption and equipment efficiency audit to identify areas of energy waste, optimize operational performance, and implement sustainable practices that reduce costs and environmental impact Annual report, 2024 by U.S. department of Energy highlights that energy assessment on SME resulted in $14.8 million in cost savings with a reduction of 53,000 metric tons of CO2 emission per year (U.S. Department of Energy, FY 2024). Plant energy use at the production and operations of the plant by reviewing and monitoring the maintenance and energy-related operation of all process units, lines, and equipment and, thus, the total efficiency of the whole production process, economic efficiency, and reduction of greenhouse gas emissions. Aided by the distributed energy resources, their standardization and digital technologies make the bidirectional control and interface dynamic, and thus, the open-ended demand for energy and the flexibility of curtailment become the results. In it, there are innumerable issues, and the chances of saving a great amount of costs (Micieta et al., 2019; Saxena et al., 2020). Sustainability 4.0 is a resilient programming that exploits the closeness of the company and the feeder suppliers. No matter the size, manufacturing companies might add a competitive edge by employing the ideas and practices of sustainable development, including smart grid and lean manufacturing that embrace both profitability and sustainability. Efficient manufacturing is the path being promoted by companies. The process focuses entirely on developing eco-friendly strategies and sustainable technology (Song and Moon, 2017; Kliestik et al., 2020).

Companies compete more with environmental problems, but with the new sustainability demands speeding up, they have gained importance in productivity. The current gaps are in world efficiency and the inclusion of sustainability viability. Waste recycling as the solution for industries is one of the methods that can help companies reduce waste and positively impact the environment. Manufacturers can also see which things are the best things to change in the workflow so that the re-usable, the repairable, and the recyclable materials can be properly segregated (Godina et al., 2020; Enyoghasi and Badurdeen, 2021; Hughes et al., 2022). Manufacturing is one of the largest economic drivers in developing countries, but the emissions of pollution and ecological damage are also serious problems connected to it. Besides, transitioning to a circular economy might mitigate most companies' energy usage and waste generation. To consumers and companies who are now much more aware of the concepts, the base of circular and waste starts to change; thus, the new circular economic principle of reuse, reduction, and recycling evolves, as we said (Furstenau et al., 2020; Rojek et al., 2020).

5.2.5 Sustainability in the Digital Age

Sustainability is a multidimensional concept that emphasizes the harmonious integration of economic growth, environmental stewardship, and social responsibility within industrial processes. Compared to previous industrial revolutions, Industry 4.0 aims to create a balanced approach where technological advancements meet SDGs. Digital advancement of technologies and CPS enables conjunction. With AI, we can handle continuous monitoring and predictive analytics. Also, with digital technologies, data-driven decision-making minimizes resource consumption and environmental effects immensely for industries. Sustainability in Industry 4.0 aims at ecological issues; it also includes economic sustainability with higher and efficient production and social sustainability. Industries aligning their business with sustainable production practices reduce carbon footprints and contribute to the green circular economy by making the reuse of discarded materials and recycling possible (Javaid et al., 2022).

5.2.6 The Importance of Integrating Sustainability with Digital Transformation

Integrating sustainability with digital transformation is important. It ensures that the technological advancements of Industry 4.0 are leveraged for the long-term well-being of society and the planet, and not just for increased efficiency and profitability. Advanced technologies allow companies to enhance their supply chains, reducing energy consumption and waste. Additionally, IoT devices can track and monitor resource usage in real-time, allowing for more precise control over production processes and minimizing the use of natural resources (Harikannan, Vinodh, and Gurumurthy, 2021).

Moreover, adding environmental consciousness to digital transformation projects increases companies' capacity to confront challenges such as scarcity of resources, stringent regulations, and varying consumer demands and helps them focus on their sustainability. As sustainability becomes more of a focus for both consumers and governments, the companies that have ingrained these practices into their operations will be more suitable to meet these targets and thus will be ahead in the competition. Thus, the foundation mobilizes green funds for future projects while companies gradually advance to sustainable products, services, and business models. The result of such a collaboration is the improvement of the environment, society, and the community's health (Sangwan and Bhatia, 2020).

5.3 INDUSTRY 4.0 AND SUSTAINABLE PRACTICES: CASE STUDIES

5.3.1 Siemens' Digital Factory

Siemens has been at the forefront of applying advanced technologies of Industry 4.0 in its manufacturing processes. At its Amberg plant in Germany, Siemens has created a fully digitalized factory where products communicate with production machines, and each step is optimized in real time. The result is a significant reduction in energy use, waste, and production time, contributing to a more sustainable manufacturing process.

5.3.2 Tesla's Gigafactories

One of the best examples of how companies drive sustainability through Industry 4.0 is Tesla's Gigafactories, which handles large-scale manufacturing. These factories utilize advanced robotics, AI, and renewable energy to produce electric vehicles and efficient batteries. The amalgamation of these technologies may prove to be revolutionary as they may reduce the environmental impact and accelerate the world's transition to sustainable energy.

5.3.3 Procter & Gamble's Supply Chain Transparency

Another example that emerges is Procter & Gamble (P&G), which has leveraged blockchain technology that amplifies its supply chain's transparency, security, and sustainability. By handling the voyage of raw materials to finished goods, P&G makes sure that sustainability standards are upheld all through its supply chain.

5.4 CHALLENGES AND OPPORTUNITIES

The shift toward Industry 4.0 has traction in terms of sustainability. There are, however, uncountable difficulties that are faced as the technologies are being implemented. The narrow spaces of the companies may be unable to afford the costs. In addition, the wide linkage of the Industry 4.0 systems brings up the issue of data security on one hand and privacy on the other. Another of the big problems is a lack of a qualified workforce. There is a growing need for the ability to handle digital technologies in the labor market. Even though matching the human labor force with the Industry 4.0 engagement is likely to be a long-drawn process, there are many positive sides to the process – the rise of new technologies and the implementation of different projects open ways for research and development. Besides, interaction with other industries, governments, and universities promotes collaboration to find environmental solutions using the latest technologies. When installed, this equipment enables the country or the company to build up its economic stability and sustain its existence to remain competitive globally (Lee et al., 2018). The most one will see looking forward is not only the sustainability of modern technologies and their fusion with the traditional business but also the empowerment of the business through its progress in the sustainability of modern technologies. The future of the growth of the companies will be governed by digital technologies, which strengthen their capacity to be more sustainable and make them more efficient; therefore, they will be able to equip and join the global market. Digital revolution ushered wave to automation and computerization and is known as Industry 4.0 that drive significant progress in communication, commerce and connectivity (Groumpos, 2021).

5.4.1 Barriers and Encounters in Implementation of Industry 4.0 for Sustainable Development

5.4.1.1 Technological Challenges

Integration of Legacy systems with fourth industrial technologies is another major technological problem for executing Industry 4.0 for sustainability. Combining old systems with new technologies. Many industries still rely on outdated equipment and processes not designed to interface with IoT devices, AI systems, or CPS. Integration of these legacy systems, along with Industry 4.0 and related technologies, can be complex and costly, requiring significant investments in infrastructure, training, and system upgrades (Müller, Kiel, and Voigt, 2018). Additionally, the interoperability of different technologies and platforms poses a challenge, as many companies use exclusive systems inconsistent with Industry 4.0 solutions. Overcoming encounters is necessary for achieving the full capability of Industry 4.0 in driving sustainability.

5.4.1.2 Cybersecurity Concerns and Data Privacy

Maintaining security, data privacy, and cybersecurity is also challenging. IoT, CPS, and other digital technologies allow industries to be more sophisticated and digital. Maintaining and building security for these industries is a top priority, to protect sensitive data and keep them away from unauthorized malfeasance (Javaid et al., 2022). This provides the companies with information and freedom to compete in the marketplace. Companies shall abide by data protection protocols and employ robust security measures against unauthorized data breaches and access (Vuksanović Herceg et al., 2020).

5.4.2 Economic and Financial Barriers

5.4.2.1 High Initial Costs and ROI Concerns

One of the main issues with the introduction of 4.0 industry technologies as a means of sustainability is the fact that this will demand a high initial investment, which in turn can become an obstacle for many companies, including medium and small industries. These expenses will include procuring new machines, software, infrastructure, and staff training. The investment return new technologies carry may not have an immediate effect, so it would be hard to justify the expense (Ali and Johl, 2023). On top of that, the economic benefits, such as the reduction of consumption of energy and waste, which are the outcomes of the sustainability initiatives, require some time to manifest. Companies must carefully evaluate the expected return on investment (ROI) and outline long-term plans to ensure the sustainability of their investments in the technologies of Industry 4.0 (Harikannan, Vinodh, and Gurumurthy, 2021).

5.4.2.2 Economic Implications for Small and Medium Enterprises (SMEs)

SMEs encounter many problems when embracing Industry 4.0. Due to a lack of financial backup and the need for expert hands to implement and handle these technologies, these companies become a stumbling block. Moreover, they will be unable to compete with large companies in the high-tech area due to their limited financial resources (Sangwan and Bhatia, 2020). The SME owner would be better off with the government, industry partnerships, or the financing of the smart factory to upgrade these small companies in the digitalization era. It is another thing they have to solve by creating scalable solutions developed by the private sector that the SMEs can use. This is because they have to contribute to the alteration of the factories to the digital revolution and the sustainability of their company (Ghobakhloo, 2020).

5.4.3 Regulatory and Policy Challenges

5.4.3.1 Need for Supportive Policies and Regulatory Frameworks

In order for Industry 4.0 to be put into practice for sustainability in the right way, there are a few vital factors that need to be met – government policies and regulations that are favorable to the incorporation of digital technologies, while also taking into account the potential risks. Government authorities must outlay standards and incentives to companies to ensure sustainability and abide by regulations (Vuksanović Herceg et al., 2020). Legislation modifications will encourage green energy, green technologies, and offer tax breaks to companies and businesses (Müller, Kiel, and Voigt, 2018). Standardization is a very important issue for engineers who want to build different systems and use them with each other. Such systems make integrating Industry 4.0 solutions in companies more comfortable. Collaboration with international cooperation can help solve worldwide issues such as microclimate change, resource scarcity, and cybersecurity. To this end, governments, industries, and international organizations can produce common standards and regulations that anchor the proper realization of technologies of Industry 4.0 (Javaid et al., 2022).

5.4.4 Social and Ethical Considerations

5.4.4.1 Workforce Displacement and Skills Gap

As common as Industry 4.0 technologies are, workforce analysis is the way to go regarding job displacement and the skills gap. Automation, AI, and robots could be the substitutes for some of the work, especially those that need to repeat jobs, which consequently brings about unemployment and social inequality (Ali and Johl, 2023). On the flip side, the job market is increasing because companies demand workers with high-end technical knowledge in data analysis, cybersecurity, and machine learning. The skills gap comes to the front as the most important aspect of the labor force to meet the demand of Industry 4.0 and the digital economy (Harikannan, Vinodh, and Gurumurthy, 2021).

5.4.4.2 Ethical Implications of AI and Automation in Sustainability

Using AI and automation in sustainability also raises ethical concerns. For instance, applying AI systems in the management process may give rise to the issues of transparency, accountability, and the possibilities for bias through AI. Ensuring that intelligent systems are designed ethically and applied in a fair and accountable manner will not discredit trust of the public and prevent the satisfaction of sustainable goals (Müller, Kiel, and Voigt, 2018). Additionally, there are ethical concerns regarding digital technologies' environmental effects. The invention and clearance of electronic devices, the electricity usage of data centers, and the digital infrastructure's environmental impact are all important issues that need to be solved to certify that Industry 4.0 acts as a tool for sustainability instead of worsening the circumstances (Ghobakhloo, 2020).

5.5 INDUSTRY 4.0 PROGRESSION TOWARD FIFTH INDUSTRIAL REVOLUTION

5.5.1 Integration of Human-Centric Technologies for Sustainability

Conceptualizing the idea of Industry 5.0 from Industry 4.0 is one main component of introducing human-centric technologies that are sustainable, socially responsible, and improve human health. Industry 5.0 is generally highly sought-after because of low automation and high human involvement; it acknowledges creative human, intuitive, and ethical thinking in a digital environment (Javaid et al., 2022). Industry 5.0 is executing production at the consumer level to satisfy individual preferences, hence offering greater flexibility and sustainability. Collaborative robots (cobots) and AI-driven design tools are some of the technologies that reduce industrial waste and usage of resources by making products customized to individual demands (Harikannan, Vinodh, and Gurumurthy, 2021).

5.5.2 The Shift Toward Personalization and Mass Customization

Consumers are increasingly clamoring for manufacturers' mass personalizations and customizations. The modes of corporation technology, like 4.0, such as 3D Printing, AI, and Big Data analytics, allow companies to supply mass personalized products quickly and cheaply. At the same time, the waste and other inefficiencies that traditionally come with production are not experienced (Ghobakhloo, 2020). Through personalized and individualized production techniques, companies are reducing the environmental pollution that comes with factory production, refining resource utilization, and improving customer satisfaction. This switch to personalization and customization is lubricating the emergence of new sustainable and profitable business models that are eco-friendly and customer-oriented (Müller, Kiel, and Voigt, 2018).

5.6 EMERGING TECHNOLOGIES AND INNOVATIONS

5.6.1 Quantum Computing, Blockchain, and Their Potential Impact on Sustainability

Emerging technologies like quantum computing and blockchain are predicted to make a big difference in the digital age in the fight for environmental sustainability. Quantum computing, a tool that has the potential to process large calculations at super-high speeds, could be the major force of change in fields like materials science, energy optimization, and climate modeling. These greatest improvements may bring about advanced solutions for more orderly and cleaner industrial processes while lowering energy consumption and environmental impact (Matthess et al., 2022). Blockchain, a technology famous for its decentralized and transparent character, enables supply chain improvement in transparency, fraud reduction, and ethical sourcing. To ensure a secure supply chain and an unchangeable record of transactions, blockchain can raise trust and accountability and thus support a more circular, sustainable economy (Hofmann et al., 2019).

5.7 THE FUTURE OF INDUSTRY 4.0 AND SUSTAINABILITY

In the modern era, sustainability and Industry 4.0 will have infinite relationship. As technologies are better utilized, the capabilities for sustaining them will, in addition, be advanced, bringing increased efficiency to industries, mitigation of environmental antagonistic effects, and sustainable development processes in the world economy. Unifying AI, IoT, and renewable energy systems will be the decisive factor in transitioning to these. Emerging technologies and innovations were shaping the future of Industry 4.0 and sustainable manufacturing. Potential societal and environmental benefits of pervasive Industry 4.0 principles adoption. Counsels for legislators, industry professionals, and researchers to foster a more sustainable and resilient manufacturing sector in the digital age. Future research on sustainable manufacturing has numerous avenues as Industry 4.0 develops. Academics can investigate the creation of novel technologies and approaches to improve manufacturing processes' environmental performance, and resource efficiency. Multidisciplinary research endeavors are also required to tackle the intricate socio-technical problems of Industry 4.0 and sustainability. This manufacturing is significantly affected by material science with more resilient, eco-friendly, and recyclable, bio-based materials (Javaid et al., 2022).

5.8 STRATEGIC RECOMMENDATIONS FOR BUSINESSES

Industry 4.0 projects to be aligned with sustainability is more like combining sustainability, Industry 4.0 processes, and activities efficiently and effectively. To achieve this, the companies must take an overall approach, accommodating sustainable and environment-friendly practices with digital technologies is a breakthrough for business operations. As per Ali and Johl (2023), this leads to establishing explicit sustainability goals, sponsoring the required facilities and education, and creating alliances with market players who have a similar bent to sustainability. Companies must concentrate on their ongoing innovation, and development activities, which will inspire staff members to develop fresh, utilizing technological ideas to manifest sustainability (Sangwan and Bhatia, 2020).

Long-term plans for sustainable planning are critical for fully manipulating the benefits of Industry 4.0 technology to the highest level. They must establish detailed roadmaps to set a plan, policy, and time frames for Industry 4.0 projects. Such a roadmap will have to mention the exact goals and performance indicators with which companies can trace whether they are on the right track or need to make any adjustments (Müller, Kiel, and Voigt, 2018). Also, the considerations should include the social and environmental effects of their projects in the 4.0 era. Thus, their strategies should align with the Goals of Sustainable Development (SDGs) (Ghobakhloo, 2020).

5.9 CONCLUSION

Industry 4.0 is a great solution in the manufacturing field because IT and digital technologies are used for resource optimized utilization, minimize the unsafe impact on the environment, and make the work more effective. The only way to reach the environmental goal and, at the same time, be more resilient in the future is by implementing the above principles and incorporating sustainability into the industry. In any event, this objective is only feasible provided the involved parties join forces to tackle the issues that arise and make the most of the digital transformation. All the research examined in this part of the study indicates the huge benefits of Industry 4.0 in sustainability, specifically in the supply chain. It can be recalled that through IT technology and the implementation of ecological practices, job holders can make the needed strides in resource efficiency, pollution elimination, and even attain a better position in the global arena.

Nevertheless, countering the potential challenges and obstacles that the transformation may bring about will result from the combined efforts of universities, producers, and policymakers committed to sustainable manufacturing through the Industry 4.0 revolution. Industry 4.0 is a pattern shift in the digital age, evident through its relationship with sustainability. Thanks to the lean and green manufacturing practices made possible with advanced digital technologies, industries can optimize resources, reduce waste, and positively impact the environment while boosting the economy. Nevertheless, the extended potential of Industry 4.0 can be exploited in the direction of sustainability only if the leading problems, such as universal goals for the lower costs of implementation, the seizure of the consumer data center, and the gap between retraining of the old and the education of the new, are settled.

As industries, governments, and academia get together to conquer these obstacles, Industry 4.0 stands far ahead of the curve for the manufacturing sector. In the digital era, companies obtain the highest efficiency at the highest speed, achieved through the optimized use of resources and energy consumption, lesser environmental damage, and, therefore, solid operational efficiency. Even so, materializing these ideas will elicit a true effort from all counterparts, including policymakers, industry leaders, and technology providers. Tackling issues such as high implementation costs, data security, and the skills gap is the only way to harness Industry 4.0 for sustainable manufacturing fully. Only the close interrelations between industries, governments, and academia will eliminate the obstacles and craft the way for the fourth evolution of industry to take the lead in a sustainable future actively. As cutting-edge technologies fill the industrial transformation gap, they also create new value and solve many urgent challenges.

REFERENCES

Abubakr, M., Abbas, A. T., Tomaz, I., Soliman, M. S., Luqman, M., & Hegab, H. (2020). Sustainable and smart manufacturing: An integrated approach. *Sustainability*, *12*(6), 2280.

Ali, K., & Johl, S. K. (2023). Driving forces for industry 4.0 readiness, sustainable manufacturing practices and circular economy capabilities: Does firm size matter? *Journal of Manufacturing Technology Management*, *34*(5), 838–871.

Appolloni, A., Jabbour, C. J. C., D'Adamo, I., Gastaldi, M., & Settembre-Blundo, D. (2022). Green recovery in the mature manufacturing industry: The role of the green-circular premium and sustainability certification in innovative efforts. *Ecological Economics*, *193*, 107311.

Brozzi, R., Forti, D., Rauch, E., & Matt, D. T. (2020). The advantages of Industry 4.0 applications for sustainability: Results from a sample of manufacturing companies. *Sustainability*, *12*(9), 3647.

Charnley, S., Carothers, C., Satterfield, T., Levine, A., Poe, M. R., Norman, K., … & Martin, K. S. (2017). Evaluating the best available social science for natural resource management decision-making. *Environmental Science & Policy*, *73*, 80–88.

Çınar, Z. M., Nuhu, A. A., Zeeshan, Q., Korhan, O., Asmael, M., & Safaei, B. (2020). Machine learning in predictive maintenance towards sustainable smart manufacturing in Industry 4.0. *Sustainability*, *12*(19), 8211.

Ejsmont, K., Gladysz, B., & Kluczek, A. (2024). Industry 4.0 and sustainability. *Encyclopedia*. Retrieved from https://encyclopedia.pub/entry/1478 (Accessed August 21, 2024).

Enyoghasi, C., & Badurdeen, F. (2021). Industry 4.0 for sustainable manufacturing: Opportunities at the product, process, and system levels. *Resources, Conservation and Recycling*, *166*, 105362.

Esquivel-Elizondo, S., Hormaza Mejia, A., Sun, T., Shrestha, E., Hamburg, S. P., & Ocko, I. B. (2023). Wide range in estimates of hydrogen emissions from infrastructure. *Frontiers in Energy Research*, *11*, 1207208.

Furstenau, L. B., Sott, M. K., Kipper, L. M., Machado, E. L., Lopez-Robles, J. R., Dohan, M. S., & Imran, M. A. (2020). Link between sustainability and Industry 4.0: Trends, challenges, and new perspectives. *IEEE Access*, *8*, 140079–140096.

García-Muiña, F. E., Medina-Salgado, M. S., Ferrari, A. M., & Cucchi, M. (2020). Sustainability transition in Industry 4.0 and smart manufacturing with the triple-layered business model canvas. *Sustainability*, *12*(6), 2364.

Ghobakhloo, M. (2020). Industry 4.0, digitization, and opportunities for sustainability. *Journal of Cleaner Production*, *252*, 119869.

Godina, R., Ribeiro, I., Matos, F., Ferreira, B. T., Carvalho, H., & Peças, P. (2020). Impact assessment of additive manufacturing on sustainable business models in Industry 4.0 context. *Sustainability*, *12*(17), 7066.

Gregori, F., Papetti, A., Pandolfi, M., Peruzzini, M., & Germani, M. (2017). Digital manufacturing systems: A framework to improve social sustainability of a production site. *Procedia CIRP*, *63*, 436–442.

Groumpos, P. P. (2021). A critical historical and scientific overview of all industrial revolutions. *IFAC-PapersOnLine*, *54*(13), 464–471.

Gupta, R., Hirani, H., & Shankar, R. (2023). Sustainable solid waste management system using technology-enabled end-of-pipe strategies. *Journal of Environmental Management*, *347*, 119122.

Harikannan, N., Vinodh, S., & Gurumurthy, A. (2021). Sustainable industry 4.0–an exploratory study for uncovering the drivers for integration. *Journal of Modelling in Management*, *16*(1), 357–376.

Hofmann, E., Sternberg, H., Chen, H., Pflaum, A., & Prockl, G. (2019). Supply chain management and Industry 4.0: Conducting research in the digital age. *International Journal of Physical Distribution & Logistics Management*, *49*(10), 945–955.

Hughes, L., Dwivedi, Y. K., Rana, N. P., Williams, M. D., & Raghavan, V. (2022). Perspectives on the future of manufacturing within the Industry 4.0 era. *Production Planning & Control*, *33*(2–3), 138–158.

Ito, K., Fujiwara, T., Cho, H., Ihara, T., Matsuoka, K., & Takaoka, M. (2024). Waste-to-energy plant evaluation of combustion stability using an automatic crane system with AI-based waste identification. *Journal of Material Cycles and Waste Management*, *26*(4), 1997–2008.

Javaid, M., Haleem, A., Singh, R. P., Suman, R., & Gonzalez, E. S. (2022). Understanding the adoption of Industry 4.0 technologies in improving environmental sustainability. *Sustainable Operations and Computers*, *3*, 203–217.

Jayashree, S., Reza, M. N. H., Malarvizhi, C. A. N., & Mohiuddin, M. (2021). Industry 4.0 implementation and triple bottom line sustainability: An empirical study on small and medium manufacturing firms. *Heliyon*, *7*(8), e07753.

Jin, M., Tang, R., Ji, Y., Liu, F., Gao, L., & Huisingh, D. (2017). Impact of advanced manufacturing on sustainability: An overview of the special volume on advanced manufacturing for sustainability and low fossil carbon emissions. *Journal of Cleaner Production*, *161*, 69–74.

Kagermann, H., Lukas, W., & Wahlster, W. (2011). Industrie 4.0–Mit dem Internet der Dinge auf dem Weg zur 4. industriellen Revolution [Industry 4.0: With the internet of things towards 4th industrial revolution]. *VDI Nachrichten*. Retrieved from https://www.dfki.de/fileadmin/user_upload/DFKI/Medien/News_Media/Presse/Presse-Highlights/vdinach2011a13-ind4.0-Internet-Dinge.pdf (Accessed June 19, 2020).

Kagermann, H., Wahlster, W., & Helbig, J. (2013). *Recommendations for implementing the strategic initiative Industrie 4.0: Final report of the Industrie 4.0 Working Group*. National Academy of Science and Engineering. Retrieved from https://www.din.de/blob/76902/e8cac883f42bf28536e7e8165993f1fd/recommendations-for-implementing-industry-4-0-data.pdf (Accessed March 30, 2020).

Kliestik, T., Nica, E., Musa, H., Poliak, M., Mihai, E. A., & Mihai, E. A. (2020). Networked, smart, and responsive devices in Industry 4.0 manufacturing systems. *Economics, Management, and Financial Markets*, *15*(3), 23–29.

Klymenko, O., Lillebrygfjeld Halse, L., & Jæger, B. (2021). The enabling role of digital technologies in sustainability accounting: Findings from Norwegian manufacturing companies. *Systems*, *9*(2), 33.

Krupnova, T., Rakova, O., Lut, A., Yudina, E., Shefer, E., & Bulanova, A. (2020). Virtual reality in environmental education for manufacturing sustainability in Industry 4.0. In *2020 Global Smart Industry Conference (GloSIC)* (pp. 87–91). IEEE.

Kumar, S., Suhaib, M., & Asjad, M. (2020). Industry 4.0: Complex, disruptive, but inevitable. *Management and Production Engineering Review*, *11*(2), 43–51.

Lee, J., Davari, H., Singh, J., & Pandhare, V. (2018). Industrial artificial intelligence for industry 4.0-based manufacturing systems. *Manufacturing letters*, *18*, 20–23.

Liao, M., & Yao, Y. (2021). Applications of artificial intelligence-based modeling for bioenergy systems: A review. *GCB Bioenergy*, *13*(5), 774–802.

Matthess, M., Kunkel, S., Xue, B., & Beier, G. (2022). Supplier sustainability assessment in the age of Industry 4.0 – Insights from the electronics industry. *Cleaner Logistics and Supply Chain*, *4*, 100038.

Micieta, B., Binasova, V., Lieskovsky, R., Krajcovic, M., & Dulina, L. (2019). Product segmentation and sustainability in customized assembly with respect to the basic elements of Industry 4.0. *Sustainability*, *11*(21), 6057.

Müller, J. M., Kiel, D., & Voigt, K. I. (2018). What drives the implementation of Industry 4.0? The role of opportunities and challenges in the context of sustainability. *Sustainability, 10*(1), 247.

Ojstersek, R., & Buchmeister, B. (2020). The impact of manufacturing flexibility and multicriteria optimization on the sustainability of manufacturing systems. *Symmetry, 12*(1), 157.

Ordieres-Meré, J., Prieto Remón, T., & Rubio, J. (2020). Digitalization: An opportunity for contributing to sustainability from knowledge creation. *Sustainability, 12*(4), 1460.

Paravizo, E., Chaim, O. C., Braatz, D., Muschard, B., & Rozenfeld, H. (2018). Exploring gamification to support manufacturing education on Industry 4.0 as an enabler for innovation and sustainability. *Procedia Manufacturing, 21*, 438–445.

Rakic, S., Pavlovic, M., & Marjanovic, U. (2021). A precondition of sustainability: Industry 4.0 readiness. *Sustainability, 13*(12), 6641.

Ramirez-Peña, M., Mayuet, P. F., Vazquez-Martinez, J. M., & Batista, M. (2020). Sustainability in the aerospace, naval, and automotive supply chain 4.0: A descriptive review. *Materials, 13*(24), 5625.

Riley, C., Vrbka, J., & Rowland, Z. (2021). Internet of Things-enabled sustainability, big-data-driven decision-making processes, and digitized mass production in Industry 4.0-based manufacturing systems. *Journal of Self-Governance and Management Economics, 9*, 42–52.

Rojek, I., Mikołajewski, D., & Dostatni, E. (2020). Digital twins in product lifecycle for sustainability in manufacturing and maintenance. *Applied Sciences, 11*(1), 31.

Sangwan, S. R., & Bhatia, M. P. S. (2020). Sustainable development in industry 4.0. In *A roadmap to industry 4.0: Smart production, sharp business, and sustainable development* (pp. 39–56).

Saxena, P., Stavropoulos, P., Kechagias, J., & Salonitis, K. (2020). Sustainability assessment for manufacturing operations. *Energies, 13*(11), 2730.

Song, Z., & Moon, Y. (2017). Assessing sustainability benefits of cyber manufacturing systems. *International Journal of Advanced Manufacturing Technology, 90*(5), 1365–1382.

Stock, T., & Seliger, G. (2016). Opportunities of sustainable manufacturing in Industry 4.0. *Procedia CIRP, 40*, 536–541.

Tukker, A. (2015). Product services for a resource-efficient and circular economy – A review. *Journal of Cleaner Production, 97*, 76–91.

U.S. Department of Energy. (FY 2024). https://www.energy.gov/sites/default/files/2024-12/MESC_FY24_Annual_Report_web.pdf?utm_source=chatgpt.com

Varela, L., Araújo, A., Avila, P., Castro, H., G. Putnik. (2019). Evaluation of the relation between lean manufacturing, industry 4.0, and sustainability. *Sustainability, 11*(5), 1439.

Varlamov, G. B., Mitchenko, I. O., Jianguo, J., Weijie, Z., & Zongyan, W. (2022). Modern technologies for increasing the energy and environmental efficiency of energy production. https://doi.org/10.20535/1813-5420.3.2022.271992

Vuksanović Herceg, I., Kuč, V., Mijušković, V. M., & Herceg, T. (2020). Challenges and driving forces for industry 4.0 implementation. *Sustainability, 12*(10), 4208.

Xu, L. D., Xu, E. L., & Li, L. (2018). Industry 4.0: State of the art and future trends. *International Journal of Production Research, 56*(8), 2941–2962.

Zhou, Z., & Zhang, L. (2022). Sustainable waste management and waste to energy: Valuation of energy potential of MSW in the Greater Bay Area of China. *Energy Policy, 163*, 112857.

Part 2

Sustainability in Digital Supply Chain

6 Ensuring Security Across Digital Supply Chain

Ritu Sharma, Seema Garg, and Col Sharad Khattar

6.1 INTRODUCTION

Supply chains have changed from being straightforward linear systems to intricate, worldwide networks that are important in a society that has gone digital. Global supply chains now include the input and acquisition of raw materials as well as the ultimate delivery of producers' completed goods to customers. However, if every link in the supply chain functions as planned, shipments must arrive on time due to this intricate structure. New possibilities for effectiveness, creativity, and international cooperation have arisen as a result of this development. But technology has also brought forth new difficulties, especially with regard to safety. Digital supply chains can help with that. The Big Data generated by every stage of the process is captured by digital supply chains using digital technologies. Data analytics is then used to make sure logistics professionals have the actionable insights they need to properly plan, manage, and strategize supply chains.

As more and more companies depend on digital technology for operational management and optimization, there exists the need to secure the digital supply chain has grown. From the beginning point, i.e., sourcing to manufacturing to distribution and finally delivery, data, information, and products are all part of the digital supply chain. Suppliers, producers, logistics providers, and consumers are just a few of the many parties involved, and they are all linked via various digital systems and platforms. While everything is great for real-time insights, predictive analytics, and quick decision-making, digitalization has the risk of exposing companies to a plethora of cyberattacks. The next important thing is securing a digital supply chain against malicious actions by hackers. New vulnerabilities in computers, networks, and other electronic equipment are discovered daily by cybercriminals, using which they can access facilities, steal data, or halt operations. Disastrous consequences include huge financial losses as well as irreparable damage to reputation and legal liabilities. Such outcomes include cyber catastrophes, including ransomware attacks on manufacturing facilities or thefts of customer data in transit.

The interconnected supply chain structure leads to a serious consequence due to a security breach. The consequence of a logistics provider breach is as follows: delay in delivery, loss of consumer confidence, and disintegration in distribution. Digital supply chain security is not merely concerned with individual enterprises but with the whole ecosystem. They must adopt a comprehensive security strategy, which involves people, processes, and technology, to address and improve the impacts of

DOI: 10.1201/9781003477242-8

these issues. With the entrance of every stakeholder into an open relationship, the identification of new risks and preventive action will become easier. That the regulation compliance factor has a close bearing on the level and degree of the security measures undertaken by the digital supply chain proves it. It all sounds simple, yet building resilience at the state level is managing risk and sustaining trust in an increasingly interconnected and digitalized world.

6.1.1 Background of the Study

Digitalization is the factor driving changes in modern business. The supply chain has changed from a static and linear model to a dynamic network model induced by the latest technologies in the modern world. The change opened up new avenues for possible worldwide interaction, innovations, and advantages of efficiencies and optimizations in production, distribution, and consumption of products and services. On the other hand, it also poses challenges, especially in terms of security. A digitalized supply chain is when data analytics and digital technologies are applied to engineering efficiency in rapid response to changing environments while informing possible choices. The data generated from current supply chains is fundamentally the way data management in warehouses and further analysis to develop actionable insights.

The digital supply chain uses new age technology like the Internet of Things (IoT), machine learning, cloud computing, artificial intelligence, and Big Data analytics to facilitate real-time visibility, predictive insights, or agile decision-making, as opposed to the traditional supply chain that depended primarily on paper-based documentation and manual processes. While on the one hand, digitalization contributes numerous positive advances to human existence, it also brings new threats and vulnerabilities. The most alarming among them is cybercriminal attacks. With increasing incidences of supply chain cyberattacks, cybercriminals now tend to break into supply networks and steal critical data, interrupt operations, or demand ransom payments. This has been opened up by the access of software, network, or device weaknesses. In fact, security breaches might leave far-reaching consequences since they are linked to the whole structure of supply chains. Downstream parties would probably be multiple and affected by a single breach in the chain, causing extensive interruptions. These could be delivery delays, shattered consumer confidence, and monetary losses for every single one concerned because a logistics company suffered a data breach, for instance. Therefore, care for the digital supply chain should also extend beyond individual corporations to the ecosystem as a whole.

The security concerns in the digital supply chains are already challenging enough without bringing in the matter of regulatory compliance. Security for customer information has become increasingly harder with the enactment of strict data protection legislation worldwide. These legislative measures were enacted to minimize the incidence of cyberattacks on customer information; however, they have added more layers of complexity to companies operating at an international level.

Increased awareness of threats requires making electronic supply chain security a top priority. Organizations should further take-in up-to-struggling stance toward cybersecurity and have in place enabling strict safeguards to successfully identify, prevent, and respond to cyberattacks. It is mandatory to achieve a secure and resilient digital supply chain, which is possible only through the consistent training of all staff, up-to-dates on technology, as well as regulations compliance programs in this increasingly complicated and ever-changing business climate.

6.1.2 Cybersecurity Threat Landscape in the Digital Supply Chain

In today's interconnected economy of the globe, vast online networks of outside suppliers, contractors, and associates are being increasingly relied upon by organizations of all kinds. Thanks to advances in technology, these intricate digital supply chains can now travel literally across oceans and nations. New models for business require sharing sensitive and proprietary information in an ever-growing collection of digital channels and platforms but also bring great flexibility and economy. A digital supply chain integrates many stakeholders and technologies—from suppliers to logistics companies to end-users. Connectivity increases visibility and efficiency, but unfortunately also increases the areas of vulnerability where bad actors can take advantage of. In this respect, the cybersecurity threat landscape keeps evolving dynamically, as adversaries continually change their tactics. Unapproved access to the private supply chain information, such as that pertaining to intellectual and proprietary data as well as customer information, causes losses, including financial and reputational, while most typically causing interruptions in the services. Strong cybersecurity policies and controls have to be prioritized across all digital service providers of organizations if they wish to address the increasingly emerging issues related to security, governance, and compliance. Software patch management, data encryption, and rigorous access limitations are often thought of as being restrictive or outdated, but in reality, these are among the most basic of practices. Together, they are the important foundations needed to build secure and resilient environments able to withstand the disruptions of these ever-changing digital ecosystems. In an example of how data encryption solutions really operate, take the traveling sensitive and secret information as it goes from one server to another and through linked devices and apps in the digital supply chain. Privileged access management and hardening access control would also come to the rescue to prevent unauthorized disclosure in the case of a breach. Software fixes provide regular blanket sealing of holes that hackers constantly scout for. Expanding on this essential foundation with more and more sophisticated cybersecurity controls and business processes would be required by organizations to manage risk further in the ever-changing world of digital supply chains. Network micro-segmentation and zero-trust architectures that tightly control access and migration through digital domains limit the damage in case of a breach. The same is true with multilayered security techniques, which eliminate vulnerabilities. Proactive protection and safeguard of key digital assets, especially in tumultuous periods like mergers, acquisitions, and divestitures, avert unauthorized access or data breaches. Irrespective of whether systems and contracts are severed, moved, or decommissioned, it's during special times like these, that unauthorized access or breaches fail to occur. Similarly, resilience can be termed as the good capability to recognize, respond to, and recover from occurrences.

- Incident response teams are trained through attack simulations, cyber wargaming, and incident response team exercises. By promoting a security-conscious workplace, businesses can prepare their employees to provide a crucial human component to technological security measures.
- Organizations can no longer afford to rely exclusively on static security in their cyber strategy thanks to the rapid evolution of threat actors and the regulatory environment.

6.1.3 Importance of Data Security in Digital Supply Chain

Nowadays, it's more important than ever to keep personal information safe and private online. Organizations and people are continually dealing with large quantities of sensitive data, as we read above, due to the growing dependence on digital systems and the internet. The smooth transmission of information between manufacturers, suppliers, logistics companies, and consumers is essential to the digital supply chain. Data security and privacy have grown crucial as businesses digitize and use the latest technologies like AI, machine learning and the IoT. Neglecting to safeguard confidential data can lead to serious problems leading to monetary losses, reputation damage, and interruptions in business operations. Data security and privacy must be prioritized and protected due to the increasing sophistication and frequency of data breaches and privacy violations.

The importance of data security and privacy in sustaining digital supply chains is nothing but establishing faith, as it is more important than ever. The stakeholders, clients, and partners expect that their data will be handled with utmost confidentiality and care; thus, organizations have made sure that strong data security and privacy barriers are embedded in protecting these critical pieces of information, and instilling confidence will develop long-lasting partnerships between partners and consumers. More so, significant private information traverses within digital supply chains, such as customer records, proprietary or secret information, financial information, and trade secrets. Many possible consequences range from reputational damages, monetary losses, and legal liabilities due to any breaches or infringement of privacy.

Data protection and privacy are the cornerstones of building a resilient and effective digital supply chain: they protect sensitive information, ensure compliance, and build confidence among stakeholders. Organizations can ill afford to deprioritize any of those in this age of increasing cyber threats coupled with very strict regulations because they all lead to greater optimum resource use, sustaining customer trust, and long-term success.

6.2 REVIEW OF LITERATURE

It is due to the interconnected nature of operations in the modern world that security problems have become more complex in the digital supply chain. A multilayered approach that leverages technology and collaboration is the most effective way to overcome challenges in today's world. Organizations must invest in strong security measures as well to develop a culture of vigilance toward their supply chains in order to withstand all the imminent threats that evolve in this ever-changing world.

Application of Distributed Ledger Technology (DLT) toward supply chain management (SCM) well depicts how these bottlenecks can be largely alleviated through decentralization, transparency, and data management efficiencies. The same innovations have a tremendous potential to take the economy away from a linear one to a fully circular one and to improve the reliability of data-driven actions, as well as the credibility of SC bodies. In this work, we shall thoroughly analyze 111 publicly available papers on the efficiency and applicability of DLT in SC. Researchers in the field of supply chain (SC) security management, both now and in the future, can use this guide to better comprehend the integration of digital technologies such as DLT. We classed these works based on standard trustworthiness criteria, including immutability, transparency, traceability, and integrity (Asante et al., 2021).

A new era in SCM solutions is upon us, and it's called "Supply Chain 4.0." This new paradigm merges traditional manufacturing with IT and telecommunications. There are new operational and cyber risks associated with supply chain 4.0, despite the fact that its main objective is to improve production systems inside supply chains by leveraging global reach, increasing agility, and emerging technology to boost efficiency, timeliness, and profitability. There are problems with the industrial and IT operations of supply chain 4.0 due to a lack of security, inadequate interoperability, and semantic standards. Blockchain, smart contracts, AI applications, cyber-physical systems, the IoT, and the Industrial Internet of Things are the technologies that support supply chain 4.0. There are cybersecurity concerns caused by all of these technologies, whether used alone or in combination. This study sheds light on the similarities and differences between the military supply network 4.0 and the industrial supply chain, outlining the features, benefits, drawbacks, dependencies, and underlying technology of each. Consideration of connection, sensing, and system convergence is all part of this evaluation of cyber dangers and possibilities for field research. The discussion also includes both existing and future semantic models that are relevant to concerns of standardization, development, and safety assurance as they pertain to the introduction of new technology into military supply chains 4.0. To determine the current status quo and the implications for potential future research routes, this is explored holistically as well as from technology-specific viewpoints (Sobb et al., 2020).

Distributors, manufacturers, buyers, and sellers all work together in what are known as digital supply chains (DSCs). As a platform for the proliferation of IoT technologies, DSCs have become more and more integrated with IoT breakthroughs like field sensors and real-time status monitoring. New cyber danger windows have opened up everywhere as the IoT has become more ubiquitous, expanding the boundaries of networks. Cyber-Supply Chain Risk Administration (C-SCRM) is a relatively new field that combines expertise in cybersecurity, supply chain administration, and enterprise risk management. Its goal is to stop cyberattacks that target DSCs in order to steal money, intellectual property, or disrupt business operations. Unfortunately, there is currently little proof that C-SCRM methods effectively limit breaches. Our research, which spans over a decade, is the first of its kind to statistically analyze how policies and practices outlined in the Cybersecurity Framework, developed by the United States National Institute for Standards and Technology, or NIST, and widely used as a de facto global standard for C-SCRM, affect an organization's breach profile. More effective control of particular breach types was substantially connected with certain framework activity categories as sets of policies/ practices inside those activity areas, according to our study. In order to secure worldwide DSCs and gain control of IT network vulnerabilities, our research provides the groundwork for an evidence-based strategy (Boyson et al., 2022).

Many industries have been affected by the fast expansion of ICT. This has coincided with a rapid transition in the supply chain toward what is known as a "digital supply chain," in which electronic and digital technology permeate the whole process. Among the many advantages brought forth by this development are shorter supply chain lead times, more profits, and less losses. On the other hand, the end-to-end supply chain is now more vulnerable to a wider range of security threats due to the widespread use of these technologies. Our comprehensive analysis of supply chain security is presented here. In this article, we'll go over some of the most common supply chain security concerns and threats. After that, we'll go over some of the security solutions and countermeasures that researchers in both academia and business have suggested to lessen the impact of these dangers. Lastly, we provide some suggestions and guidelines for establishing a safe supply chain (Hammi et al., 2023).

It is crucial to educate cybersecurity experts on these topics and to have a good grasp of these types of assaults, attack pathways, security obstacles, and mitigation strategies. This chapter provides a systematic overview of the security threats, challenges, and relevant cybersecurity measures in DSCs. It then goes on to discuss the knowledge and skills which cybersecurity professionals should have in order to efficiently manage cyber risks in DSCs. Finally, it identifies areas that require further research (Nygård & Katsikas, 2022).

In terms of total operating cost, data security, and auditability, traditional centralized storage and processing of data options have obvious drawbacks. While there is a way to securely and easily transfer data between users, it is difficult to keep up with who has accessed the information as well as how it may have changed during its lifespan. This is one of the main problems with current systems. Inadequately handled by current regulatory and legal compliance regimes is the capacity to electronically control data exchange inside and between various organizational units within the SC. Here we introduce Cydon, a distributed database management system that runs custom distributed apps with a new search and retrieve technique that relies on metadata properties. For every data flow in an environment and every degree of data access and modification within a SC, Cydon uses a smart distributed database to provide an unchangeable audit trail and transaction history. Cydon seems to maintain a "always-on" chain of custody, distribute encrypted data securely over several nodes to prevent single points of failure, and provide authorized users rapid access to dispersed, encrypted data (Asante et al., 2021).

In order to determine how cyber risk and supply chain vulnerability affect digital operations within the UAE pharmaceutical sector, this study analyzed existing empirical data. Approach, methodology, and data analysis were carried out using 243 replies from 14 pharmaceutical manufacturing businesses in Dubai. Normality, instrument validity, and regression analysis were performed on the data. Using a descriptive and analytical study methodology and a practical sample technique, the effects of cybersecurity and SC risk upon digital operations were investigated. The results confirmed the positive and statistically significant relationship between cyber-security and the risk to the supply chain posed by digital activities.

6.2.1 Constraints and Implications of the Study

The research framework was constructed using three variables that were exclusively assessed in relation to the Dubai pharmaceutical business. Expanding the scope of future studies to include more industrial sectors and regions is essential. Implications for practice: Customer-focused organizations facing high supplier concentration and uncertain digital transformation levels should benefit from diversifying their clientele in order to better manage supply chain risk. In order to gain and keep customers' confidence, pharmaceutical companies may need to put more emphasis on technology-based manufacturing companies (Solfa, 2022).

In order to better understand the way organizations can implement a CSCRM strategy that transcends the technical, internal workings of individual companies and transcends the dyad, this chapter aims to investigate supply chain managers' perspectives on the components of CSCRM and the associated level of alignment. The ultimate goal is to foster better alignment, which can enhance cyber resilience in the supply chain. Approach, methodology, and design: More than a hundred Italian companies participated in an exploratory assessment of the fast-moving consumer products (FMCG) sector. To assess the disparities in attitudes across the many participants in the fast-moving consumer goods (FMCG) supply chain, one-way examination of variance was used to analyze the results. Findings: While there is some consensus among FMCG supply chain participants on what constitutes a cyber threat or risk event, the research shows that logistics suppliers may steer the CSCRM process in a more "supply chain-oriented" direction. The importance of people as critical components in enhancing supply chain cyber resilience is also emphasized by the study (Creazza et al., 2022).

As a result, getting to a safe trading environment requires some serious overhauls to the present supply chain architecture. In this post, we will go over the parts of the supply chain that are most important for security and look at the security problems with the current design of the supply chain. To address the weaknesses in the current supply chain architecture, many new technologies have been considered, including blockchain, ML, and physical unclonable functions (PUFs). This chapter reviews recent research that shows how the sector is becoming more open to new technologies, including the Internet of Everything (IoT), blockchain, and machine learning. A large-scale use of blockchain technology is still in the future, even if many companies have already implemented IoT applications and AI systems. Research has also shown that authentication systems based on PUF have made significant strides in the last decade. Nevertheless, there is currently no appropriate reference model for incorporating them into intricate supply networks (Hassija et al., 2020).

A complicated, international, and sometimes unreliable supply chain is used to create modern automobile systems and IoT devices. At least two things may happen in this supply chain: (1) one participant can provide sensitive data and information (hereafter "assets") that have to be kept secret from the others and (2) another participant can have limited access to the assets that other participants have brought. On top of that, there can be some bad apples in the supply chain. It is critical to prevent these organizations from compromising or accidentally disclosing any sensitive information stored on the device. Managing security in the face of supply chain difficulties is an important, though often ignored, part of the security architecture of contemporary electronic systems. This chapter reviews the current state of the art and delves into the security issues plaguing automotive and IoT systems caused by the complexity of the supply chain.

Businesses are more vulnerable to cyberattacks as a result of this adoption, which may lead to data leaks, operational disruptions, and the erosion of reputation, intellectual property, and financial assets, among other negative outcomes. Therefore, it is crucial for the network of supply chains to have an efficient cybersecurity and threat management plan. This plan should include learning everything you can about the critical assets and resources within a traceability structure, as well as the cyber risks that might affect them and how to protect yourself from them. Over one hundred relationships between assets, threats, and countermeasures pertinent to supply channel traceability are laid out in this chapter, which adds to a thorough threat modeling study on supply network traceability systems. Reproducibility, extensibility, and falsifiability characterize our analysis. An organized asset-centric threat estimation approach is used to achieve reproducibility. The STRIDE threat model is adopted to describe common threats. The model is extensible through the use of a layered architecture over supply chains, which the analyst can adapt to a concrete implementation. The relationship (asset, threat, countermeasure) is falsifiable by providing the sources that generated it. While much of the chapter focuses on technology, it does a quick review of safe traceability across supply chains with human and process components just to be thorough (Syed et al., 2022).

The idea was proposed of a virtual SC twin, which is a computer model that depicts the states of the network at any particular instant in real time. We go into the details of the digital twins' design and deployment as they pertain to SC disruption risk management. Finding the connections between risk data, disruption simulation, and outcome evaluation is made possible by combining model-based and data-driven methods. Digital twins are urgently needed to map supply networks and provide visibility, as shown by the shocks and modifications experienced by SCs during and after the COVID-19 pandemic. In order to guarantee end-to-end visibility as well as business continuity in multinational corporations, this study's findings add to the literature and practice for SC risk management by improving both proactive and reactive decision-making through the utilization of SC visualization, analysis of historical disruption data, and real-time disruption data (Ivanov & Dolgui, 2019).

Business continuity planning skills that are believed to provide firm-level advantages are documented in this chapter using a framework. Here, SCM serves as the backdrop. Therefore, the following question is intended to be addressed in this chapter: In order to guarantee the continuation of operations, which areas of competence development should companies prioritize in order to safeguard their supply chains? First, we define the ideas around the focus context, which include supply chain security, the safety of SCM, or the development the firm-level competencies. Then, we will address this question. Afterward, a theoretical structure is presented that unifies the methods, tools, and personnel required to build capacities for managing the security of the supply chain. This will aid in the preparation for future company operations' continuity plans.

There is a dearth of empirical studies on Big Data analytics, despite the fact that it is receiving more and more attention in management circles. The authors use the Delphi method of research since there is a dearth of relevant literature on the topic of Big Data research as well as SCM. Findings: The results of the presented Delphi survey add to the current body of knowledge by illustrating the growing trend of digital business transitions and by highlighting 43 possibilities and threats associated with the advent of massive data sets. Analytics from both the corporate and supplier chain vantage points. Restrictions and consequences of the study: Researchers now have a starting point with these frameworks to build upon when they conduct future studies at the intersection of SCM and Big Data analytics (Kache & Seuring, 2017).

The distributed ledger system that initiated the bitcoin boom has been the subject of increased interest from both academics and practitioners due to the skyrocketing value of bitcoin and the cryptocurrency craze. The possible applications and commercial repercussions of blockchain technology are often the source of controversy. Blockchain technology is a decentralized, distributed ledger system that records transactions involving digital assets directly between users, rather than relying on trusted third parties like governments or banks. So, it may lessen the chances of hacking, privacy breaches, political instability, financial institution instability, regulatory costs, contractual conflicts, and compromised security that come with intermediaries' actions. Discover the secrets of blockchain technology as you learn to use it to make your supply chain more resilient in the face of rising risks and uncertainties in this enlightening essay (Min, 2019).

They identify many obstacles that may hinder the widespread use of blockchain technology in supply chain applications. These include issues with scalability, investment costs in relation to perceived advantages, difficulties in sharing data and ensuring interoperability, and the absence of relevant standards and legislation. The number of documented uses of blockchain technology is growing, even if its acceptance is still at its early stages. As a first step, we categorize the reported supplier chain applications by industry. Scholars and practitioners may get a better grasp of blockchain's potential utility in modern supply chains from this chapter (Ahmed & MacCarthy, 2022).

Companies nowadays have access to resources and items in unprecedented quantities, varieties, delivery times, and transportation modes, thanks to supply chains that have evolved from local trading units to worldwide physical and virtual marketplaces. Natural catastrophes and pandemic breakouts are two examples of the irrepressible conditions that have caused crises in the global economy in recent decades. Among other things, traditional SCR depended on diversified suppliers, partial order allocations, and enhanced safety stock levels. The digital alteration of its pieces into SCR modeling and control, on the other hand, may lead to the next generation of robust supply chain operations. This document has a dual purpose. To begin, it explains how the SCR may be enhanced via digital transformation, sophisticated analytics, automation, and the use of cyber-physical systems and security (CPSS) technologies from the era of industry 4.0. The second benefit is that it lays out a plan for using AI and the IoT to build a supply chain with digital resilience (DSCR) that is strong, up-to-date, and effective. We hope that by engaging in this conversation, organizations will be better able to address the digital capabilities needed to increase transparency and control, boost revenue, decrease expenses, and fortify their supply chains against disruptions brought about by digitalization, automation, and AI (Al-Banna et al., 2022).

6.3 DISCUSSION

The flow of products, services, and data in linked companies across the globe today owes its importance to the digital supply chain for goods. The risk of cyber infiltrations, however, has risen with the proliferation of digital solutions and technology. This chapter extensively unfolds the relevance of cybersecurity to the digital supply chain—from cyber reconnaissance, how important it is to take precautionary steps, as well as the implications for securing this crucial component of modern business operation. Issues may arise as a result of ongoing cyberattacks directed against supply chains. Damaging effects of these attacks may include breaching sensitive information, disrupting or halting operations, and losing money that touches on the entire chain. Cybercriminals exploit an interrelated environment of the supply chains in launching ransomware attacks, making it a threat to the electronic goods supply chains based on the premise of instant and huge payments. Tough measures for cybersecurity are required as the ransom demands increase when various firms are affected. The digital supply chain is an essential modern supply chain for linking products, services, and information throughout the world for easy movement across organizations. The risk of cyberattacks has increased together with the penetration of digital solutions and technologies, however. This chapter raises issues pertaining to the need for cybersecurity within the digital supply chain, particularly in relation to cyberattacks, why preventive measures should be taken, and how to keep this vital part of contemporary business activity secure. Existing cyberattacks directed at the supply chain can also pose some problems. Such attacks usually lead to the manipulation of sensitive data and disruption or stoppage of operational activities, and financial losses cover the entire chain. Cybercriminals take advantage of the interdependent supply chain environment by conducting ransomware attacks, causing threats to the supply chains of electronic goods, which can collect sudden and huge payments from suppliers to keep their operations functional. There is an urgent requirement for secure cybersecurity since the risk of paying ransom for the majority of firms increases.

6.3.1 Recommendations

The study's results provide the following suggestions for strengthening digital supply chain security like strong cybersecurity measures must be incorporated, encouragement of teamwork, and the free exchange of data should be done by encouraging open dialogue and cooperation on security-related matters, organizations may help foster a culture of trust and faith. Also, the transparency among all the involved supply chain partners. Efforts to meet regulatory requirements to reduce legal risks and make sure everyone is following the rules, training programs, and compliance evaluations should be done on a regular basis. And to improve the management of supplier risks to reduce interruptions in the supply chain and maximize compliance with security standards, it is important to set explicit security criteria and standards for suppliers and to implement frequent monitoring and supervision procedures.

Money should be well spent on raising staff knowledge and awareness to better protect themselves against social engineering, phishing, and other forms of cybercrime, businesses should invest heavily in education and awareness campaigns for their employees and to better equipped to identify and react to security risks in the digital supply chain if they participate in regular security awareness workshops, practice phishing attacks, and get continuing education on new threats. Also to keep an eye on things and assess your security strategy regularly to stay ahead of potential security threats, organizations should evaluate and analyze their security posture on a regular basis. Digital supply chain security events may be detected and mitigated early with the use of security analytics, feeds of threat intelligence, etc. incident response frameworks.

Any business can be better prepared for the increasingly digital and linked business world by adopting these practices and recommendations, which will help them secure their digital supply networks and reduce cybersecurity risks, and be more resilient to new attacks.

6.4 CONCLUSION

Supply chain security is both a challenge and a need in today's connected digital world. This study emphasizes the significance of strong cybersecurity defenses, adherence to legal requirements, and cooperative risk-reduction strategies. Digital change creates new vulnerabilities even as it fosters creativity and efficiency. Organizations can create robust and reliable supply chains by putting a high priority on data protection, investing in cutting-edge technologies like blockchain and artificial intelligence, and encouraging collaborations among stakeholders. Businesses may successfully navigate the changing threat landscape and sustain dependable operations in a world that is quickly digitizing with proactive measures and ongoing adaptation.

BIBLIOGRAPHY

Ahmed, W. A., & MacCarthy, B. L. (2022). Blockchain in the supply chain–A comprehensive framework for theory-driven research. *Digital Business*, 2(2), 100043.

Al-Banna, A., Franzoi, R. E., Menezes, B. C., Al-Enazi, A., Rogers, S., & Kelly, J. D. (2022). Roadmap to digital supply chain resilience. In *Computer aided chemical engineering* (Vol. 49, pp. 571–576). Elsevier.

Asante, M., Epiphaniou, G., Maple, C., Al-Khateeb, H., Bottarelli, M., & Ghafoor, K. Z. (2021). Distributed ledger technologies in supply chain security management: A comprehensive survey. *IEEE Transactions on Engineering Management*, 70(2), 713–739.

Bhargava, B., Ranchal, R., & Othmane, L. B. (2013, February). Secure information sharing in digital supply chains. In *2013 3rd IEEE international advance computing conference (IACC)* (pp. 1636–1640). IEEE.

Boyson, S., Corsi, T. M., & Paraskevas, J. P. (2022). Defending digital supply chains: Evidence from a decade-long research program. *Technovation*, 118, 102380.

Chen, Y., & Paulraj, A. (2004). Towards a theory of supply chain security. *Journal of Operations Management*, 25(4), 303–317.

Christopher, M. (2016). *Logistics & supply chain management*. Pearson UK.

Creazza, A., Colicchia, C., Spiezia, S., & Dallari, F. (2022). Who cares? Supply chain managers' perceptions regarding cyber supply chain risk management in the digital transformation era. *Supply Chain Management: An International Journal*, 27(1), 30–53.

Giunipero, L., & Eltantawy, R. (2004). Securing the upstream supply chain: A risk management approach. *International Journal of Physical Distribution & Logistics Management*, 34(9), 698–713.

Grant, D. B., Wong, C. Y., & Trautrims, A. (2017). *Sustainable logistics and supply chain management: principles and practices for sustainable operations and management*. Kogan Page Publishers.

Hammi, B., Zeadally, S., & Nebhen, J. (2023). Security threats, countermeasures, and challenges of digital supply chains. ACM Computing Surveys, 55(14s), 1–40.

Hassija, V., Chamola, V., Gupta, V., Jain, S., & Guizani, N. (2020). A survey on supply chain security: Application areas, security threats, and solution architectures. *IEEE Internet of Things Journal*, 8(8), 6222–6246.

Ivanov, D., & Dolgui, A. (2019). Viability of intertwined supply networks: Extending the supply chain resilience angles towards survivability. A position paper motivated by COVID-19 outbreak. *International Journal of Production Research*, 58(10), 2904–2915.

Kache, F., & Seuring, S. (2017). Challenges and opportunities of digital information at the intersection of Big Data Analytics and supply chain management. *International Journal of Operations & Production Management*, 37(1), 10–36.

Lee, H. L. (2004). The triple-A supply chain. *Harvard Business Review*, 82(10), 102–112.

Li, S., Rao, S. S., Ragu-Nathan, T. S., & Ragu-Nathan, B. (2005). Development and validation of a measurement instrument for studying supply chain management practices. *Journal of Operations Management*, 23(6), 618–641.

Min, H. (2019). Blockchain technology for enhancing supply chain resilience. *Business Horizons*, 62(1), 35–45.

Monczka, R. M., Handfield, R. B., Giunipero, L. C., & Patterson, J. L. (2015). *Purchasing and supply chain management*. Cengage Learning.

Nygård, A. R., & Katsikas, S. (2022, August). SoK: Combating threats in the digital supply chain. In *Proceedings of the 17th international conference on availability, reliability and security* (pp. 1–8).

Pagell, M., & Wu, Z. (2009). Building a more complete theory of sustainable supply chain management using case studies of 10 exemplars. *Journal of Supply Chain Management*, 45(2), 37–56.

Prajogo, D., & Olhager, J. (2012). Supply chain integration and performance: The effects of long-term relationships, information technology and sharing, and logistics integration. *International Journal of Production Economics*, 135(1), 514–522.

Rodrigue, J. P., Comtois, C., & Slack, B. (2009). *The geography of transport systems*. Routledge.

Saberi, S., Kouhizadeh, M., Sarkis, J., & Shen, L. (2017). Blockchain technology and its relationships to sustainable supply chain management. *International Journal of Production Research*, 55(14), 4177–4194.

Sarkis, J., Zhu, Q., & Lai, K. H. (2011). An organizational theoretic review of green supply chain management literature. *International Journal of Production Economics*, 130(1), 1–15.

Sheffi, Y. (2005). *The resilient enterprise: Overcoming vulnerability for competitive advantage*. MIT Press.

Simchi-Levi, D., Kaminsky, P., & Simchi-Levi, E. (2008). *Designing and managing the supply chain: Concepts, strategies, and case studies*. McGraw-Hill/Irwin.

Sobb, T., Turnbull, B., & Moustafa, N. (2020). Supply chain 4.0: A survey of cyber security challenges, solutions and future directions. *Electronics*, 9(11), 1864.

Solfa, F. D. G. (2022). Impacts of cyber security and supply chain risk on digital operations: evidence from the pharmaceutical industry. *International Journal of Technology, Innovation and Management (IJTIM)*, 2(2), 18–32.

Stevenson, W. J., & Hojati, M. (2007). *Operations management* (Vol. 8). McGraw-Hill.

Syed, N. F., Shah, S. W., Trujillo-Rasua, R., & Doss, R. (2022). Traceability in supply chains: A cyber security analysis. *Computers & Security*, 112, 102536.

Tang, C. S. (2006). Perspectives in supply chain risk management. *International Journal of Production Economics*, 103(2), 451–488.

Tracey, M., & Vonderembse, M. A. (2000). Using information sharing to increase supply chain visibility. *Supply Chain Management: An International Journal*, 5(5), 243–254.

Trkman, P., McCormack, K., De Oliveira, M. P. V., & Ladeira, M. B. (2010). The impact of business analytics on supply chain performance. *Decision Support Systems*, 49(3), 318–327.

van Hoek, R. I. (2001). From reversed to forward logistics: The fifteenth fisher memorial lecture. *Journal of Marketing Management*, 17(5–6), 463–477.

Waters, D. (2003). *Logistics: An introduction to supply chain management.* Palgrave Macmillan.

Zhao, X., Xie, J., & Leung, S. C. (2014). On supply chain coordination for a deteriorating inventory item with price-and stock-dependent demand under consignment stock agreement. *International Journal of Production Economics*, 156, 1–10.

Zsidisin, G. A., & Ritchie, B. (2009). *Supply chain risk: A handbook of assessment, management, and performance.* Springer Science & Business Media.

7 Ranking of Barriers to Adoption of Blockchain Technology in Supply Chain

Akshat Kumar Rai and T. Bangar Raju

7.1 INTRODUCTION

Blockchain technology has been described as an information technology revolution. Its ability to provide a secure, transparent, and immutable ledger has found varied applications across numerous sectors. However, despite the apparent benefits, its integration into supply chain management has been met with significant resistance. The present introduction delves into the obstacles impeding the implementation of blockchain technology inside the supply chain, with particular attention to their causes and consequences. The supply chain industry is the backbone of the global economy and involves the transportation of goods from manufacturers to consumers by a chain of retailers, wholesalers, and logistics companies. Blockchain offers higher levels of traceability, lower loss of revenue due to counterfeit and gray markets, increased levels of regulatory compliance, and operational efficiencies. Although the proponents of technology championed the prospects, the adoption rate of blockchain within supply chains has not yet lived up to the initial hype. The supply chain is, in its most basic sense, complex and involves especially the manufacturers, suppliers, distributors, and retailers. All players in the chain must work and coordinate in a synchronized manner in order that there is a continuous availability in the chain of goods and services. What blockchain technology offers is a decentralized yet shared and fixed medium for these negotiations to occur in a manner that is hard for an individual party to manipulate. By so doing, this capability has the potential of eliminating fraud, enhancing materials' traceability and compliance, in addition to minimizing administrative expenses by up to 50 percent. But the migration from traditional system to a blockchain-based system is a complex process and has technical, organizational, and legal complexities.

First and foremost, there is more knowledge and experience to overcome when supply chains try to implement blockchain technology. Thanks to its fairly recent nature, there are still many things that must be discovered before the full potential of blockchain can be realized. Most firms require more talent knowledgeable in blockchain to help them assess how to deploy and orchestrate a blockchain system. This

 DOI: 10.1201/9781003477242-9

gap also reaches a number of relevant stakeholders that have to trust the technology and base their essential operations on it.

The visibility of supply chain management operations is very important. Blockchain opens up fully the visibility by creating a record of each record or transfer of goods on a distributed ledger. It enables all stakeholders, including manufacturers, retailers, wholesalers, and consumers track the status of a given product, creating credibility for the product. Second, there is a technical factor which is a barrier to successful implementation of blockchain technology; it includes the following: the present blockchain solutions sometimes require support to quickly process large transactions at an exponential rate, which is essential for extensive supply chains.

Further, while we have various types of blockchains that have to work together for more efficiency, many of the current blockchain systems are isolated and have little compatibility. In addition, another major factor is that there are regulatory risks involved in the sector. Because of the decentralized and intrinsic nature of the blockchain, new difficulties for the regulators in the spheres of compliance, as well as client privacy and asset security, appear. As it now stands—the legal requirements for blockchain technology adoption in many parts of the world remain either non-existent or immature, which does not allow companies to use an emerging technology that may not be fully compliant with the legal guidelines sometime in the future. One more barrier is the resistance of organizations to changes; to change is also essential. The integration of blockchain solutions in a supply chain motivates substantial shifts in the current processes and structures. Most organizations do not embrace such changes because they either disrupt their internal operations, the new technology is too expensive, or there are definite odds of NOT getting any returns. This resistance can be extreme, especially in large organizations that may already be set with very long launched information systems that are well grounded. One noteworthy factor is cost consideration; this factor cannot be ignored whatsoever. The first and recurring costs related to transitioning to or implementing blockchain systems may be tremendously costly. This comprises the cost of procuring the technology, the cost of incorporating this into the total system of the organization, and the subsequent cost of maintaining this technology. These expenses can be a challenge to many small and medium-sized firms, or any firm that may not immediately grasp or envision the benefits of blockchain technology. One of the problems with blockchain is that it is transparent. While there exist strengths for transparency to ease traceability or even fraud, restrictions may be encountered because of the secrecy of commercial information. This is the fine line that companies need to tread as they seek to capture blockchain's value without courting new risks.

However, cultural factors cannot be overlooked. Transitioning to a blockchain system requires one to rethink how they handle and process data in a decentralized way rather than a centralized way. The process of getting to this type of change will therefore require time, because stakeholders will only have to compromise to shift from their highly controlling and less collaborative ways of working. Lastly, since this technology is still relatively new, it is easier for the supply chain to identify one or two areas where it demonstrates success. An absence of blatant and visibly proven success thus works against the practical rationale for change and shift to blockchain technology.

Consumer interest in blockchain is increasing as blockchain can improve supply chain operations by reducing costs and increasing consumer trust. But this is where the challenges with the road toward full implementation come in. Challenges include the complexity of the technology, the problems of scale, and the amount of capital required to implement the solution. Furthermore, the need to maintain standardization of these blockchain solutions, issues of privacy, and regulation all contribute to the conservative approach taken toward integration of blockchains in supply chain applications. This is why the strategic approach to identify and confront these issues and obstacles based on their relevance and solubility level of difference that present a company is needed. That kind of approach is useful for grading and selecting concerns and creating detailed approaches. By systematically addressing those challenges, the companies can fully unlock the opportunities of blockchain, thus creating a new wave of supply chain disruption.

7.1.1 Overview of Blockchain Technology in Supply Chain

Blockchain technology is gradually taking its place as the newest innovation that supply chain management stands to gain, with the potential to provide value such as transparency, security, and efficiency improvements. With development in blockchain further and its expansion in different sector, these advantages are coming to revolutionize the global supply chain management significantly.

Increased transparency is one of the major advantages associated with the use of blockchain. The system uses an immutable record to capture all activities and the flow of products, making it easier to track them. For example, Walmart uses blockchain technology in the food chain business to quickly identify the source of a food item in cases of contamination while promoting transparency from the farm to the store.

It also disallows fraud and counterfeiting in the supply chain because of the security that blocks any change on the records, thereby enhancing the efficiency of the global supply chain. An example is De Beers; it utilizes blockchain technology to regulate the shipment of diamonds from the mines to the retailers, and the diamonds are conflict-free, natural. This, in turn, helps prevent the production and trade of counterfeit and conflict diamonds. This further enhances compliance and quality control by implementing these processes through smart contracts on blockchain. For instance, in the pharmaceutical industry, blockchain helps to allow compliance with regulations such as the Drug Supply Chain Security Act in the United States, where the user tracks drugs through production, distribution, and delivery to ensure they are not counterfeit.

It also makes work easier and is cost-effective since it eliminates or rather reduces the work which requires manual input and paperwork. Maersk and IBM's Trade Lens Blockchain is used in the management of maritime shipping transactional paperwork, thus minimizing paperwork, enhancing the processing time frame, besides tapping costs. Asset identification, location, status, and condition can be tracked with ease through the use of blockchain, making the tracking in real time possible. Corporate, such as Everledger in particular, uses blockchain to keep records of the origin of luxury products as a means of giving them credibility.

Although, there are certain advantages of using the blockchain-based implementation in supply chain, a number of challenges has been identified, such as scalability issue, interoperability with existing systems, and regulatory issues. However, it is predicted that the used cases of blockchain might be improved more in the future by integrating with IoT and AI solutions focusing on SCM.

Blockchain implementation is gradually seeping into different sectors, including retailing, manufacturing, and healthcare, hence showing more acceptance in supply chain use. This is more so considering that blockchain is today increasingly and significantly playing the role of molding the future of supply chain business. The study is expected to find out that technological and financial factors are challenging but organizational and cultural factors are key to hindering the blockchain implementation within supply chain organizations. The results are intended to help fuel further debate on blockchain by demonstrating how managerial and strategic approaches should also accompany technological progressions.

7.2 LITERATURE REVIEW

The application of blockchain is now being highlighted for its potential for true change in supply chain and mainly for providing transparency, efficiency, along with the traceability of the chain. There are, however, significant challenges of technology, organization, culture, and strategy when applying blockchain with existing supply chain processes, despite the improvements that blockchain offers. In the following literature, this chapter derives a comprehensive review of a wide array of research to shed light on these challenges. It provides a conceptual framework upon which solutions can be fashioned out to address these challenges and optimize blockchain in supply chain setting.

Interest in blockchain in supply chain management comes from its capacity to transform critical aspects of supply chains, visibility, efficiency, and mutuality of trust. This review is based on recent studies that employ methodological approaches which includes Decision-Making Trial and Evaluation Laboratory (DEMATEL), Analytic Hierarchy Process (AHP), and fuzzy logic. These methodologies assist in demystifying the relations and interactions between the barriers to blockchain adoption, thus bringing out the lens that practicality has.

Due to its powerful and secure processing of a high number of transactions and data, blockchain technology is intricate and difficult to interface with current IT systems. Khan et al. (2023a) notes that many supply chain systems require assistance in the adoption and integration with blockchain and leading to massive enhancements or even the overhaul of current systems. This is politically undesirable, economically costly, and organizationally disruptive. Another vital technological hindrance highlighted is technology incompatibility, where blockchain is not compatible with other systems of blockchain and other existing technologies. Since blockchain gives organizations no way to communicate or exchange data with anything else that does not adhere to the exact same standards, organizations have to be uniform in order to gain these benefits. In addition, as stated by Yadav and Singh (2020), there is an importance of data decentralization and safety into the blockchain. It can only be effectively addressed through well laid down standard operating procedures that will in the process define security as well as compatibility of the systems.

In the case of organizations, the most influential hindrance to change is known as resistance to change. When adopting blockchain solutions, major overhauls of business processes and organizational structures are always required, and staff and management may not always be ready for them. Saberi et al. (2019) mention that trust in relation to the issues of its management hampers its implementation, accompanied by resistance, and ignorance of the opportunities offered by blockchain. The requirement for highly qualified staff increases because not everyone is aware of blockchain and supply chain management; thus, the companies face a problem in finding qualified employees. Such a vanilla skill gap can hold hostage projects, create project overruns, or cause project failure, which means the organization needs to invest in training and development to establish those skills.

Also, although blockchain is believed to be more secure, this transition involves such key security and privacy concerns as foundational privacy and security concerns that must be addressed. From the leadership point of view, Bag et al. (2020) identified absence of management vision as one of the main challenges. Leadership is highly effective in a way that supports the improvement of innovation and technology in organizations. We wish to assert that blockchain projects can only achieve the critical mass and financing required for their implementation with the support of leaders. This lack of vision can also result in a strategic mismatch between the organization's goals and objectives, and the underlying blockchain technology's capabilities in this technological age, leading to minimized innovation. It emerged that organizational culture plays a major role as a factor that affects the decision to adopt new technology. One of the main challenges is the culture that is not friendly to change or risk because it may either slow the adoption of blockchain or totally stop it. This often causes an opposition to change because they are not familiar with the new technology or are unfamiliar with change in the established procedures and techniques. From the financial aspect, the key issue is the cost of implementing blockchain technology. Implementing blockchain is also very expensive since organizations must spend on technology and personnel to improve the existing systems or purchase new hardware, and seek or develop blockchain software.

Ongoing costs related to the maintenance and management of blockchain systems can also be significant. Sahoo and Thankur (2023) highlight the financial barriers for Indian MSMEs, where limited budgets may restrict the ability to adopt new technologies requiring substantial upfront investment. The uncertainty of the return on investment (ROI) from blockchain projects is another financial barrier, as the long-term benefits of blockchain, while promising, are still to be fully proven across all industries. By design, blockchain is a distributed platform that creates the need for various players, such as suppliers, customers, regulatory agency, and technology solution providers, to be engaged proactively. As it will be seen, the application of blockchain requires the consensus and collaboration of these players to deliver a technology that can address multiple requirements and standards set within the industry and legislation. In their work, Bae et al. (2024) delineate cooperation and non-cooperation in blockchain investment, and noted that cooperation brings out the best when it comes to adoption and performance. Another challenge due to the diversity of stakeholders is that the stakeholders have different motivations and views due to the different roles they play, hence making the adoption process difficult. Sustaining these objectives calls for an effective corporate governance structure responsive to the stakeholder concerns of the diverse shareholders.

To analyze the impact of this technology in supply chain management (SCM), Treiblmaier (2018) drew on economic theories. Future empirical research can build on this study by suggesting research questions and a framework to follow. By applying the PEEST framework—which posits that political, economic, environmental, social, and technological factors are interconnected—Wang et al. (2023) adopt a more multistakeholder approach to identify the barriers to blockchain adoption, identifying 27 specific barriers. Their findings also remind leaders involved in the adoption process to consider various factors to effectively address these challenges. Utilizing the research method developed by Al Amin et al. (2023), by Integrated Fuzzy TOPSIS–ISM model helps tackle blockchain challenges in sustainable supply chains by comparing the significance of different options, revealing the threshold effects of supply chain barriers, and discussing potential strategic directions based on the categorization of sub-barriers.

In the study of Surucu-Balci and his team, maritime supply chain is the subject of the investigation, where the role of blockchain technology and its impacts are discussed. Their research applies a systematic literature review to outline the DIS capabilities improved by blockchain and the main challenges and facilitators of its use. Challenges discussed include concerns with scalability, infrastructural characteristics, and the nature of the industry, which may slow down the use of blockchain. The same study also shows how blockchain technology can cause a shift in maritime operations and management by enhancing its transparency, traceability, and efficiency; the barriers highlighted should therefore be addressed with specific strategies.

Hackius and Petersen see one intriguing view using an online survey conducted among the professionals related to logistics (Hackius & Petersen, 2017). This is what their research focuses on—an analysis of how blockchain is viewed with regard to logistics and supply chains. In the overwhelming mood of clients, it showcases a relatively positive attitude but differentiated level of passion corresponding to the hierarchical level, industry sector, and experience in blockchain. This survey raises the possible advantages of blockchain solutions, including increased clarity and speed, and identifies the main impediments, including the lack of a clear legal framework and the necessity to involve multiple stakeholders. What the research points out is that, although there is awareness of what blockchain entails, practical issues as well as industry preparedness have important roles in defining its implementation path.

In their study about the impact of blockchain technology on sustainability, Güner, Son-Turan, and Ayan (2022) employ both the quantitative and qualitative approaches. Their work shows how the use of blockchain can solve the problems typical for the industry such as the lack of supply chain transparency, poor labor conditions, or products origin. Some of the advantages that have been discovered from the research include the fact that blockchain is likely to bring about positive impacts in these sectors, but implementing it under the current circumstances has some limitations, which have been imposed by the volatility of the industries in question.

Some of the critical barriers enumerated are as follows: Technological issues, that is, the ability to interface blockchain with existing systems, lack of structure and standards concerning blockchain that would enable interoperability, and an issue to content and form that addresses issues of decentralizing and securing data with the use of blockchain. Popular organizational barriers include resistance to change, trust management, lack of necessary skills, privacy/security concerns, absence of engaging leadership, and cultures rigid against innovation. Implementation costs and maintenance costs, as well as the unpredictability of the overall ROI, are major financial issues related to the establishment of information systems. Moreover, due to decentralized characteristics of the blockchain, it requires engaging and incorporating multiple stakeholders to attain consensus regarding the implementation of multiple solutions, while maintaining comprehensive, yet fair governance mechanisms to address varied interests.

Blockchain has the potential to revolutionize supply chains but there are different organizational, cultural, technological, and strategical challenges which are preventing it from doing so. When these barriers are properly understood and evaluated with references to actuality of supply chain cases, the managers and policymakers can understand how to utilize the blockchain technology for improving the availability, credibility, authenticity as well as efficiency of the supply chain networks.

7.3 RESEARCH METHODOLOGY

The research design and methodology used in the proposed study adopt the combination of quantitative and qualitative data collection methods to fully assess barriers in SCM.

7.3.1 Data Collection

The information for this research was gathered through both an extensive analysis of the existing literature and an online survey, questionnaire including questions posed to experts in the field. This method, therefore, seeks to identify the factors which are likely to lead to resistance to the use of blockchain technology in the supply chain. The identified barrier is then systematically subjected to Fuzzy Analytical Hierarchy Process (AHP) and DEMATEL method.

7.3.2 Analysis Method

Fuzzy AHP: To cope with the inherent uncertain and vague character of the impact assessment of the different barriers.

DEMATEL: To explain the interconnections between the established barriers.

7.3.2.1 Fuzzy Analytical Hierarchy Process (AHP) Method

Analytic Hierarchy Process (AHP) is a decision support technique that originated to finish problems by fragmenting, categorizing, and prioritizing them in terms of hierarchical redress. The Fuzzy Analytical Hierarchy Process (AHP) is an improvement of the conventional AHP; a systematic approach to decision-making. An expert known as Thomas L. Saaty initiated it in the 1970s. While AHP employs numerical differences directly to compare the comparative significance of factors, Fuzzy AHP introduces fuzzy arithmetic into dealing with fuzziness and imprecision present in the decision-making process.

Breakdown of how the fuzzy analytical hierarchy process (AHP) method works:

1. **Problem Structuring**: Similar to AHP, Fuzzy AHP begins by the formulation of the decision problem and the construction of a hierarchical model made up of the goal, criteria (and sub-criteria), and the alternatives.
2. **Pairwise Comparisons with Fuzzy Logic**: While the AHP scale employs definite figures, Fuzzy AHP adopts what are known as linguistic variables, which include slightly important, important, and very important, which are translated into fuzzy numbers. These fuzzy numbers are used to develop pairwise comparison matrices where an element is the relative measure of the extent that one element is more important than another with respect to the criterion or sub-criterion.
3. **Synthesizing the Comparisons**: The fuzzy pairwise comparison matrices are applied to determine the relative weights of the criteria and the sub-criteria. This entails different approaches that include the fuzzy arithmetic approach, fuzzy geometric mean, and the centroid technique.
4. **Consistency Check**: The reliability of the judgments provided in the pairwise comparison is checked. If the consistency is not good then the above said process will have to be altered.
5. **Aggregating the Priorities**: The relative weights are summed up on the levels of the hierarchy to identify the ranking of the alternatives in the general respect.

7.3.2.2 Decision-Making Trial and Evaluation Laboratory (DEMATEL) Method

The Decision-Making Trial and Evaluation Laboratory (DEMATEL) is an integrated approach that sustains the analysis and modeling of a complicated problem, especially the one compounds with human factors and interactions. This approach is used in management science, operations research, and systems engineering.

Critical aspects of the DEMATEL method include:

1. **Identifying Factors**: The first step just requires finding the components or parameters of the problem or system under consideration. These are factors that concern areas which are either impacted on by the problem or have an impact on the problem.
2. **Developing an Interaction Matrix**: Each of the factors is studied as to its impact on all the other factors. This is usually done by way of opinion of some expert or by computation. Such a result in the form of the interaction matrix is achieved where each cell is an indication of the extent to which one factor affects the other.
3. **Normalizing the Matrix**: Later on, the interaction matrix is normalized so that the sum of all the influences will always be less than 1. This is needed for the sake of mathematical and analytical simplicity as well as for the expediency of the data handling.
4. **Calculating the Impact Scores**: From the resulting normalized matrix, impact scores for each factor can be obtained. These scores tell about the

extent of impact that the total factors of the system have. Factors are categorized into two groups: The first type, consisting of factors that impact those mentioned above, known as the cause or dispatching factors, while the second type consists of factors impacted by the other factors more, known as the effect or receiving factors.

5. **Creating a Causal Diagram**: The last step entails producing a causal chart from impact scores that were obtained from the data collected. This diagram provides information about the relations between factors and which are essential or minor. It facilitates the analysis of structure and activity related to the occurrence of the system or problem.

7.4 RESEARCH PROBLEM AND OBJECTIVE

When adopting blockchain in supply chains, several challenges arise. Firstly, people still did not understand the innovation and the benefits that came with it, which escalated to doubt and a slow adoption. Secondly, the integration of blockchain in existing supply chain technologies is a technical complexity, which brings high costs and implementation issues. In addition, constantly evolving regulatory fronts regarding blockchain increase business risks; regarding data privacy and cross-border transactions, for instance, they hold back. There are issues related to efficiency and the impact that blockchain will have on the environment in large scale, especially on the energy consumption of the network. The following are the various challenges that affect the implementation of blockchain in supply chains, thereby preventing its full uptake despite the positive impact it can particularly have on the supply chain. Education, progressive technology, increased transparency in regulation laws and policies, and the employment of environmentally sound strategies to overcome such challenges are the ways to make green procurement more acceptable.

1. **Lack of Understanding of Blockchain Technology and Its Potential Benefits**:
   - **Problem**: The general public, business entities in particular, requires better awareness of what blockchain is and its uses. This technology, upon which currencies such as Bitcoin are constructed, is usually viewed solely through the prism of payments.
   - **Impact on Adoption**: Where there is no knowledge of blockchain, it may create skepticism or resistance when it comes to investing in things that deal with blockchain. Having expounded on the possible performance improvement, cost reduction, intelligibility, and security solutions realized through blockchain, the stakeholders can look into blockchain for supply chain.
2. **Technical Challenges in Integrating Blockchain with Existing Supply Chain Systems**:
   - **Problem**: Blockchain technology stands out as a different system to traditional database systems. A major technological problem is how

to integrate it into architecture of the current supply chain frameworks that may require restructuring or development to interface with blockchains.
   - **Impact on Adoption**: These integration challenges may result in higher cost, longer time to implement the solution, and technical issues which may discourage businesses to consider blockchain solutions. Integration with other systems in the organization is important to ensure an easy running of the system and its acceptance by the users.
3. **Regulatory and Compliance Issues**:
   - **Problem**: In general, the choice of the legal framework of blockchain is still in progress. Other areas of concern also include protection of data, making cross-border payments, and legal position of smart contracts.
   - **Impact on Adoption**: One of the reasons companies are reluctant to embrace certain technologies is that the regulations governing their use may be up in the air. This is even more so an issue as legal systems differ from one jurisdiction to another and by legal precedent, no organization can function beyond the legal sphere.
4. **Scalability Concerns and the Environmental Impact of Blockchain Technologies**:
   - **Problem**: While some of the blockchain shared networks and in particular those based on proof-of-work similar to Bitcoins are known to be power-consuming. Besides, as the number of transactions grows, many technologies implemented in the blockchain context require assistance to operate effectively.
   - **Impact on Adoption**: Another interesting implication of blockchain technologies is that environmental organizations can consider it as an issue. Maintaining scalability issues also means, as the supply chain grows, it may require technical support to manage the increased load, leading to scalability issues affecting the efficiency of the system.

Objective:

- Identify and rank barriers toward blockchain adoption in the supply chain.

Before delving into the analysis, it is essential to outline the core objectives of evaluating the adoption barriers for blockchain in SCM. By reviewing relevant literature, obstacles impeding blockchain's integration were identified (Table 7.1).

Table 7.2 illustrates the barriers and its definition.

7.5 RESEARCH GAP

In the existing literature on the adoption of blockchain technology in supply chains, a notable research gap is apparent, highlighting the need for further exploration. This gap specifically includes the following areas:

TABLE 7.1
List of Main Criteria

Main Criterion	Description
Technological barriers (TB)	These encompass the technical and operational obstacles to blockchain adoption within supply chains.
Regulatory and legal barriers (RLB)	These challenges arise from the uncertain and complex legal and regulatory environment surrounding blockchain technology.
Organizational and cultural barriers (OCB)	These barriers refer to internal and external resistance to adopting new technologies like blockchain within organizations and across the supply chain.
Economical and financial barriers EFB)	Economic and financial barriers concern the costs and benefits of implementing blockchain technology in supply chains.

1. **Technology-Organization-Environment Framework**: While studies have used this framework to examine the barriers to blockchain adoption in sustainable SCM, there might be a gap in exploring how different framework components interact and influence blockchain adoption in various contexts.
2. **Interrelationships Among Barriers**: Some research has unraveled the interrelationships among barriers in SCM, particularly highlighting market-based risks, financial issues, inadequate knowledge, and contractual risk. Further research could explore how these barriers are interrelated in different supply chains or industries and how they impact adoption.
3. **Specific Barriers in Different Supply Chain Types**: Different types of supply chains, such as manufacturing, green supply chains, or construction, face unique barriers. For instance, transaction-level uncertainties, usage in the underground economy, managerial commitment, scalability challenges, and privacy risks are significant in manufacturing supply chains. In contrast, green supply chains face barriers like a lack of management vision and cultural differences among partners. Examining these barriers in more detail and comparing them across various supply chain types could be a fruitful area of research.
4. **Business Awareness and Familiarity**: A significant barrier is the lack of business awareness and familiarity with what blockchain can deliver for future supply chains. Research could focus on developing frameworks or models that enhance business leaders' and decision-makers' understanding and familiarity with blockchain technology.
5. **Impact of Emerging Technologies**: Research could explore how integrating blockchain with other emerging technologies like AI, IoT, and Big Data analytics influences the barriers to adoption and the overall effectiveness of blockchain in supply chains.

While extensive research exists on the barriers to blockchain adoption in supply chains, focusing on the interplay of these barriers in different contexts, comparative studies, business awareness, and integration with other technologies can provide new insights and uncover significant research gaps in this field.

TABLE 7.2
List of Barriers

Criterion	Name of Barrier	Description
S1	High implementation costs	Significant upfront and ongoing operational costs for setup and maintenance.
S2	Scalability issues	Difficulty in handling large volumes of transactions quickly and efficiently.
S3	Interoperability with existing systems	Challenges in integrating blockchain with existing supply chain systems.
S4	Technical complexity	Complexity of blockchain concepts and technologies.
S5	Lack of technical expertise	Shortage of skilled professionals in blockchain and supply chain management.
S6	Energy consumption	High energy requirements for certain blockchain implementations.
S7	Data standardization	Need for standardized data formats and protocols.
S8	Technology maturity	Blockchain may not yet meet all operational needs.
S9	Data veracity	Ensuring accuracy and truthfulness of data entered into the blockchain.
S10	Regulatory uncertainty	Lack of clear, consistent legal and regulatory frameworks for blockchain.
S11	Legal and compliance issues	Complex legal landscape related to blockchain transactions and data storage.
S12	Intellectual property concerns	Sharing and protecting IP on blockchain platforms.
S13	Compatibility with regulatory standards	Ensuring compliance with existing and future regulations.
S14	Cross-border transactions	Navigating legal and operational challenges of international transactions.
S15	Resistance to change	Organizational reluctance toward adopting new technologies.
S16	Lack of awareness and understanding	General unawareness of blockchain's benefits.
S17	Integration with suppliers and partners	Difficulty in achieving adoption among all supply chain participants.
S18	Trust issues	Building trust to share data on a blockchain.
S19	Ownership and control concerns	Questions about who owns and controls the data on the blockchain can arise, especially when multiple parties are involved in the supply chain.
S20	Loss of intermediaries	Resistance from those whose roles may be threatened by blockchain.
S21	Lack of proven ROI	Uncertainties about the return on investment from blockchain.
S22	Fragmentation of blockchain solutions	Competing platforms hindering widespread adoption.
S23	Risk of obsolescence	Rapid technological advancements potentially render solutions obsolete.
S24	Environmental concerns	Addressing environmental impacts of blockchain operations.
S25	Monitoring and reporting	Developing systems for monitoring and reporting within blockchain frameworks.

7.6 DATA ANALYSIS

1. **Define the Problem and Build Hierarchy**: As per Figure 7.1 it defines the hierarchical structure of overall goal and criteria which is divided into overall goal, main criteria & sub criteria
2. **Select Fuzzy Numbers**:

TABLE 7.3
TFNs of Linguistics Comparison Matrix

Linguistic Variable	Assigned TNF
Not significant (NS)	(0.5, 1, 1.5)
Slightly significant (SS)	(1.5, 2, 2.5)
Moderately significant (MS)	(2.5, 3, 3.5)
Very significant (VS)	(3.5, 4, 4.5)
Extremely significant (ES)	(4.5, 5, 5.5)

3. **Application of Fuzzy AHP**:

Table 7.3 represents a comparison table using Triangular Fuzzy Numbers (TFNs) represents the pairwise comparisons between criteria or alternatives, where each cell in the matrix contains a TFN that reflects the decision-maker's fuzzy judgment of the relative importance of the two items being compared.

This stage entails applying the fuzzy AHP technique to rank the identified barriers to blockchain implementation. The seven experiments analyze the significant criteria and sub-criteria using triangular fuzzy numbers (TFNs) from the table above. Tables 7.4–7.8 show the fuzzy pairwise comparison and decision matrix of the primary and sub-criteria barriers and their estimated weight and ranking.

According to the fuzzy synthetic extent values for the four main criteria are calculated and presented in Table 7.9.

Since the method for reaming criteria is the same as the method for calculating the weight of other criteria, it is not covered in this description. Table 7.9 displays the weights of the other criteria as well as the outcomes of pairwise comparisons of the criterion and sub-criteria (Tables 7.10–7.12).

7.7 RESULTS AND DISCUSSION

This research has pinpointed 25 obstacles to the adoption of blockchain, identified through an extensive review of literature and expert insights. These challenges are divided into four principal categories. Although it can be challenging to determine how much of an impact these obstacles will have on supply chains when adopting blockchain, there are a lot of advantages. These advantages include improved traceability, transparent transactions, and the facilitation of secure, real-time tracking of goods in the supply chain.

This study used a multi-criteria decision-making approach, specifically the analytical hierarchy process in a fuzzy environment (fuzzy AHP), to identify the most

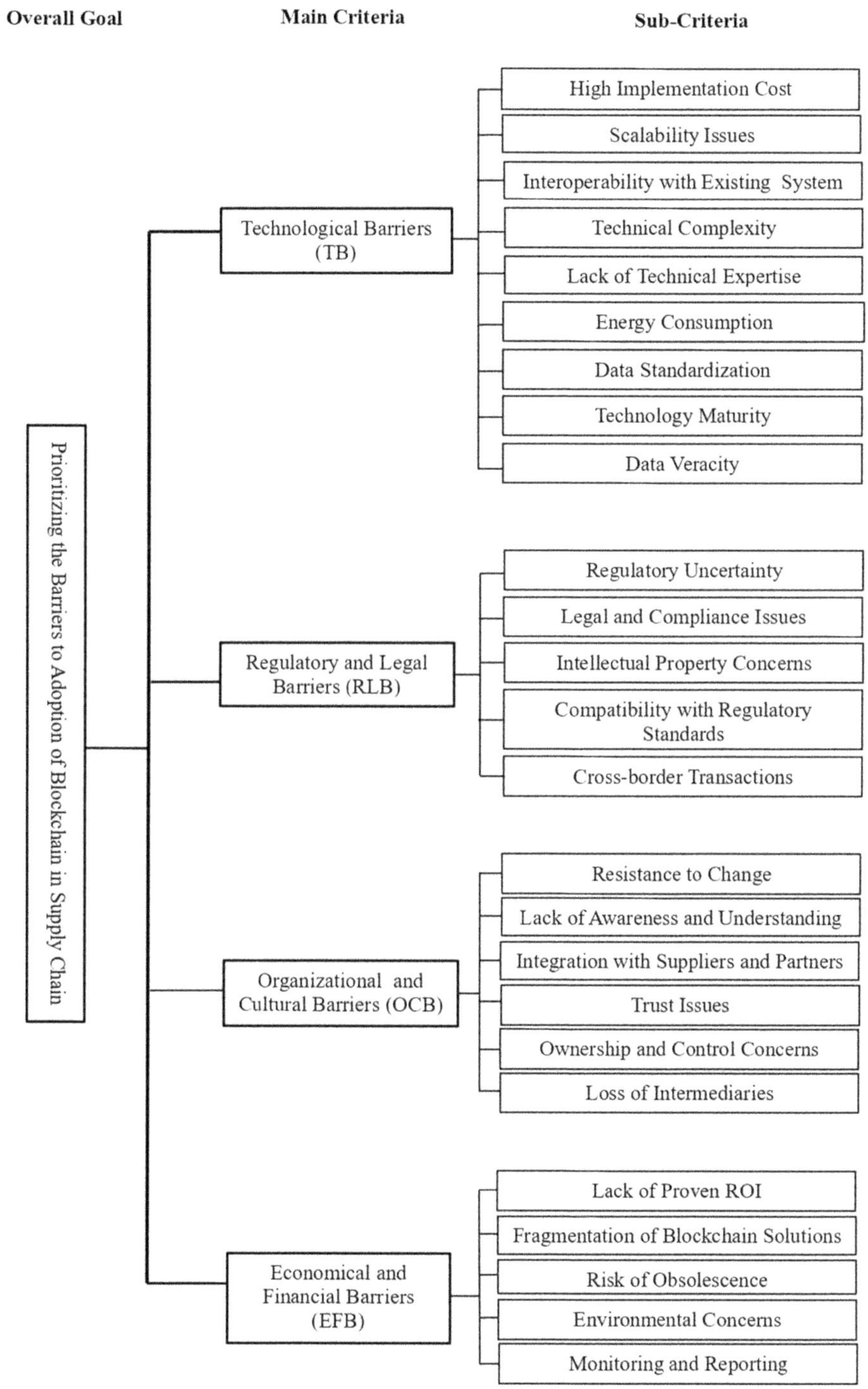

FIGURE 7.1 Hierarchical structure of overall goal and criteria by author.

Source: Authors own creation.

TABLE 7.4
The Main Criteria Computed Fuzzy Evaluation and the Decision Matrix

	TB	RLB	OCB	EFB	Weight	Rank
TB	(0.5, 1, 1.5)	(2.5, 3, 3.5)	(2.5, 3, 3.5)	(2.5, 3, 3.5)	0.42	1
RLB	(0.4, 0.3, 0.3)	(0.5, 1, 1.5)	(3.5, 4, 4.5)	(2.5, 3, 3.5)	0.25	2
OCB	(0.4, 0.3, 0.3)	(3.5, 4, 4.5)	(0.5, 1, 1.5)	(2.5, 3, 3.5)	0.23	3
EFB	(0.4, 0.3, 0.3)	(0.4, 0.3, 0.3)	(0.4, 0.3, 0.3)	(0.5, 1, 1.5)	0.08	4

important barriers to blockchain deployment in the supply chain. This will help decision-makers prioritize the barriers that have been identified. Furthermore, the research underscores the importance of addressing these barriers to facilitate the successful integration of blockchain technology into supply chain operations. Through overcoming these obstacles, businesses can enhance their supply chain operations, augment their competitiveness, and effectively leverage the benefits provided by blockchain technology.

Table 7.4 presents the primary criteria (barriers) outcomes, where a higher weight value indicates more significant barriers. According to the table, the ranking order for the primary barriers to blockchain implementation is as follows: Technological Barriers are the most critical, followed by Regulatory and Legal Barriers, Organizational and Cultural Barriers, and Economic and Financial Barriers. This ranking suggests that barriers related to technology pose the greatest challenge, followed by issues in regulation and legal frameworks, organizational culture, and financial constraints.

The fuzzy AHP analysis results for the organizational barriers sub-criterion, detailed in Table 7.5, reveal that high implementation costs are the most significant sub-criterion impeding blockchain adoption in the supply chain, receiving the highest ranking. This raises some questions, such as the deployment of blockchain requires massive capital investment, thus proving to be difficult, especially for SMEs. Firms need to consequently ensure that they critically evaluate the advantages and disadvantages of adopting blockchain technology. The evaluation of the barriers reveals that scalability challenges are the second biggest concern, and yet, a blockchain that can handle the expansion of supply chain needs to be deployed. This underlines the need to fund the growth of blockchain by more efficient structures capable of processing more transactions and competing demands. Thirdly, it refers to the flow of information, where it is pointed out that for blockchain to be effective in an organization, it has to mesh well with the other existing systems. This goes ahead to highlight the need to ensure that all blockchain solutions created align with the current technological architectures. Technical complexity appears fourth, and this illustrates inherent characteristics of blockchain as a hindrance. This puts more emphasis on the fact that technology should be unbandaged in a way that makes the users and other stakeholders get a clear understanding, hence embrace the technology in question. The fifth prior rated barrier, which presupposes a lack of more technical analysis, also points to the scarcity of qualified specialists who could have

TABLE 7.5
Calculated Pairwise Comparison and Decision Matrix of Sub-criteria (Technological Barriers)

	S1	S2	S3	S4	S5	S6	S7	S8	S9	Weight	Rank
S1	(0.5, 1, 1.5)	(2.5, 3, 3.5)	(2.5, 3, 3.5)	(2.5, 3, 3.5)	(3.5, 4, 4.5)	(2.5, 3, 3.5)	(2.5, 3, 3.5)	(2.5, 3, 3.5)	(3.5, 4, 4.5)	0.26	1
S2	(0.4, 0.33, 0.29)	(0.5, 1, 1.5)	(2.5, 3, 3.5)	(2.5, 3, 3.5)	(3.5, 4, 4.5)	(2.5, 3, 3.5)	(2.5, 3, 3.5)	(2.5, 3, 3.5)	(2.5, 3, 3.5)	0.19	2
S3	(0.4, 0.33, 0.29)	(0.4, 0.33, 0.29)	(0.5, 1, 1.5)	(2.5, 3, 3.5)	(2.5, 3, 3.5)	(3.5, 4, 4.5)	(2.5, 3, 3.5)	(2.5, 3, 3.5)	(2.5, 3, 3.5)	0.15	3
S4	(0.4, 0.33, 0.29)	(0.4, 0.33, 0.29)	(0.4, 0.33, 0.29)	(0.5, 1, 1.5)	(2.5, 3, 3.5)	(2.5, 3, 3.5)	(2.5, 3, 3.5)	(3.5, 4, 4.5)	(3.5, 4, 4.5)	0.12	4
S5	(0.29, 0.25, 0.22)	(0.29, 0.25, 0.22)	(0.4, 0.33, 0.29)	(0.4, 0.33, 0.29)	(0.5, 1, 1.5)	(2.5, 3, 3.5)	(2.5, 3, 3.5)	(2.5, 3, 3.5)	(2.5, 3, 3.5)	0.08	5
S6	(0.4, 0.33, 0.29)	(0.4, 0.33, 0.29)	(0.29, 0.25, 0.22)	(0.4, 0.33, 0.67)	(0.4, 0.33, 0.29)	(0.5, 1, 1.5)	(1.5, 2, 2.5)	(2.5, 3, 3.5)	(2.5, 3, 3.5)	0.07	6
S7	(0.4, 0.33, 0.29)	(0.4, 0.33, 0.29)	(0.4, 0.33, 0.29)	(0.4, 0.33, 0.29)	(0.4, 0.33, 0.29)	(0.67, 0.5, 0.4)	(0.5, 1, 1.5)	(2.5, 3, 3.5)	(2.5, 3, 3.5)	0.06	7
S8	(0.29, 0.25, 0.22)	(0.4, 0.33, 0.29)	(0.4, 0.33, 0.29)	(0.29, 0.25, 0.22)	(0.4, 0.33, 0.29)	(0.4, 0.33, 0.29)	(0.4, 0.33, 0.29)	(0.5, 1, 1.5)	(2.5, 3, 3.5)	0.04	8
S9	(0.29, 0.25, 0.22)	(0.4, 0.33, 0.29)	(0.4, 0.33, 0.29)	(0.29, 0.25, 0.22)	(0.4, 0.33, 0.29)	(0.4, 0.33, 0.29)	(0.4, 0.33, 0.29)	(0.4, 0.33, 0.29)	(0.5, 1, 1.5)	0.03	9

TABLE 7.6
Calculated Pairwise Comparison and Decision Matrix of Sub-criteria (Regulatory & Legal Barriers)

	S10	S11	S12	S13	S14	Weight	Rank
S10	(0.5, 1, 1, 5)	(2.5, 3, 3.5)	(2.5, 3, 3.5)	(2.5, 3, 3.5)	(2.5, 3, 3.5)	0.40	1
S11	(0.4, 0.33, 0.29)	(0.5, 1, 1.5)	(3.5, 4, 4.5)	(3.5, 4, 4.5)	(2.5, 3, 3.5)	0.28	2
S12	(0.4, 0.33, 0.29)	(0.29, 0.25, 0.22)	(0.5, 1, 1.5)	(2.5, 3, 3.5)	(2.5, 3, 3.5)	0.15	3
S13	(0.4, 0.33, 0.29)	(0.29, 0.25, 0.22)	(0.4, 0.33, 0.29)	(0.5, 1, 1.5)	(2.5, 3, 3.5)	0.10	4
S14	(0.4, 0.33, 0.29)	(0.4, 0.33, 0.29)	(0.4, 0.33, 0.29)	(0.4, 0.33, 0.29)	(0.5, 1, 1.5)	0.07	5

TABLE 7.7
Calculated Pairwise Comparison and Decision Matrix of Sub-criteria (Organizational & Cultural Barriers)

	S15	S16	S17	S18	S19	S20	Weight	Rank
S15	(0.5, 1, 1.5)	(3.5, 4, 4.5)	(3.5, 4, 4.5)	(3.5, 4, 4.5)	(3.5, 4, 4.5)	(3.5, 4, 4.5)	0.42	1
S16	(0.3, 0.3, 0.2)	(0.5, 1, 1.5)	(2.5, 3, 3.5)	(3.5, 4, 4.5)	(3.5, 4, 4.5)	(2.5, 3, 3.5)	0.24	2
S17	(0.3, 0.3, 0.2)	(0.4, 0.33, 0.29)	(0.5, 1, 1.5)	(3.5, 4, 4.5)	(3.5, 4, 4.5)	(3.5, 4, 4.5)	0.17	3
S18	(0.3, 0.3, 0.2)	(0.29, 0.25, 0.22)	(0.4, 0.33, 0.29)	(0.5, 1, 1.5)	(3.5, 4, 4.5)	(3.5, 4, 4.5)	0.11	4
S29	(0.3, 0.3, 0.2)	(0.29, 0.25, 0.22)	(0.4, 0.33, 0.29)	(0.4, 0.33, 0.29)	(0.5, 1, 1.5)	(2.5, 3, 3.5)	0.07	5
S20	(0.3, 0.3, 0.2)	(0.4, 0.33, 0.29)	(0.4, 0.33, 0.29)	(0.4, 0.33, 0.29)	(0.29, 0.25, 0.22)	(0.5, 1, 1.5)	0.05	6

adequate and profound knowledge about the blockchain systems as an issue. This underlines why it is important to invest in training and development or hire a third-party service to help the company implement blockchain effectively. The sixth threat—the energy consumption—is an issue of the environmental impact that blockchain may pose although in systems that require huge processing power. This necessitates the need to embrace green and efficient blockchain solutions. The last one is data standardization, where an adequate standard is necessary for the convertibility and interoperability of the blockchain systems in the supply chain. It makes it crucial to develop benchmarks that are acceptable to everyone in the industry. Eighth, technology maturity, which indicates worries of immaturity of blockchain technology and the risk associated with its future sustainability. This is, in a nutshell, an

TABLE 7.8
Calculated Pairwise Comparison and Decision Matrix of Sub-criteria (Economical & Financial Barriers)

	S21	S22	S23	S4	S5	Weight	Rank
S21	(0.5, 1, 1.5)	(3.5, 4, 4.5)	(2.5, 3, 3.5)	(2.5, 3, 3.5)	(2.5, 3, 3.5)	0.41	1
S22	(0.3, 0.3, 0.2)	(0.5, 1, 1.5)	(3.5, 4, 4.5)	(2.5, 3, 3.5)	(2.5, 3, 3.5)	0.25	2
S23	(0.4, 0.33, 0.29)	(0.3, 0.3, 0.2)	(0.5, 1, 1.5)	(3.5, 4, 4.5)	(2.5, 3, 3.5)	0.16	3
S24	(0.4, 0.33, 0.29)	(0.4, 0.33, 0.29)	(0.3, 0.3, 0.2)	(0.5, 1, 1.5)	(2.5, 3, 3.5)	0.10	4
S25	(0.4, 0.33, 0.29)	(0.4, 0.33, 0.29)	(0.4, 0.33, 0.29)	(0.4, 0.33, 0.29)	(0.5, 1, 1.5)	0.07	5

TABLE 7.9
Fuzzy Synthetic Extent Values of Criteria

Criteria		Calculation		Results
TB	(1.67, 2.28, 2.83)	x	(1/4.40, 1/5.55, 1/6.49)	(0.379, 0.410, 0.436)
RLB	(1.15, 1.41, 1.61)	x	(1/4.40, 1/5.55, 1/6.49)	(0.261, 0.254, 0.248)
OCB	(1.15, 1.41, 1.61)	x	(1/4.40, 1/5.55, 1/6.49)	(0.261, 0.254, 0.248)
EFB	(0.42, 0.44, 0.43)	x	(1/4.40, 1/5.55, 1/6.49)	(0.095, 0.079, 0.66)

indication of the need for firms to look at technology progress and maturity more closely before leaping into this type of adoption. Hence, the ninth threat of effective blockchain adoption is the issue of data credibility that refers to accuracy of data stored in the blockchain systems. This goes a long way in highlighting mechanisms for integrity checks of the data before it is recorded permanently on the blockchain, to enhance the quality of blockchain systems.

Table 7.6 outlines the analysis in this research also produced the ranking of the sub-criteria under the category Regulatory and Legal barriers, and the study found that the limited regulatory frameworks received the highest score. This means that the lack of an extensive regulatory framework for blockchain is a key issue that companies must find ways to overcome for proper adoption. This reemphasizes the need for the governments to find ways of passing policies that do not only affect the companies that want to use blockchain but also policies that are friendly and easily understandable. The second major one is legal and compliance, which includes the problem of the existence of lack of a legal environment and the necessity to act within the given law. This barrier stresses the need to be capable of maneuvering in the legal structures to enable blockchain adoption. Legal issues are mentioned third, indicating major challenges associated with the issues of IP protection and integration within the blockchain platform. Therefore, there is a focus on strong strategies and legislation, which will protect the IPR during the blockchain deployment. The fourth

TABLE 7.10
Final Ranking of the Blockchain Implementation Barriers

Category	Weight	Sub-category	Local Weight	Global Weight	Rank
Technological barriers (TB)	0.4163	High implementation costs	0.2619	0.1091	21
		Scalability issues	0.1917	0.0798	18
		Interoperability with existing systems	0.1495	0.0622	14
		Technical complexity	0.1204	0.0501	13
		Lack of technical expertise	0.0828	0.0345	9
		Energy consumption	0.0663	0.0276	5
		Data standardization	0.0566	0.0235	4
		Technology maturity	0.0397	0.0165	2
		Data veracity	0.0311	0.0129	1
Regulatory and legal barriers (RLB)	0.2534	Regulatory uncertainty	0.4	0.101	23
		Legal and compliance issues	0.28	0.072	22
		Intellectual property concerns	0.15	0.039	15
		Compatibility with regulatory standards	0.1	0.025	10
		Cross-border transactions	0.07	0.017	6
Organizational and cultural barriers (OCB)	0.2524	Resistance to change	0.42	0.106	25
		Lack of awareness and understanding	0.24	0.06	19
		Integration with suppliers and partners	0.17	0.043	17
		Trust issues	0.11	0.027	12
		Ownership and control concerns	0.07	0.017	7
		Loss of intermediaries	0.05	0.012	3
Economic and financial barriers (EFB)	0.077	Lack of proven ROI	0.41	0.032	24
		Fragmentation of blockchain solutions	0.25	0.019	20
		Risk of obsolescence	0.16	0.013	16
		Environmental concerns	0.1	0.008	11
		Monitoring and reporting	0.07	0.005	8

least cited reason is compatibility with current regulations, which hints at the choice of blockchain solutions compatible with current regulations. This serves to call for the need to advance in the production of blockchain technologies that go further than the legislated prescriptions, which will be efficient and effectively implementable. In the fifth spot, cross-border transactions decipher the issues and the legal impendent that surround international transactions. This barrier makes it clear that blockchain solutions must be able to service international systems in a way that is not restricted

TABLE 7.11
Calculation of Consistency Ratio (CR)

Category	Ci	Ri	$CR = \frac{Ci}{Ri}$
Technological barriers (TB)	0.110	1.46	0.075
Regulatory and legal barriers (RLB)	0.098	1.12	0.087
Organizational and cultural barriers (OCB)	0.094	1.24	0.076
Economic and financial barriers (EFB)	0.102	1.12	0.091

Note: A CR of less than or equal to 0.1 is generally acceptable, indicating that the matrix has an acceptable level of consistency. If the CR is above 0.1, the judgments may be inconsistent.

TABLE 7.12
Determining Cause and Effect Relationships

Category	Ri	Ci	Ri + Ci	Ri − Ci	Identity
Technological barriers	1.5579	0.1790	1.7369	1.3790	Cause
Regulatory and legal barriers	0.9722	0.5402	1.5124	0.4320	Cause
Organizational and cultural barriers	0.4938	1.0194	1.5131	−0.5256	Effect
Economic and financial barriers	0.1885	1.4739	1.6624	−1.2854	Effect

Note: The row sum (Ri) and column sum (Ci) calculations determine cause-and-effect relationships within categories. The Ri − Ci value identifies the relationship between the criteria. The negative value criteria are identified as the effect group, and vice versa. These calculations help determine if a potential in the decision framework is a cause or an effect.

by local regulations but also legal in each area for cross-border trade to be made easier. Such insights highlight the need to foster polices and laws that will encourage the use of blockchain solutions in supply chains. It is recommended that blockchain initiatives reach proper compliance with data protection legal requirements and principles or adhere to current laws and regulations when it comes to intellectual property and regulations across different geographies while transacting. A proper solution to these barriers can avoid legal hurdles and even going international compliance issues in order to conduct efficient and effective operations smoothly.

Table 7.7 presents the hierarchy of sub-criteria of the organizational and cultural barriers; the study reveals that the most critical barrier to change is Resistance to Change. This barrier concerns the notion that the associated costs and the efforts needed to provide an organizational sustainable blockchain are high, and clearly shows that organizational resistance can significantly hinder the implementation of blockchain in the supply chain. They are ranked second, and the need for more awareness signifies the scarcity of profound understanding of blockchain technology. This means that if implementation of blockchain technology is to be improved, business organizations should support training and educational departments to produce skilled personnel. The third strategic imperative identified is Integration with Suppliers and Partners to stress the importance of synchronization of blockchain frameworks both at the company and at the level of interaction with the partners, that is critical for supply chains. Coming in fourth place, the Trust Issues indicate that various problems concerning the efficacy, integrity, and safety of the blockchain need to be solved by the businesses. This is important to achieve increased acceptance among stakeholders. Ownership and Control Concerns, ranked fifth, underlines that problems of ownership and control must be defined and regulated within the context of blockchain systems. Hence, involvement of all the supply chain members is essential to make blockchain deployment effective, and where needed, fairly distribute the benefits. Last is the Loss of Intermediaries, ranked sixth, concerns replacing traditional mediators in the chain. It identifies key themes, challenges, and opportunities of this shift for businesses with a focus on role transformation and new sources of value for intermediaries impacted by blockchain technology within the supply chain networks.

Table 7.8 is the kind of Economic and Financial barriers sub-criteria ranking, and according to the ranking responses received, the most prominent challenge is Lack of Proven ROI (Return on Investment). This also explains why there must be definitive values that reflect the worth of blockchain technology investment, since the plan is likely to adopt the technology in environments where many organizations are wary of new costs. The second most significant sub-criterion, Fragmentation of Blockchain Solutions, notes the problem that there are different blockchain platforms. Ineffective interfaces also prevent end-to-end integration and compatibility, which is vital in decision-making in blockchain within the supply chain platforms. It means that businesses face a vast number of blockchain platforms in search of solutions that can be integrated across business processes. The third risk factor is thus Risk of Obsolescence, which also underlines the need to avoid using a form of blockchain technology that has limited flexibility in change within a given fast-evolving technological space. Lastly, businesses ought to understand or consider the different long-term impacts of their chosen blockchain solutions since they are likely to become outdated in due time. Environmental Concerns are placed at the fourth level to represent the fact that rising consciousness of the effects of blockchain technologies on the environment is critical when those technologies are energy-intensive. Firms are thus urged to explore the likelihood of producing sustainable blockchain projects and reduce their negative environmental impact. In conclusion, the roles and responsibilities of this phase are ranked fifth, stressing the need for special focus on more enhanced modes through which monitoring and reporting of blockchain activities will be communicated. This proclaims the requirement to sustain sight as well as compliance with stipulated guidelines while making it possible for blockchain systems to operate effectively in the supply chain.

7.8 CONCLUSION

This chapter discusses the most prominent challenges that prevent the broader adoption of blockchain technology in the context of logistics and SCM industries. Thus, the present research, while following a comprehensive, methodologically proven approach based on the Fuzzy AHP and DEMATEL, is able to systematically and exhaustively detect and categorize multiple factors hindering the blockchain adoption. The present analysis underscores the difficulty of incorporating blockchain into supply chain networks and suggests approaches to addressing these issues.

The major propositions of this study revolve around analyzing and ranking barriers under technological, regulatory/legal, organizational, and financial categories. All are neatly packed and embodied in each category, that make it arduous to implement blockchain technology in a cohesive and efficient manner.

Technological Barriers: Based on the research, technological challenges are flagged as the biggest buts to implementation of the blockchain technology. This category is characterized by problems like technical aspects of blockchain systems, the problem of scalability, and compatibility with existing IT systems. These are important barriers as they affect the ability of supply chain actors to adopt blockchain, as well as the proper running of the blockchain in the intended applications. For instance, the problems of scalability imply the present-day technological constraints of the blockchain systems in completing numerous transactions per day in a timely manner, necessarily sufficient in the sphere of global supply chains.

Regulatory and Legal Barriers: Legal and regulatory factors constitute a major moving factor, exacerbated by the fact that the legal systems governing the application of the technology are still evolving. The decentralized and unchangeable nature of blockchain presents one difficulty for regulatory solutions since usual legal environments have to be prepared for that. People might refrain from adopting blockchain in their organizations because they might realize that they may not meet new laws created in the future, affecting their functionality, in case there is no set standard and rules governing blockchain.

Organizational Barriers: The chapter also pointed out the role of resistance to change within organizations. This resistance can be due to the fact that critical stakeholders do not comprehend the need for blockchain technology or because they are apprehensive about the change of practice. It does so because organizational culture is crucial here; producers who adhere to conventional values may regard blockchain as a disruptive rather than a facilitating factor.

Financial Barriers: One of the main implementation challenges is important initial investment, which increases costs related mainly to the purchases of new technology and training of the workforce. SMEs, for instance, who need a greater degree of risk margin to pursue a hunch on a technology most are unsure of being viable, might find it hard to do so.

The chapter proposes several strategic approaches to mitigate these barriers:

1. First, awareness and understanding campaigns about blockchain and its application within an organization at all the hierarchical levels may help in easing resistance to change and accepting of innovation.
2. Second, technology providers have sound and scalable blockchain solutions which are by technology offering today's technical hurdles.
3. Third, getting regulators involved for purposes of influencing policies enabling the right use of this technology could reduce legal and regulatory risks.

Additionally, the study avers that for this technology to be fully integrated in supply chain systems, there is a need for collective efforts of the technology developers, supply chain operators, and policymakers to collectively champion the cause of this technology. This includes managing the technical and financial stronghold and ensuring there is a compatibility of the strategic vision of the blockchain plan with other company visions.

In conclusion, although blockchain faces major challenges, the obstacles that have been identified are not insurmountable. The application of the blockchain will provide transparency, increase efficiency, and create better security measures through improved planning and organization of SCM. The conclusions made from this chapter provide novel and important understandings of the phenomenon of blockchain in SCM and present a clear direction for both scholars and practical applications in the future development of this emerging technology.

BIBLIOGRAPHY

Akhtar, P., Azima, N., Ghafar, A., & Din, S. U. (2021). Barricades in the adoption of blockchain technology in supply chain management: Challenges and benefits. *Transnational Marketing Journal*, *9*(1), 3–16.

Al Amin, M., Nabil, D. H., Baldacci, R., & Rahman, M. H. (2023). Exploring blockchain implementation challenges for sustainable supply chains: An integrated fuzzy TOPSIS–ISM approach. *Sustainability*, *15*(18), 13891.

Amazon. (2023). *Blockchain for Supply Chain: Track and Trace 2*. Amazon Web Services, Inc.

Asadi, M., Hashemkhani Zolfani, S., Pamucar, D., Salimi, J., & Saberi, S. (2023). The appropriation of blockchain implementation in the supply chain of SMES based on fuzzy LMAW. *Engineering Applications of Artificial Intelligence*, *123*, 106169.

Ayan, B., Güner, E., & Son-Turan, S. (2022). Blockchain technology and sustainability in supply chains and a closer look at different industries: A mixed method approach. *Logistics*, *6*(4), 85. MDPI.

Bae, S. H., Saberi, S., Kouhizadeh, M., & Sarkis, J. (2024). Evaluating blockchain technology adoption in multi-tier supply chains from an institutional entrepreneurship theory perspective. *Transportation Research Part C: Emerging Technologies*, *160*, 104510.

Bag, S., Viktorovich, D. A., Sahu, A. K., & Sahu, A. K. (2020). Barriers to adoption of blockchain technology in green supply chain management. *Journal of Global Operations and Strategic Sourcing*, *14*(1), 104–133.

Bai, Y., Liu, Y., & Yeo, W. M. (2022). Supply chain finance: What are the challenges in the adoption of blockchain technology? *Journal of Digital Economy*, *1*(3), 153–165.

Chavalala, M. M., Bag, S., Pretorius, J. H. C., & Rahman, M. S. (2022). A multi-method study on the barriers of the blockchain technology application in the cold supply chains. *Journal of Enterprise Information Management*, *37*(2), 745–776.

Deloitte. (2017, September 20). *Using Blockchain to Drive Supply Chain Transparency and Innovation*. Deloitte.

Fosso Wamba, S., Queiroz, M. M., & Trinchera, L. (2020). Dynamics between blockchain adoption determinants and supply chain performance: An empirical investigation. *International Journal of Production Economics*, *229*, 107791.

Gaur, V., & Gaiha, A. (2020, May). *Building a Transparent Supply Chain*. Harvard Business Review.

Hackius, N., & Petersen, M. (2017). Blockchain in logistics and supply chain: Trick or treat? *Proceedings of the Hamburg International Conference of Logistics (HICL)*, *23*, 3–18.

Hatibaruah, K. (2023, June 9). *Overcoming Barriers to Blockchain Adoption in Supply Chain Management: Challenges and Solutions*. www.linkedin.com

Kafeel, H., Kumar, V., & Duong, L. (2023). Blockchain in supply chain management: A synthesis of barriers and enablers for managers. *International Journal of Mathematical, Engineering and Management Sciences*, *8*(1), 15–42.

Kaur, J., Kumar, S., Narkhede, B. E., Dabić, M., Rathore, A. P. S., & Joshi, R. (2022). Barriers to blockchain adoption for supply chain finance: The case of Indian SMEs. *Electronic Commerce Research*, *24*(1), 303–340.

Khan, S., Haleem, A., Husain, Z., Samson, D., & Pathak, R. D. (2023a). Barriers to blockchain technology adoption in supply chains: The case of India. *Operations Management Research*, *16*(2), 668–683.

Khan, S., Kaushik, M. K., Kumar, R., & Khan, W. (2022). Investigating the barriers of blockchain technology integrated food supply chain: A BWM approach. *Benchmarking: An International Journal*, *30*(3), 713–735.

Khan, S., Khan, M. I., & Singh, R. (2023b). Modelling the barriers of blockchain technology implementation in supply chain. *Journal of Industrial Integration and Management*, *8*(03), 367–385.

Kouhizadeh, M., Saberi, S., & Sarkis, J. (2021). Blockchain technology and the sustainable supply chain: Theoretically exploring adoption barriers. *International Journal of Production Economics*, *231*, 107831.

Mathivathanan, D., Mathiyazhagan, K., Rana, N. P., Khorana, S., & Dwivedi, Y. K. (2021). Barriers to the adoption of blockchain technology in business supply chains: A total interpretive structural modelling (TISM) approach. *International Journal of Production Research*, *59*(11), 1–22.

Peng, G., et al. (2021, September). An application of fuzzy analytic hierarchy process in risk evaluation model. *Frontiers in Psychology*, *12*, 22.

Putra, D., & Sobandi, M., et al. (2018, October 1). Fuzzy analytical hierarchy process method to determine the quality of gemstones. *Advances in Fuzzy Systems*, *2018*, 1–6,

Saberi, S., Kouhizadeh, M., Sarkis, J., & Shen, L. (2019). Blockchain technology and its relationships to sustainable supply chain management. *International Journal of Production Research*, *57*(7), 2117–2135.

Sadhya, V., & Sadhya, H. (2018). Barriers to adoption of blockchain technology. *AMCIS 2018 Proceedings*.

Sahoo, P. S. B. B., & Thankur, V. (2023). The factors obstructing the blockchain adoption in supply chain finance: A hybrid fuzzy DELPHI-AHP-DEMATEL approach. *International Journal of Quality & Reliability Management*, *41*(9), 2292–2310.

Saurabh, S., & Dey, K. (2020). Blockchain technology adoption, architecture, and sustainable agri-food supply chains. *Journal of Cleaner Production*, *284*, 124731.

Singh, A., Prasath Kumar, V. R., Dehdasht, G., Mohandes, S. R., Manu, P., & Rahimian, F. P. (2023b). Investigating barriers to blockchain adoption in construction supply chain management: A fuzzy-based MCDM approach. *Technological Forecasting and Social Change*, *196*, 122849.

Singh, R., Khan, S., Dsilva, J., & Centobelli, P. (2023a). Blockchain integrated IoT for food supply chain: A grey based Delphi-DEMATEL approach. *Applied Sciences*, *13*(2), 1079.

Surucu-Balci, E., Iris, Ç., & Balci, G. (2024). Digital information in maritime supply chains with blockchain and cloud platforms: Supply chain capabilities, barriers, and research opportunities. *Technological Forecasting and Social Change*, *198*, 122978.

Treiblmaier, H. (2018). The impact of the blockchain on the supply chain: A theory-based research framework and a call for action. *Supply Chain Management: An International Journal*, *23*(6), 545–559.

Vafadarnikjoo, A., Badri Ahmadi, H., Liou, J. J. H., Botelho, T., & Chalvatzis, K. (2021). Analyzing blockchain adoption barriers in manufacturing supply chains by the neutrosophic analytic hierarchy process. *Annals of Operations Research*, *327*(1), 129–156.

Wang, Z.-J., Chen, Z., Xiao, L., Su, Q., Govindan, K., & Skibniewski, M. J. (2023). Blockchain adoption in sustainable supply chains for Industry 5.0: A multistakeholder perspective. *Journal of Innovation & Knowledge*, *8*(4), 100425.

Yadav, S., & Singh, S. P. (2020). Blockchain critical success factors for sustainable supply chain. *Resources, Conservation and Recycling*, *152*, 104505.

Yadav, V. S., Singh, A. R., Raut, R. D., & Govindarajan, U. H. (2020). Blockchain technology adoption barriers in the Indian agricultural supply chain: An integrated approach. *Resources, Conservation and Recycling*, *161*, 104877.

8 Role of Corporate Social Responsibility in Circular Economy: A Bibliometric Mapping Research Trends

Madhuri Yadav, D. D. Tewari, and Sukanta Kumar Baral

8.1 INTRODUCTION

The global economy has continuously been faced with dual challenges, such as resource scarcity and environmental degradation; these are a direct result of years of following a linear economy style that strictly adheres to the principle of taking, making, and disposing. In answering this, the term "circular economy" has emerged as a conversion model that advocates for a more efficient use of resources, less waste, and the looping of materials in the disposal system so as to enable further development. It definitely targets the transforming, redefinition of the economic production and consumption process with regards to reuse, recycling, and regeneration as a means, and eventually transition toward a resilience-based economy. However, that potential will be realized only when the cooperation of all stakeholders takes an approach that resonates with environmental, economic, and social norms. Here, CSR becomes a key player in grounding organizational practices and strategy in CE principle. CSR as a construct talks about the moral obligations of businesses to do so in a manner that contributes to society and the environment, becoming positive instead of moving only on the legal path. That is, it includes such voluntary actions which are directed toward social, economic, and environmental matters, and so positions corporations in the frame as really critical agents of sustainable change. The intersection of CSR and CE is at a point where both are premised on a long-term value approach, include stakeholder engagement, and capitalize on innovation toward sustainability goals. By integrating CSR principles into their practices, companies can themselves become major forces in facilitating circularity, including sustainable supply chain management, eco-design of products, and waste reduction strategies, as well as reputational and competitive advantage to themselves.

DOI: 10.1201/9781003477242-10

The coming together of corporate social responsibility (CSR) and the circular economy (CE) has attracted much attention in academia and policy domains around the world on its potential to create transformative changes in the utilization and sharing of resources. This intersection defines the necessity for a systematic understanding of how CSR initiatives could impact the adoption and scalability of CE practices across industries and regions. Due to the growing global pressures like resource depletion, environmental degradation, and socio-economic inequalities, the principles of sustainability have now become the pivot for outlining the future of industries and societies. CE has become the transformational framework for addressing all such concerns by doing so through the efficient use of resources, waste reduction, and ecosystem regeneration. At the same time, CSR evolved from merely a voluntary business ethic into a strategic necessity in aligning corporate objectives with those of the environment and society. This intersection of CSR and CE depicts a promising path toward sustainable futures through a more in-depth analysis of research trends in these areas. Bibliometric mapping is a systematic approach to the analysis of the scholarly literature in a field and could provide some very useful insights into the development, trends, and gaps in this nascent area. CSR and CE are two very complementary concepts with respect to sustainable development. CSR measures the ethical practices a corporation implements alongside societal welfare and environmental concerns beyond mere compliance. Meanwhile, CE is an economic model intending to de-couple economic growth from resource consumption by recycling, reusing, and remanufacturing. It entails the adoption of integrating CSR and CE that allows holistic approaches to businesses in minimizing environmental footprints to stakeholder value dynamics. Activities such as circular practice can be implemented for sustainable supply chains, eco-conception, and waste management systems, which would align with core CSR efforts toward achieving environmental goals but contribute to social well-being through fair labor practices and inclusive growth. Some examples could be promoting recycling schemes in the community through CSR initiatives or stimulating micro-enterprises that adopt circular activities and thus build local economies while going global in sustainability-to-be-issues dimensions. Of course, this is a new heading for a business organization that has engaged in the activity of traditional sustainable practices. This bibliometric mapping is in itself an overview of existing knowledge and, more so, it is a guide for future research and practice. Today, these studies largely, and in some cases, I may add, integrate, the field academic research and practice with policy into what would ultimately contribute to the global agenda of sustainability by identifying trends, gaps, and new opportunities. The broadened meaning of this conceptualization lays a very firm basis in which CSR can be understood in terms of its transforming potential for advancing the CE both theoretically and practically. It positions CSR as a key enabling agent in the transitions that will gradually make a society regenerative, inclusive, and sustainable in terms of the economic future it will shape. A systematic mapping of the existing research landscape, this study highlights CSR as an important instrument in propelling the CE principles into actualization. The utility of this research can be distinguished by several dimensions:

1. **Academic Insights**: This bibliometric analysis synthesizes current studies and hence offers scholars an understanding of how research on CSR in CE has changed and which aspects are still pending. It visualizes the underexploited areas, mainly guiding future research directions toward more interdisciplinary investigations of emerging themes such as digital transformation, green technologies, and stakeholder-driven CSR initiatives. Interdisciplinary adoption of bibliometric mapping is a highly replicable methodology for further study of other domains adopting similar approaches.
2. **Policy Relevance**: The insights afforded from this study are directed toward devising frameworks and incentives that can be employed by the policymakers to have businesses integrate CSR approaches with CE objectives. Geographic and sectoral trends help ensure that specific policies address certain local issues and opportunities pertaining to sustainable development. The contribution of CSR-driven CE initiatives to, and the impact on attainment of, the United Nations Sustainable Development Goals (SDGs) has also been depicted in this study.
3. **Business Applications**: This research makes the business case for corporate leaders adopting CSR initiatives supporting circular practices, such as waste reduction, resource efficiency, and sustainable innovation in products. This study illustrates how CSR-driven CE strategies can create simultaneous trust and brand reputation; improve relations with customers, investors, and the wider community; and help enterprises identify emerging opportunities through which to differentiate themselves through innovative circular business models that respond to increasing consumer demand for sustainability.
4. **Societal and Environmental Benefits**: This advocacy makes people aware of sustainable practices among different societal stakeholders by highlighting linkages between CSR and CE, as integrating CSR into CE projects can have significant environmental benefits, such as less waste, lower carbon emissions, and conservation of natural resources, ultimately contributing to a more sustainable and just economy.

CSR is the pillar that supports the CE and aligns the corporate objectives with sustainability principles. This may lead to trust as a stakeholder in giving an opportunity for the innovations that businesses need and in adopting a circular model for reuse, remanufacturing, or sustainable supply chain in this role. The 2021 research from the Ellen MacArthur Foundation found that 65% of companies in Europe are considering their CSR policy as one important aspect for all sorts of CE practices, including waste reduction and sustainable product design. The "People & Planet Positive" strategy of IKEA joins CSR with CE in that by 2030 its product lines will use recyclable or renewable inputs for sourcing, thus making a major contribution to reducing furniture waste.

A CE embraces change within itself, such that it is all wasted and consumption efficiency across its entirely including lifecycles production, distribution, and consumption stages. It includes reducing resource inputs and reusing resources, and recycling (Kirchherr et al., 2017). CSR stands for Community Supported Restoration, which means that an organization is responsible for the sustainability of the economy of the society. They should be putting that up with the communities, identifying their most pressing needs, and intertwining those with corporate strategy and corporate goals (Baral, 2017). It is companies that are conscious about building better society and really know how to carry their responsibility toward the environment, the employees, and so many others (López-Concepción et al., 2022). Carroll's four-pyramid model encompasses four distinct categories of corporate responsibilities: philanthropic, ethical, legal, and economic (Tian et al., 2011). CSR and CE have different definitions and concepts, but strategic CSR, the circular economy, and the environment fall under "sustainability." They encompass various aspects to achieve long-term social and economic well-being (Esken et al., 2018). Also, both concepts can be complementary for attaining sustainable developments toward the well-being of society; organizations and corporations have the opportunity to embrace and incorporate sustainable practices into their conduct and operational frameworks (Leandro & Paixao, 2018). CE is not assigned to a single person or time (Fortunati et al., 2020). Factory-based economy that aims to be repaired by design includes various principles, such as the use of renewable resources and a focus on promoting reusability and minimizing waste through reduction (MacArthur, 2013). It is pivotal in realizing SDGs and promoting sustainable practices among individuals, organizations, and businesses (Leandro & Paixao, 2018). It can replace the linear production and consumption model, where items are created, used, and discarded (Merli et al., 2018). The analysis reveals a recent surge in research output concerning CSR and CE, reflecting an escalating global awareness of sustainable business practices.

Nonetheless, circular economies emphasize sustainability, which includes three essential components: economic, social, and environmental (Upadhyay et al., 2021). These pillars are also integral to the UN SDGs for 2030, highlighting the significance of environmentally conscious, socially responsible, and economic development (Purvis et al., 2019). The CE includes reusing, repairing, and recycling, leaving waste behind as a thing of the past (Stoyanova, 2019). Nowadays, both scholars and academicians are using the terms "CSR" and "CE." Bibliometric analysis measured the popularity level of the two terms and thus provided insight into the data analysis. It is useful for identifying trends and patterns for research in the area of CSR and CE relationships. In this regard, data collection involved the Scopus database and employed some specific keywords, such as "CSR," "corporate social responsibility," and "CE," or "circular economy." Following the Boolean operator (*), we received 198 articles that relate to both CSR and CE from the Scopus database. Bibliometric analysis findings unveil the current situational state of research, more so the CSR and CE. In general, CSR refers to responsibility on the part of companies toward society's well-being, while CE focuses on reuse and less wastage. Both of these bear the same goal of change for the betterment of the society through their initiatives.

8.2 LITERATURE REVIEW

CSR and CE: New avenues for achieving sustainable development. By adopting CSR principles into circular business practices, these organizations can address environmental, economic, and social challenges using the two concepts together. The present literature review introduces the unique development of research on this subject, the key themes and issues, and bibliometric mapping and its role in advancing the understanding of scholarship from a bibliometric perspective.

CSR and CE thus evolve as totally entangled thematic objects in the scope of discourses pertaining to the emerging, increasingly well-reputed, potentially hydra-headed nature of their combined promise for sustainability. Bibliometric analyses of emerging research demonstrate significant increases in the number of articles dealing with this subject since the adoption of the Paris Agreement and the United Nations Sustainable Development Goals (SDGs) at the end of 2015. A bibliometric analysis by García-Barragán et al. (2019) examines the publications related to CE, indicating an increase in their number in a year, with CSR seen as a conceptual core driving more than 25% of these articles. The initial literature mainly discussed sustainability in general terms. Recently, areas where CSR and CE meet have been found, such as green supply chains, eco-innovation, and social equity. CSR initiatives play a crucial role in promoting CE practices within supply chains. These include green procurement, reverse logistics, and product lifecycle management.

Scopus data (2010–2023) shows a 40% increase in keyword co-occurrences such as "CSR practices," "circular business models," and "stakeholder collaboration" in CE research. Bai et al. (2021) studied electronics companies in China, finding a 30% reduction in waste and 20% cost savings attributed to CSR-driven CE initiatives.

CSR catalyzes innovation in business models by encouraging practices such as leasing, sharing economies, and product-as-a-service models. Kirchherr et al. (2018) analyzed European firms and found that CSR was a key driver in adopting circular strategies, with 35% of firms implementing take-back schemes due to stakeholder pressure.

Effective CSR strategies rely on collaboration with various stakeholders, including customers, suppliers, and communities, to co-create circular solutions. Andersen et al. (2022) demonstrated that community-driven recycling programs under CSR frameworks achieved a 25% higher material recovery rate compared to traditional programs.

CSR is often shaped by regulatory frameworks that encourage or mandate sustainable practices aligned with CE. The European Union's Circular Economy Action Plan has driven CSR-CE integration in industries like packaging and electronics, with over 70% of firms citing regulatory compliance as a motivator (Ellen MacArthur Foundation, 2021).

Europe leads global research on CSR and CE due to strong policy frameworks and public awareness. Asia focuses on industrial applications, particularly in manufacturing and electronics. There are limited studies, but there is emerging interest in using CSR to address resource scarcity and waste management challenges. A bibliometric study by Kumar et al. (2024) found that 60% of publications from Europe focus on regulatory aspects, while 40% from Asia focus on practical applications.

Industries such as technology, fashion, and food are prominent in CSR-CE research. Studies reveal CSR-driven circular practices like recycling and upcycling textiles. Emphasis on technology in the e-waste management and sustainable product design arena. It is expected that H&M will be at 100% circular production by 2030, which represents the point at which corporate ethics and circular commitments collide (Khan et al., 2021).

For example, several bibliometric approaches are widely intended in research dissemination mapping, impact work identification, and specific thematic gaps exploration. A bibliometric analysis was performed by Zeng et al. (2019) to illustrate an increment of co-citation of both concepts, CSR and CE, as presented in scholarly articles, particularly in journals such as *Journal of Cleaner Production and Sustainable Production and Consumption.*

Most empirical studies are case studies or surveys analyzing the practical dimensions of CSR on CE adoption. For instance, Bai et al. (2021) have found that nearly 80% of the sample surveyed regarded CSR as the major element in the adoption of CE in organizations.

While most of the research in this area has occurred in developed countries, it has left a significant gap in understanding how CSR-CE varies in developing economies. In their bibliometric review, Kumar et al. (2024) showed that only 15% of publications deal with issues that relate to Asia or Africa, or both. This indicates an urgent need for all Asia-related research or studies conducted in an African context. Though some aspects of environmental and economic sides have seen terms being thrown about and studies being done, social dimensions like inclusivity or fair sharing of resources, as far as CSR-created CE is concerned, remain largely unexplored. Another largely unexplored new, emerging area is the social impact that organizations that benefit from CSR-CE are able to create and the role those emerging technologies, such as blockchain, IoT, or even AI, can play in enhancing CSR-CE practices. Both CSR and CE can be used as complementary concepts for attaining sustainable development toward the well-being of society, organizations, and corporations to adopt and integrate sustainable practices into their behaviors and operations (Leandro & Paixao, 2018).

Carroll's definition encompassed both the economic and legal responsibilities that society had for businesses, considering them within the larger framework of CSR (Carroll, 2016). The interchangeability of sustainability and CSR gained traction primarily due to John Elkington, who introduced the term "triple bottom line" and connected it with the notion of sustainability. The Triple Bottom Line (TBL), driven by sustainability principles, offers a framework for evaluating a business' performance and the organization's success through three dimensions (Alhaddi, 2015). It emphasizes the integration of economic and social aspects with the environmental agenda. Elkington defined profit, people, and the planet. He defined social, economic, and ecological dimensions as profit, people, and earth (Elkington, 1997). Nevertheless, the objective of any business aspiring to contribute to the betterment of society, communities, and specific stakeholder groups involves fulfilling the economic, legal, ethical, and philanthropic responsibilities (Carroll, 2016). It is used for the policies, procedures, and actions, which may include transparency in reporting, implemented by companies to enhance the societal and welfare conditions of their stakeholders and the broader community (Zaman et al., 2022).

However, the idea of the CE needs a singular, definitive term (Del Baldo & D'Anghela, 2020). Nonetheless, the CE term was brought by Pearce and Turner in

1990, which explored the connections between economic activities and the environment. CE operates on "reduce, reuse, and recycle." It can be effectively implemented through five key areas of action: acquisition (take), production (make), distribution, utilization (use), and the recovery process (Dey et al., 2022). This ultimately redefines economic growth toward overall well-being in society. CE is shifting, over time, from dependence on these resources to renewable energy sources. It is a model to support development of economic, environmental, and social well-being. It has as its three foundational components: reduced waste and pollution, extending the life of products, and restoring natural systems (Fortunati et al., 2020). It is a progressive business model that converts CSR into action. It works on the elements of the three Rs: recycle, reduce, and reuse. CE is effectively reducing waste production and using resources efficiently (Mazzucchelli et al., 2022). CSR endeavors are multidimensional, promoting business profits, protecting the environment, consumer and employee welfare, and many related themes (Hong et al., 2023). It is an industry strategy that deliberately designs such regeneration or restoration coupled with sustainable resource management and practices (Geissdoerfer et al., 2017).

8.3 OBJECTIVES OF THE STUDY

1. To identify the research trends and patterns in the association between CSR and the CE.
2. To analyze the existing literature on the role of CSR in the context of the CE.

8.4 DATA AND METHODOLOGY

For many decades now, people have utilized the bibliometric techniques for the quantitative evaluation of the publications written. When assessing research, they must determine the number of co-authors on those papers or the extent to which people value international cooperation. Bibliometric methods are now rapidly becoming important, with rankings of journals and institutes (Ellegaard & Wallin, 2015). The present study incorporated bibliometric and content analysis to present the current state of the literature in CSR and CE. Data collection involved the Scopus database until 2023 while concentrating on the keywords "CSR," "corporate social responsibility," and "circular economy." The research aimed to cover an 11-year span between 2012 and 2023. This resulted in the retrieval of 198 published articles. The data were then arranged according to credible databases. These databases were majorly centered on Scopus since it has a broad collection of papers, authors, and journals. The following information was examined: the year published, journal, subject area, authorship information, countries, and keywords, among others. The study analyzed using the VOS viewer tool that creates networks for every specific variable and enables word combination creation (Meseguer-Sánchez et al., 2021). Despite this, it has several methods to present the different maps and zoom, scroll, and search, which allow analysis and exploration of the details of the map (Van Eck & Waltman, 2011). These networks are established based on association including co-citation, bibliographic, and co-authorship (Alsharif et al., 2020) (Figure 8.1).

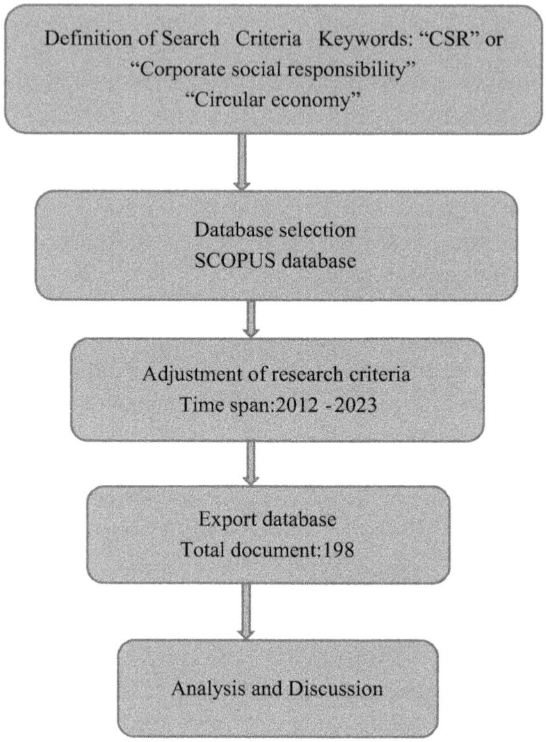

FIGURE 8.1 Flow chart of methodology.

Source: Framed by the authors.

8.5 RESULTS AND DISCUSSION

Table 8.1 depicts the evolution of essential characteristics from 2012 to 2023 in articles on CSR and the CE. To make the analysis more accessible, the 11 years are split into two-year sections. The volume of scientific articles produced over this time has increased exponentially significantly. It is seen that two articles were released in 2012, while 93 articles were published in 2023. Only two articles were published during the initial two-year period (2012–2013). However, in the most recent 2022–2023, the number of publications surged to 93, marking a remarkable increase of 4650%.

Over the last 11 years, there has been a remarkable increase in various aspects of the study. The study's timeline has been divided into six segments spanning two years. The frequency of authorship has notably surged. In 2012–2013, only four authors published articles, whereas in 2022–2023, a staggering 300 authors contributed to publications. The scope of countries addressing the analyzed issue has also expanded significantly, growing from two in 2012–2013 to 98 in 2022–2023. Additionally, citations have experienced substantial growth over the same period. In 2012–2013, 167 articles were cited, while in 2020–2021, this number soared to 1815, resulting in 3,592 citations. Moreover, journal numbers have increased significantly, jumping from just one in the initial two years (2012–2013) to an impressive 58 in 2022–2023.

TABLE 8.1
Article Publication from 2012 to 2023

Year	Articles	Authors	Citations	TC/A	Journals	Countries
2012–2013	2	4	167	83.5	1	2
2014–2015	3	6	30	10	1	3
2016–2017	7	14	164	23.4	3	8
2018–2019	25	72	1002	40	18	32
2020–2021	68	202	1815	26.6	77	52
2022–2023	93	300	414	4.39	58	98

Source: Compiled by the researchers.
Note: (TC/A): An average of citations per article.

Table 8.2 depicts the five leading journals publishing the most productive and scientific journals in the CSR and CE fields. The *Sustainability Switzerland* journal has contributed a maximum of 22 articles with 523 citations. While business strategy and the environment had the highest impact index (2.87), they also had the highest citations (837). It is the only journal based in the US among the leading five journals. Closely followed were CSR and environment management with 2.134. *Sustainability Switzerland*, business strategy, and the environment have the highest H-index with 11 each. Then, the *Journal of Cleaner Production* has a 10 H-index.

Table 8.3 presents the top five productive countries in CSR and CE. However, Italy leads with 33 articles, followed closely by the UK with 31 and China with 16. Spain follows with 15. The UK boasts the maximum number of citations at 756 and

TABLE 8.2
Top Five Journals with the Highest Citation

Journal	Article	Citation	TC/A	SJR	H Index (Article)	Country
Sustainability Switzerland	22	523	23.77	0.664	11	Switzerland
Business Strategy and the Environment	17	837	49.23	2.87	11	US
Journal of Cleaner Production	17	639	37.58	1.981	10	Netherland
Corporate Social Responsibility and Environment Management	13	323	24.84	2.134	6	UK
Worldwide Hospitality Tourism Theme	9	19	2.11	0.423	3	UK

Source: Compiled by the researchers.
Note: (TC/A): An average of citation per article, SJR: Scimago Journal Rank.

TABLE 8.3
Top Five Productive Countries

Country	Article	H Index	Citation (TC)	TC/A
Italy	33	13	695	21.06
UK	31	13	756	24.38
China	16	7	387	24.18
Spain	15	9	601	40.06
India	12	4	67	5.58

Source: Compiled by the researchers.
Note: (TC/A): An average of citations per article.

the highest H-index of 13. All the top productive countries have seen a remarkable surge in scientific output during five years (2019–2023), with India also emerging as one of the leading contributors.

It can be seen in Table 8.4 that the author with the highest productivity, Morea, wrote five articles between 2012 and 2023. Following closely are Caeiro, Fortunato, Operferkuch, and Salomone, each with four articles to their names. Interestingly, Morea also holds the distinction of having the highest number of citations, totaling 164. Additionally, both Morea and Fortunati share the highest H-index, each with a score of four. It is worth noting that Morea and Fortunati are affiliated with institutions in Italy.

Figure 8.2 visually represents published articles from 2012 to 2023, year by year. It is evident from the figure that only one article was published from 2012 to 2014. Fourteen articles were published in 2018. The highest number of papers, totaling 49, were published in 2023. Additionally, 44 articles were published in 2022 and 39 in 2021. In contrast, only 11 papers were published in 2019, with an additional three in 2018.

TABLE 8.4
Top Five Leading Authors

Authors	Articles	Citation	Country	H Index	Institution
Morea, D	5	164	Italy	4	Universita degli Studi di Cagliari
Caeiro, S	4	71	Portugal	3	Universidade Aberta
Fortunati, S	4	160	Italy	4	Università degli Studi della Tuscia
Operferkuch, K	4	71	Portugal	3	Faculdade de Ciênciase Tecnologia
Salomone, R	4	71	Italy	3	Università degli Studi di Messina

Source: Compiled by the researchers.

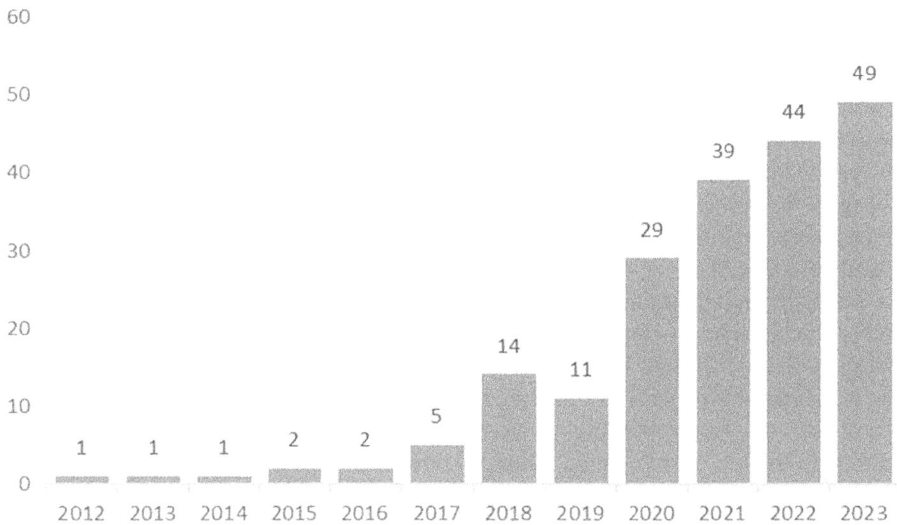

FIGURE 8.2 Publication yearly.

Source: Authors' own creation.

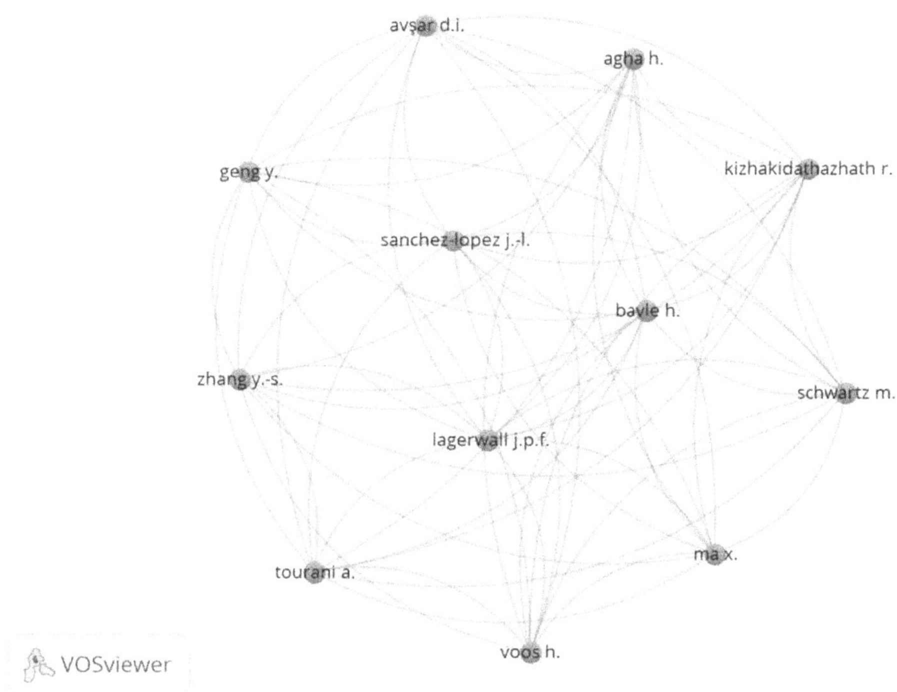

FIGURE 8.3 Co-authorship of the primary author.

Source: Authors' own creation using VOS viewer.

Based on co-authorship analysis, Figure 8.3 depicts a collaboration map among critical CSR and CE authors. The network displays significant dispersion, which may facilitate the swift expansion of this research area. Notably, among the highly productive authors, Sanchez, Lagerwall, and Bavle maintain a consistent international network of collaborations. Meanwhile, Voos, Tourani, Schwartz, and Zhang are also observed to be involved in a global network for partnership on CSR and the CE.

Figure 8.4 presents the keywords cluster based on co-occurrence. Using VOS viewer software, keyword co-occurrence was analyzed to identify the most frequently associated terms within CSR and CE. The findings indicate that the prevalent keywords in digital technologies encompass CSR, CE, recycling, business sustainability, sustainable development, and environmental impact. The analysis shows five distinct clusters. Cluster 1, emphasizes the significance of the "circular economy" and encompasses ten items. Following this, cluster 2 (centered around the concept of "corporate social responsibility" and comprises nine items). The cluster 3, associated with "sustainable development," encompasses nine things. Cluster 4 is related to "CSR" and consists of eight items. Cluster 5, connected to "Industry 4.0," comprising six items.

Figure 8.5 illustrates a collaboration map based on co-authorship among various countries. The dimensions of the circles reflect the number of associations, denote

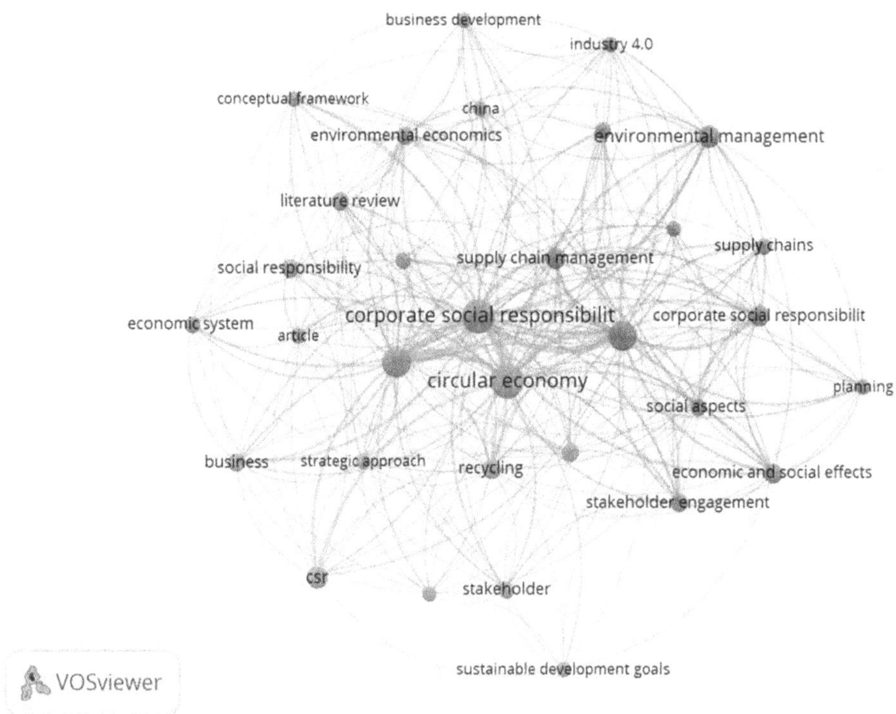

FIGURE 8.4 Keyword clusters based on co-occurrence.

Source: Authors' own creation using VOS viewer.

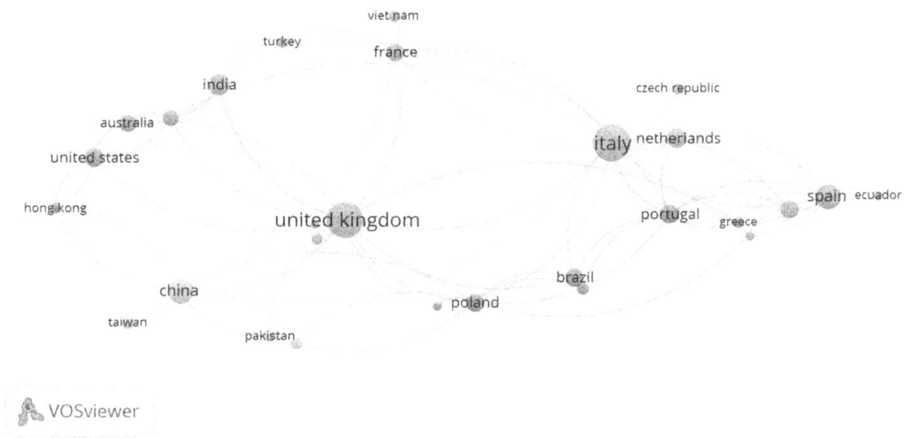

FIGURE 8.5 Co-authorship between countries.

Source: Authors' own creation using VOS viewer.

various international working groups. The other cluster, comprises three countries and is headed by the UK. Following, Italy cluster, involving three countries. The largest cluster involves five countries, led by Spain. Also, Portugal forms the cluster, consisting of five countries. India anchors the cluster, which encompasses four countries.

8.6 CONCLUSION

The study comprehensively analyzed the association between CSR and the CE over the past 11 years. Hence, a remarkable increase in scientific output can be seen over the next 11 years. The number of published articles surged from a mere two in 2012–2013 to an impressive 93 in 2022–2023, indicating a 4650% increase. The study highlighted the number of authors contributing to publications upward. In 2012–2013, only four authors were involved, while in 2022–2023, a staggering 300 authors contributed. It suggests a growing interest and collaboration in this area of study. The scope of countries addressing the topic expanded notably, growing from two in 2012–2013 to 98 in 2022–2023. The journals also experienced an exponential increase, from one in 2012–2013 to 58 in 2022–2023. Among the top five journals, *Sustainability Switzerland* emerged as the most productive in European Union member countries. Italy, the UK, and China were the leading countries in productivity. Morea emerged as the most prolific author, contributing five articles and receiving 164 citations. The study revealed a clear evolution in research output over the years, with a notable surge in articles from 2018, peaking in 2022–2023. The keyword for co-occurrence analysis indicates really important terms; those indicated by clusters are significant concepts such as CE, CSR, and sustainable development. This study, therefore, adds to the academic repository of CSR and CE and also has practical implications for various stakeholders and decision-makers across sectors. This research gives a solid foundation to build upon an economy-sustainability-linked system that bridges the gap between theory and practice. A bibliometric approach ensures that evidence has really been obtained empirically; thus, the credibility and relevance of findings about the key global challenges of the 21st century are enhanced with a promise of complex utility for academics, policymakers, business leaders, and practitioners.

8.7 EMERGING TRENDS AND FUTURE DIRECTIONS

Dynamic areas at the intersection of CSR and CE, influenced by changing technologies, emerging regulations, and social expectations. Digital transformation technologies such as blockchain, IoT, and AI provide transparency and efficiency in CSR-CE practices, such as real-time tracking of material flows and predictive waste management. Increasing obligations from governments and international organizations for CSR reporting encourage circular practices through subsidies and tax benefits. Social Inclusion Research is now also focusing on the social dimensions of CE – equal access, benefits among communities, and empowerment that are founded on commitments from CSR. Change is evident in these directions, which constitutes the ever-changeable nature of CSR and CE, thus warranting adaptive behavior and further innovation.

8.8 IMPLICATIONS FOR STAKEHOLDERS

The applicability of bibliometric mapping insights to different stakeholders may include the result of complementing the CSR strategies with CE principles, which, in turn, creates improved corporate reputation, stakeholder trust, and long-term profits. Evidence-based insights are in favor of policy formulation that can motivate CSR-influenced CE activities and remove systematic barriers to sustainability for policymakers. The result of CSR-CE integration benefits communities through job creation, resource efficiency, and environmental friendliness for the entire society.

REFERENCES

Alhaddi, H. (2015). Triple bottom line and sustainability: A literature review. *Business and Management studies*, *1*(2), 6–10.

Alsharif, A. H., Salleh, N. O. R. Z. M. D., & Baharun, R. (2020). Bibliometric analysis. *Journal of Theoretical and Applied Information Technology*, *98*(15), 2948–2962.

Andersen, M. M., Larsen, T. H., & Christensen, P. R. (2022). CSR and community engagement in circular systems. *Corporate Social Responsibility and Environmental Management*, *29*(6), 1450–1463.

Bai, C., Sarkis, J., & Dou, Y. (2021). Corporate social responsibility and sustainability in supply chains: Drivers and practices. *Journal of Cleaner Production*, *298*, 126789.

Baral, S. K. (2017). A critical study on corporate social responsibility (CSR) models in India. *Splint International Journal of Professionals*, *4*(5), 74–85.

Carroll, A. B. (2016). Carroll's pyramid of CSR: Taking another look. *International Journal of Corporate Social Responsibility*, *1*(1), 1–8.

Del Baldo, M., & D'Anghela, M. (2020). Circular economy and corporate social responsibility: A literature review. *Symphonya. Emerging Issues in Management*, 70–87. https://doi.org/10.4468/2020.1.06delbaldo.danghela

Dey, P. K., Malesios, C., De, D., Budhwar, P., Chowdhury, S., & Cheffi, W. (2022). Circular economy to enhance sustainability of small and medium-sized enterprises. In *Supply Chain Sustainability in Small and Medium Sized Enterprises* (pp. 10–45). Routledge.

Elkington, J. (1997). The triple bottom line. *Environmental Management: Readings AASES*, *2*, 49–66.

Ellegaard, O., & Wallin, J. A. (2015). The bibliometric analysis of scholarly production: How great is the impact? *Scientometrics*, *105*, 1809–1831.

Ellen MacArthur Foundation. (2021). The circular economy in action: Scaling innovation. https://www.ellenmacarthurfoundation.org/topics/circular-economy-introduction/overview

Esken, B., Franco-García, M. L., & Fisscher, O. A. (2018). CSR perception as a signpost for circular economy. *Management Research Review*, *41*(5), 586–604.

Fortunati, S., Martiniello, L., & Morea, D. (2020). The strategic role of the corporate social responsibility and circular economy in the cosmetic industry. *Sustainability*, *12*(12), 5120.

García-Barragán, J. F., Egilmez, G., & Singh, S. (2019). Sustainability and circular economy in corporate practices. *Sustainability*, *11*(9), 2726.

Geissdoerfer, M., Savaget, P., Bocken, N. M., & Hultink, E. J. (2017). The circular economy – A new sustainability paradigm?. *Journal of Cleaner Production*, *143*, 757–768.

Hong, T., Ou, J., Jia, F., Chen, L., & Yang, Y. (2023). Circular economy practices and corporate social responsibility performance: the role of sense-giving. *International Journal of Logistics Research and Applications*, *27*(11), 2208–2237.

Khan, S., Bhatti, A., & Shahid, R. (2021). Circular business models in the fashion industry: The role of CSR. *Sustainable Production and Consumption*, *27*, 1157–1165.

Kirchherr, J., Reike, D., & Hekkert, M. (2017). Conceptualizing the circular economy: An analysis of 114 definitions. *Resources, Conservation and Recycling*, *127*, 221–232.

Kirchherr, J., Reike, D., & Hekkert, M. (2018). Conceptualizing the circular economy: An analysis of 114 definitions. *Resources, Conservation, and Recycling*, *127*, 221–232.

Kumar, A., Snehvrat, S., Kumari, P., Priyadarshani, P., & Ray, P. (2024). A corporate social responsibility (CSR)-led competitiveness framework through a bibliometric analysis (2001–2023). *Journal of Asia Business Studies*, *18*(3), 737–766.

Leandro, A., & Paixao, S. (2018). Corporate social responsibility and circular economy: Two ways, same destinations? An outlook on both concepts and cases from Portugal. *Proceedings of the Congrès avniR*, Lille, Portugal.

López-Concepción, A., Gil-Lacruz, A. I., & Saz-Gil, I. (2022). Stakeholder engagement, Csr development and Sdgs compliance: A systematic review from 2015 to 2021. *Corporate Social Responsibility and Environmental Management*, *29*(1), 19–31.

MacArthur, E. (2013). Towards the circular economy. *Journal of Industrial Ecology*, *2*(1), 23–44.

Mazzucchelli, A., Chierici, R., Del Giudice, M., & Bua, I. (2022). Do circular economy practices affect corporate performance? Evidence from Italian large-sized manufacturing firms. *Corporate Social Responsibility and Environmental Management*, *29*(6), 2016–2029.

Merli, R., Preziosi, M., & Acampora, A. (2018). How do scholars approach the circular economy? A systematic literature review. *Journal of Cleaner Production*, *178*, 703–722.

Meseguer-Sánchez, V., Gálvez-Sánchez, F. J., López-Martínez, G., & Molina-Moreno, V. (2021). Corporate social responsibility and sustainability. A bibliometric analysis of their interrelations. *Sustainability*, *13*(4), 1636.

Purvis, B., Mao, Y., & Robinson, D. (2019). Three pillars of sustainability: In search of conceptual origins. *Sustainability Science*, *14*, 681–695.

Stoyanova, T. (2019). CSR strategies applied in terms of circular economy. *Economic Alternatives*, *2*, 263–274.

Tian, Z., Wang, R., & Yang, W. (2011). Consumer responses to corporate social responsibility (CSR) in China. *Journal of Business Ethics*, *101*, 197–212.

Upadhyay, A., Mukhuty, S., Kumar, V., & Kazancoglu, Y. (2021). Blockchain technology and the circular economy: Implications for sustainability and social responsibility. *Journal of Cleaner Production*, *293*, 126130.

Van Eck, N. J., & Waltman, L. (2011). VOSviewer manual. *Manual for VOSviewer version*, *1*.

Zaman, R., Jain, T., Samara, G., & Jamali, D. (2022). Corporate governance meets corporate social responsibility: Mapping the interface. *Business & Society*, *61*(3), 690–752.

Zeng, W., Zhang, L., & Zhao, S. (2019). Bibliometric analysis of CSR and circular economy intersections. *Journal of Cleaner Production*, *237*, 117650.

9 Sustainable Transformation through Organizational Ambidexterity: A Supply Chain Imperative

Muskan Khan, Arpana Kumari, Rudra Rameshwar and Bridget Irene

9.1 INTRODUCTION

Amidst the geopolitical events and uncertainty, supply chain risk management has developed into one of the most active areas within supply chain management research. These challenges developed the need to integrate technological and sustainable approaches in managing the supply chains of firms (Zavala-Alcívar et al., 2020). Managers are realizing that their company's environmental and social responsibilities are not just the responsibility of one organization, but a collaborative effort of the industry. To fulfill these social responsibilities, various nodes across the supply chain must be considered. Thus, they are endeavoring to adopt mechanisms that pursue sustainability goals. Multiple disciplines have defined the term "sustainability;" however, World Commission on Environment and Development (WECD, 1987) offers one of the most comprehensive definitions as "using resources to meet the needs of the present without compromising the ability of future generations to meet their own needs." Thus, the definition suggests focusing not only on economic development but also on the sustenance of natural resources and the society that firms serve. The present chapter discusses the significance of ambidexterity in propelling sustainable supply chain system of firms.

9.1.1 The Imperatives of Sustainability and Digitization in Supply Chains

Although over the past two decades, academic research on sustainable supply chain management (SCM) has grown, its potential to support supply chain practitioners is still under question. Previous practitioners studied purchasing ethics, safety management, carbon footprint, green purchasing, and re-manufacturing (Winter and

DOI: 10.1201/9781003477242-11

Knemeyer, 2013). However, this body of research has primarily highlighted individual contexts of sustainability, such as environmental, and not on its multiple dimensions. Sustainable SCM has been defined as

> *the management of material, information and capital flows, as well as cooperation among companies along the supply chain, while taking goals from all three dimensions of sustainable development, i.e., economic, environmental and social, into account, which are derived from customer and stakeholder requirements.*
>
> (Seuring and Müller, 2008)

Figure 9.1 Growth of publication for organizational ambidexterity in extant literature and its role in various areas of business management, like strategy, human resource management, and innovation. The Supply chain needs of firms holistically aim at all three dimensions, namely profit, people, planet. Any product and process-related decision of supplier impacts the manufacturer and other members of the supply chain. The sustainability efforts of one member influence the service agreement of all the stakeholders in product development. A more integrative approach will help companies work together, while adopting sustainable practices, considering the economic consequences for each other. Here, managerial competence, trust, transparency, and robust policies must be guaranteed from the first supplier to end-customer in the value chain. Also, the indulgence of customers in firms' sustainability practices in the supply chain will support the company's marketing efforts. However, researchers

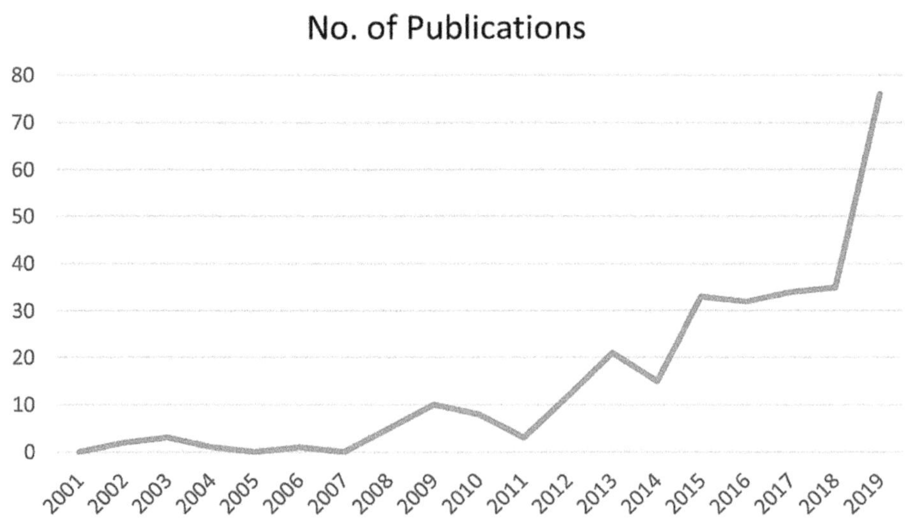

FIGURE 9.1 Growth of publication for organizational ambidexterity.

Source: Chakma et al., 2021.

see the scope of exploring the validation of metrics to support sustainable SCM, which can measure the impact of sustainable activities on economic performance.

While we talk of supply chains of firms in the globalized world, the role of digitization and technology-enablement becomes pivotal in the competitive market of Industry 4.0. Digitization with smart devices and great accuracy presents the potential of storing and processing vast amounts of data from both internal and external sources in supply chain (Schniederjans et al., 2020). Forecasting applications such as Big Data have revolutionized the supply chain in the entertainment, hotel, and retail industry (Cui et al., 2018). With a direct understanding of customers' behavior patterns, such as purchase history and payment methods, manufacturers understand the shopper's characteristics and further predict consumer behavior (Guha and Kumar, 2018). Moreover, advanced devices such as sensors and actuators provide real-time information to manufacturers and suppliers on several aspects that enable them to make timely decisions. With the proliferation of digitization, companies are facing certain challenges in terms of data security, strategic usage of devices, and staying safe and ahead in the market. The scope of detection, assessment bias, and abnormality existence in data needs to be managed. Further, the induction of artificial intelligence (AI) and Big Data as part of digitization brought both possibilities and challenges, in terms of job losses in the area of bill processing, invoicing, and customer services. Also, the advent of drone and robot vehicles is reducing the challenges of transportation and, thus, reducing the requirement of workers in remote locations. Yet, despite these opportunities, firms still face challenges of managing the complete supply chain process smoothly in terms of managing the data security, creating knowledge out of that, exploiting existing organizational potential, and exploring new ideas for performing in the continuous market shifts.

9.1.2 The Concept of Organizational Ambidexterity

The term "organizational ambidexterity" was originally used by Duncan (1976) (Gianzina-Kassotaki, 2017). Organizational ambidexterity refers to the process of achieving business efficiency by exploiting the organizational potential and creating new ideas in business with the help of an exploration approach in changing market trends (O'Reilly and Tushman, 2008). Thus, firms engage themselves in two contrary and apparently incompatible approaches of exploration and exploitation, to achieve competitive advantage and business performance (Brix, 2019; Chakma et al., 2021). Organizations' concentration only on existing strengths and utilizing those for business processes prevents them from outperforming in the market, as they lag in exploiting the new technologies and opportunities which their competitors will grab and perform. Therefore, those firms who simultaneously balance utilizing their competencies and adopting the revolutionary (exploration) changes are known as ambidextrous. These explorations may cover technological shifts and advance processes (Tushman and O'Reilly, 1996).

According to Li et al. (2016), obtaining breakthrough innovations at the team level necessitates placing a higher value on knowledge exploration than knowledge exploitation in the management of knowledge. As a result, they contend that striking a balance between exploitation and exploration is insufficient. In recent years, researchers have found relevant implications of organizational ambidexterity in multiple business processes such as organizational learning, design, entrepreneurship, and strategic management. Though quite vast research has been done in this area, yet the term "organizational ambidexterity" has not been discussed much with the sustainability approach of SCM. Witnessing the growing interest of firms in the adoption of business sustainability, the investigation on its link with organizational ambidexterity is crucial.

9.1.3 Significance of Ambidexterity for Sustainable Transformation

It is now widely acknowledged that organizational ambidexterity performs a critical role in helping businesses balance the effects of disruption with other aspects (Ivanov et al., 2014; Ojha et al., 2018). The global panorama of supply chain of firms poses multiple challenges for operationalizing sustainable practices under multiple situations in different countries. These challenges are in the form of distinct legal formalities, associated costs, product quality, and mutual understanding, and others (Koberg and Longoni, 2019). Thus, if firms in supply chains aim to achieve their sustainability goals, they ought to consider the potential influence of ambidexterity on their sustainability performance in business. The exploration and exploitation features of ambidexterity will bring effective possibilities to address the effect of disruption, due to sustainability processes in organizations. In supply chain when business intends to execute sustainability practices in coordination with all other entities, it needs to exploit its existing competencies and explore all the relevant options available out there to adopt. However, this process is not easy and risk-free for firms. They need ample support from all the associated partners in the supply chain. Multiple inconsistencies, such as lack of top management support, low team coordination, and other human and technical behaviors that impact the execution of ambidexterity (Venugopal et al., 2020). Studies on ambidexterity's importance for a sustainable supply chain are scarce (Sahi et al., 2020). Supply chain requires the resources to utilize the capacities and opportunities of firms for achieving sustainable performance. Therefore, the present study primarily focuses on mapping the importance of ambidexterity for sustainable business in the supply chain.

9.2 THEORETICAL FOUNDATION OF ORGANIZATIONAL AMBIDEXTERITY

9.2.1 The Exploration Exploitation Paradox

The literature combines and balances the two views of exploitation and exploration to conceptualize ambidexterity (He and Wong, 2004). According to Patel et al. (2013), organizational ambidexterity is a coordinated operational component pursued through investigation and utilization. Exploration creates new avenues for prosperity, whereas exploitation makes use of available resources to live (Vessey and Ward, 2013). Researchers linked exploitation with other terms, such as "knowledge exploitation." Also, exploitation is understood as "betterment, refinement, choice, production, selection, implementation and execution." However, exploration is known as "search, risk-taking, variation, experimentation, flexibility, discovery and innovation." While firms have limited resources, they face tough challenges to match changing market trends, and this causes them to get into a "failure trap". To continue performing, they need to explore the opportunities matching their requirement and market needs under the exploration stage. Many researchers have highlighted that while practicing ambidexterity, firms face the contradiction between the exploration and exploitation processes. The adoption of one impacts some features of another (Gupta et al., 2006); however, several scholars contended that both the dimensions are achievable simultaneously (Rothaermel and Alexandre, 2009). This inconsistency and disagreement among the practitioners bring a notable need to investigate more on this concept, specifically in the sustainability context of supply chain in the firms.

9.2.2 Structural and Contextual Ambidexterity

Studies indicated two methods (the structural and contextual approach) that can assist businesses in developing ambidexterity (Gibson and Birkinshaw, 2004; O'Reilly and Tushman, 2013). These approaches are proposed to manage the contradiction and inconsistency organizations face in exploitation and exploration processes. According to Birkinshaw et al. (2016), the structural approach recommends that organizations distribute their exploration and exploitation efforts among functional units that possess complementary competencies. Whereas, the contextual approach proposes that organizations establish a supportive environment with common ideals to enable their members to be ambidextrous (Birkinshaw and Gibson, 2004). This can be accomplished by creating distinct structures and by putting in place procedures that support and encourage decision-making regarding how best to divide time between two conflicting ambidexterity construct demands. Through enhanced operations and sensemaking, inter-organizational relationship systems facilitate contextual ambidexterity (Im and Rai, 2014). At the intra- and inter-organizational levels, IT spending, usage intensity, centralization, and training all contribute to structural ambidexterity (Park et al., 2020). With the relevance of both approaches in different contexts, our study aims to explore how these factors are significant in achieving sustainability in supply chain processes.

9.2.3 Antecedents and Outcomes of Ambidexterity

Ambidexterity has shown vital importance in industries. Small- and medium-scale sectors (Mu et al., 2022) and information technology sectors (Werder and Heckmann, 2019) have witnessed the consequences of ambidexterity implementation. Also, many driving factors of ambidexterity have been discussed by scholars in previous studies. García-Granero et al. (2018) stated about dynamic capabilities which help managers in taking decisions to reallocate and reshape skills in organizations. Further, organizational learning, organization design, and technological innovation are also considered as strong antecedents for ambidexterity in organization. The execution of ambidexterity has provided multiple results. With its adoption, firms have experienced enhanced individual and team performance (Mom et al., 2018), and organizational success as outcome (Good and Michel, 2013). Moreover, less autocracy, high individual initiatives, optimum human resources and financial resources, and sufficient managerial expertise decrease the challenges for organizational ambidexterity (Chang and Hughes, 2012). Also, scholars highlighted that flat and informal organizational structure makes ambidexterity easier to be administered. Researchers have studied the outcomes of ambidexterity in both positive and negative dimensions (Kauppila and Tempelaar, 2016). Enhancement in individual performance has been seen when organizations adopt organizational ambidexterity. Whereas cognitive stress and burnout are the negative consequences of this process. This happens due to switching from exploitation to exploration stages and vice versa (Kauppila and Tempelaar, 2016). Thus, multiple dimensions of antecedents and outcomes of ambidexterity bring potential scope for exploring their meaning in the supply chain sustainability context. This chapter delves into the integration of ambidexterity and sustainability practices of the supply chain process.

9.3 AMBIDEXTERITY AND SUSTAINABLE SUPPLY CHAIN PRACTICES

Organizational ambidexterity connotes a delicate balance between exploiting existing competencies and exploring novel opportunities and has emerged as a critical capability for supply chain organizations, seeking to integrate sustainability into their practices. In an era rampant with environmental challenges, resource scarcity, and evolving regulatory landscapes, supply chains are under immense pressure to reduce their ecological footprint while continuously innovating to meet dynamic consumer demands and stay competitive.

9.3.1 Exploiting Existing Competencies for Operational Efficiency

Supply chain organizations that have mastered ambidexterity leverage their established expertise and resources to drive operational efficiency and minimize waste. This exploitation facet of ambidexterity involves optimizing processes, streamlining logistics, and implementing lean practices to curtail energy consumption, emissions, and material waste (Pagell and Shevchenko, 2014). By doing so, companies can not only mitigate their environmental impact but also achieve significant cost savings and improved profitability.

A prominent example of this is Walmart, the world's largest retailer, which has capitalized on its vast supply chain network and logistics expertise to implement initiatives such as route optimization, fuel efficiency improvements, and packaging reduction strategies. These efforts have yielded substantial cost savings and environmental benefits for the company (Walmart, 2022). Walmart has been working closely with its suppliers. They want to use energy better and cut down on waste in their supply chain. It's a big deal! Also, organizations that can adapt really see how sharing ideas and working together matter in their supply chains. This includes everyone from the suppliers to the folks who help get products to stores. When companies build strong partnerships—with suppliers, manufacturers, and logistics teams—they uncover ways to improve processes and make the best use of resources. According to

Beske et al. (2014), this contributes to the development of a supply chain that is more environmentally friendly. Consider Unilever as an example. That is just what this large consumer products firm is doing! To reduce water consumption, energy expenditure, and trash, they are collaborating with their suppliers. The company's Sustainable Living Plan (Unilever, 2020a, 2020b) encompasses all of these initiatives. Thanks to everyone's hard effort, Unilever has become far more efficient and less harmful to the environment.

9.3.2 Exploring Sustainable Innovations and Circular Economy Models

Companies that excel in SCM understand the need of both focusing on efficiency and exploring new ideas. Means always seeking new methods to conduct business that are both environmentally friendly and innovative so that you can remain ahead of the competition. Spending money on research is crucial! As an example, the circular economy is one of many daring ideas that may be tried and tested (Guldmann et al., 2019).

When businesses hold the ideas of the circular economy, it's obvious that they are vigorously looking out for sustainable methods. By following these guidelines, resources could be used in a better way and waste could be cut off. For example, Dell and Philips offer fantastic programs where customers can return fragmented devices and have them fixed. In doing so, they may recuperate useful minerals while using less renewed resources (Bocken et al., 2016). This aids the globe, and it also bounces these businesses more opportunities to generate money and preserve their viable edge.

Moreover, industries are examining innovative new resources and more eco-friendly production approaches. Things resembling renewable energy and bioplastics are being recycled. Plus, they may generate environmentally helpful items using methods such as 3D printing (Ford and Despeisse, 2016). The environmental influence may be meaningfully condensed via this kind of sustainable novelty, which also unlocks new opportunities in the market and gives their goods a competitive advantage.

To recover sustainability and efficacy, smart supply chain enterprises are embracing new technology. They may diminish resource use, releases, and waste in this manner (Kamble et al., 2018). Since it can make supply chains more exposed and accountable, blockchain technology is also being used carefully. According to Kouhizadeh et al. (2021), this lets organizations to well monitor their effect and steer clear of rules-related and reputation-damaging risks.

These firms are not just refining technology, but they are also coming up with ground-breaking business models and cooperating to make sustainability a precedence. One example is industrial relationships, in which businesses in close proximity to one alternative share resources and waste products (Fraccassia et al., 2017). Another method that helps reduce resource usage and makes matters live longer is the product-service system or sharing platform, where customers share things instead of purchasing them absolute (Tukker, 2015).

9.3.3 Case Studies: Companies Leveraging Ambidexterity for Sustainability

A number of businesses have revealed that ambidexterity is an influential tool for making long-term enhancements to their supply chains. Patagonia is a fantastic instance in point. This outside wear tag is renowned for its eco-conscious products. Their know-how in material procurement and production permits them to diminish energy use and waste. Offering repair and recycling services, Patagonia has hopped on the circular economy bandwagon. Furthermore, they are pioneers when it comes to using eco-friendly materials such as organic cotton and recycled polyester (Patagonia, 2022). One more awesome thing? Among their brainstormed business strategies is the option to rent or lease outdoor equipment rather than sell it outright. Thanks to this, we can save resources and cut down on waste.

Additionally, this large food and drink corporation accomplishes some quite remarkable things. They use their supply chain know-how to improve logistics and cut greenhouse gas emissions while still innovating in sustainability. Danone has put money into regenerative farming, made plant-based dairy alternatives, and looked into biodegradable packaging options (Danone, 2021). What's more? They have teamed up with others to share by-products and resources, which means less waste and extra value from things that might have been thrown away.

IKEA, the Swedish furniture giant, is another inspiring example. Known for its flat-pack furniture, they also use their logistics skills to lower transportation emissions and waste. At the same time, IKEA looks for sustainable materials. They invest in bamboo and recycled plastics (IKEA, 2021). IKEA also offers take-back and resale programs for used furniture, showcasing how ambidexterity helps promote sustainability throughout every step of a product's life.

We must also acknowledge certain difficult obstacles that businesses may encounter, even if these examples demonstrate how ambidexterity may result in genuine, long-term transformation. It may be challenging to implement these agile tactics due to issues with resource allocation, team culture, and the sluggish pace of organizational change. Later on in this chapter, we will discuss such difficulties.

Each of these case studies exemplifies a different company's use of ambidexterity to promote environmentally friendly supply chain processes. These companies solidify their roles as sustainability pioneers by building on existing successes and investigating new avenues of operation; in the process, they benefit all parties and contribute to a more promising future for Earth.

9.4 AMBIDEXTERITY AND DIGITAL TRANSFORMATION IN SUPPLY CHAINS

To flourish in the era of Industry 4.0, supply chain organizations must experience digital transformation to endure competition. The capability to play to one's métiers while also discovering new chances is a vital skill for businesses to efficaciously traverse the digital landscape. By uniting the use of familiar technologies with the incorporation of novel concepts, ambidextrous supply chains have the possibility to unleash previously unimagined levels of competence, quickness, and sustainability.

9.4.1 Leveraging Digital Technologies for Supply Chain Optimization

Another facet of ambidexterity that can be exploited is the use of well-established technologies like RFID, Enterprise Resource Planning (ERP) systems, and advanced analytics to lift supply chain visibility, cut waste, and augment decision-making (Büyüközkan and Göçer, 2018).

Walmart, the world's largest retailer, has executed a sophisticated SCM system that mixes radio frequency identification (RFID), data analytics, and the aptitude to monitor items in real time. Waller and Fawcett (2013) has allowed the business to maximize inventory levels, diminish waste, and increase product handiness during its widespread network of shops and delivery hubs. Likewise, Unilever, a global consumer goods company, has modernized supply chain operations, condensed lead times, and increased receptiveness to changing market demands through the use of digital technology, such as advanced planning and scheduling tools (Unilever, 2020a, 2020b). Over better transportation and waste minimization, Unilever has been able to diminish its environmental influence while simultaneously aggregate operational efficiency obligations through this technology.

9.4.2 Exploring Emerging Technologies and Disruptive Innovations

New technologies such as AI, AR, and VR brought digital transformation and created avenues to integrate with ambidexterity (Dolgui et al., 2020). The size of the global generative artificial intelligence (AI) supply chain market is predicted below for the years 2023–2032, with a compound annual growth rate (CAGR) of 45.3%, as seen in Figure 9.2.

DHL and Amazon have invested in AI-powered solutions to augment warehouse operations, optimize routes, and conduct predictive analytics. This has led to cost

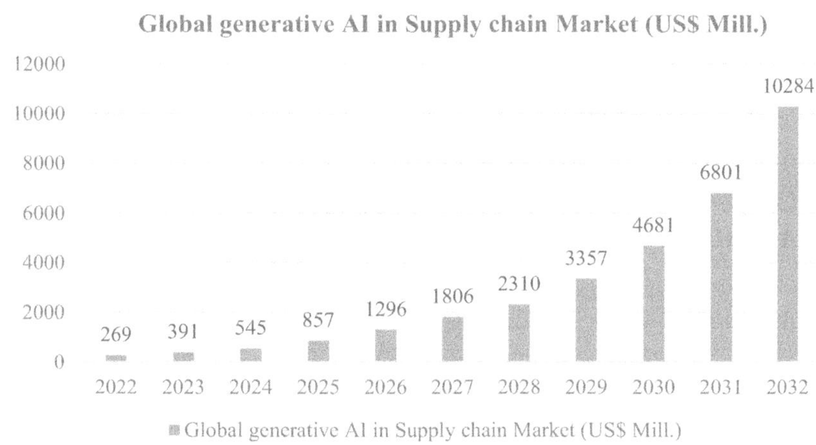

FIGURE 9.2 Prediction of the global Generative AI in Supply Chain Market size from 2023 to 2032.

Source: https://marketresearch.biz/report/generative-ai-in-supply-chain-market/

reserves and increased efficacy (Galindo and Batta, 2013). New trade models and collaboration methods are being discovered by ambidextrous supply chain organizations as a consequence of digital technology. A good instance is the beginning of sharing economies and digital platforms, which have unlocked the door for companies to discover cooperative logistics models. In these models, many companies may cartel their transportation capabilities and distribution routes to diminish costs and environmental influence (Nica et al., 2022).

9.4.3 Case Studies: Companies Embracing Ambidexterity for Digital Transformation

Ambidexterity has the prospective to hasten digital transformation in supply chains, as revealed by several firms. Amazon is a prime case; it is a vast online retailer known for its ground-breaking supply chain and logistics. Along with its existing fortes in distribution and warehouse management, Amazon has vigorously explored and developed innovative technologies such as drone delivery systems, AI, and robotics (Amazon, 2022). This ambidextrous method has allowed Amazon to elevate its supply chain operations and stay ahead of digital innovation.

Among these prominent cases is DHL, a global logistics provider. By using its expertise in logistics and transportation, DHL has heightened its prevailing operations via the use of digital technologies such as telematics, route optimization, and advanced analytics. The company is on the viewpoint for evolving technologies like the Internet of Things (IoT), augmented reality (AR), and collaborative robotics to lift operational efficiency and augment its products (DHL, 2021).

The German industrial corporation Siemens assists as a last illustration. Siemens has optimized its supply chain with the use of digital technology such as enterprise resource planning (ERP) systems and advanced scheduling and planning tools, using its expertise in manufacturing and automation. In an attempt to revamp its manufacturing procedures and new product development, the company has been observing state-of-the-art technologies containing industrial IoT, digital twins, and additive manufacturing (Siemens, 2022). This ambidextrous method has helped Siemens endure competition in the digital age while also encouraging sustainability.

9.5 OBSTACLES AND METHODS FOR DEVELOPING AMBIDEXTERITY

Ambidexterity, or the aptitude to involve in both exploitative and exploratory activities at the similar time, is a multifaceted and challenging goal that businesses must endeavor for via careful planning and execution. Even with the impending aid of ambidexterity, such as amplified creativity, adaptableness, and endurance, striking this balance will not be modest. This section investigates into the main challenges and methods for cultivating ambidexterity, with a specific prominence on resource allocation and trade-offs, cultural barriers and change management, overpowering organizational inaction and risk aversion, and plans that are predominantly appropriate to supply chain organizations.

9.5.1 Resource Allocation and Trade-offs

When it comes to allocating one's physical and mental energies wisely between exploitative and exploratory endeavors, ambidexterity poses one of the most significant challenges. When organizations divide their few resources (time, money, and human capital) among what seem to be competing priorities, they always face trade-offs (Birkinshaw and Gupta, 2013).

Spending a lot of money to make minimal changes while enhancing preexisting items, processes, and abilities is what exploitative activities are all about. However, in order to conduct research, experiment, and take risks, exploratory ventures that seek to create ground-breaking ideas and go into unexplored areas need substantial financial investments (Raisch et al., 2009).

Striking a balance between these competing demands is challenging because, on the one hand, going all in on exploration could drain resources and jeopardize the company's present revenue sources (O'Reilly and Tushman, 2013), while on the other hand, going all in on exploitation could lead to stagnation and missed opportunities. Diversifying an organization's focus between exploratory and exploitative tasks, forming teams or individuals with ambidextrous abilities, or implementing dynamic resource allocation mechanisms to adapt to changing conditions are all potential approaches to effective resource allocation (Birkinshaw and Gupta, 2013; O'Reilly and Tushman, 2013).

9.5.2 Cultural Barriers and Change Management

Due to the potential inherent differences between the mindsets, values, and behaviors needed for exploitation and discovery, cultivating ambidexterity often necessitates a substantial shift in organizational culture (Lubatkin et al., 2006).

While exploratory projects necessitate a culture that promotes creativity, risk-taking, and a tolerance for failure, exploitative activities flourish in settings that give efficiency, discipline, and incremental improvements top priority (Andriopoulos and Lewis, 2009).

Dealing with these cultural obstacles may be somewhat difficult as firmly ingrained ideas and behaviors often resist change. In this sense, effective change management strategies—which include transparent communication of the vision and reasoning for ambidexterity, therefore, creating a common awareness of its importance—as well as aggressively addressing opposition and concerns (Kotter, 2012).

Moreover, companies could have to carry out focused campaigns to foster the desired cultural traits, including setting up specialized innovation labs or incubators, encouraging cross-functional cooperation and knowledge exchange, and providing training and development chances to improve creativity and risk-taking attitudes (Andriopoulos and Lewis, 2009; O'Reilly and Tushman, 2013).

9.5.3 Overcoming Organizational Inertia and Risk Aversion

Pursuits of ambidexterity might be seriously hampered by organizational stagnation and risk aversion. Well-known companies often become firmly rooted in their current procedures, systems, and business models, which makes it difficult to adopt the innovative and exploratory attitude needed (Hannan Freeman, 1984). According to Tushman and O'Reilly (1996), this lack of movement might be caused by a combination of factors, including sunk costs in existing infrastructure, concerns about being eaten up by competitors, and the inherent risk aversion that comes with expanding and thriving businesses. Companies that are serious about overcoming these challenges should push for a culture of constant learning, new ideas, and cautious risk-taking. Fast prototyping and concept testing, skunk works projects or specialist innovation teams, and a relentless pursuit of and adoption of disruptive technologies and business models are all examples of what may fall under this category (Tushman and O'Reilly, 1996; O'Reilly and Tushman, 2013).

Good leadership is also crucial for organizations to overcome risk avoidance and stagnation. According to Smith and Tushman (2005), senior executives should make an effort to promote ambidexterity, explain its worth, and provide the resources needed for experimental ventures. They must also be willing to challenge conventional thinking, encourage calculated risks, and foster a learning environment that values success and failure equally.

9.5.4 Strategies for Fostering Ambidexterity in Supply Chain Organizations

The pursuit of ambidexterity fetches exclusive opportunities and challenges. When it derives to taming agility, receptiveness, and customer pleasure, as well as operational efficiency and cost optimization, these organizations must find an equilibrium amid the two demands (Lee, 2004).

One approach that supply chain companies may use to endorse ambidexterity is to create distinct organizational units or teams that emphasize exploratory and exploitative activities, separately. This organizational division permits each department to emphasize its own unique mission while also reducing the probability of friction and conflict (O'Reilly and Tushman, 2013). In order to facilitate this, it may be obliging to implement flexible supply chains, stand-in strong associations with key suppliers and partners, and use of agile approaches and digital technologies that allow rapid reconfiguration and adaptation (Naim and Gosling, 2011; Ivanov et al., 2019).

Having leaders with ambidextrous skills is helpful for supply chain firms at all levels. Agreeing to Mom et al. (2009) and Hajmohammad and Vachon (2016), ambidextrous leaders generate a setting that emboldens both discovery and exploitation since they can balance instant needs with longer-term purposes. Attaining ambidexterity in supply chain firms needs a long-term commitment that includes aligning strategy, structure, processes, and culture.

9.6 CONCLUSION

For primary supplies business organizations, these approaches include the committing to an organization's ambidexterity. In this approach, exploitation and exploration are achieved at the same time. To successfully attain success, there should be a delicate balance maintained between using existing competencies and searching for new horizons. Some of the available instances include resource efficacy, incorporating circular economy concepts into the structure (Ivanov et al., 2019). According to the outstanding case studies, Walmart, Unilever, and Patagonia illustrate how such an approach has been hardly practiced in supply chain operations toward sustainability and competitive advantage rationale. However, resources, organizational inertia, cultural issues, and trade-off in resource have limited a lot of focus on what else needs to be addressed (Kamble et al., 2018). In order to achieve this successfully, it is important to employ certain strategic measures like structural differentiation, promoting ambidextrous organization, engaging leaders, and so on. Foresighted, an ambidextrous supply chain possesses the potential to change the industry by delivering sustainability, pragmatism, and non-innovative exposure, and is receptive toward new trends and supporting advancements of the evolution. Thus, this will enable them to maintain a competitive edge in an always evolving industry.

9.6.1 Implications for Supply Chain Practitioners and Policymakers

Most importantly, people involved in the field need to maintain a proper balance between the aspects of exploration and exploitation within the scope of the supply chain. This includes seeking out novel ideas and "green" technologies while making sure that all the currently in use processes are geared toward achieving maximum possible efficiency (Bocken et al., 2016; Chakma et al., 2021). In such circumstances, the development of cooperative relations with suppliers and other participants of the supply chain becomes as necessary as acknowledging the possible and ample challenges in the supply chain. Likewise, however, policymakers also play a prime role in the molding of the new practices toward a more multifaceted approach. There is a need for establishing a legal environment which not only supports creativity and novel practices but also offers benefits to people putting money in green technologies, "normative influences" (Guldmann et al., 2019). Furthermore, ambidexterity enhancement policies in various organizations may include information exchange and risk-taking behavior within the organization as long as such policies come with appropriate benefits. Supply chain practitioners and policymakers may augment the resilience, creativity, and sustainability of supply chains by paralogically enhancing the dimension of borrowed strategies as a means of safeguarding their relevance and competitiveness in the contemporaneous global setting (Hajmohammad and Vachon, 2016).

9.6.2 Future Research Directions

New study is done in the field of organizational ambidexterity and its influence on the sustainability of the supply chain which is a worthwhile area of growing interest. Within the context of long-term competition and sustainability, practitioners should, in the first instance, understand how companies effectively cope with the management of the portfolio comprising of both exploitation and exploration strategies. This would require looking deeper into the dynamics of resource distribution and concessions in ambidextrous supply chain structures. Moreover, more extended study of the roles of leadership and organizational culture in promoting and embedding ambidexterity in absolute supply chain enterprises may help to understand mechanisms linked to successful ambidextrous strategies. The relationship between ambidexterity and digital supply chain transformation appears to be very interesting for future studies, in particular within the light of the influence of future technologies such as blockchain, AI, IoT, and others on the extent of ambidexterity and creation of sustainable innovations. The sustainability impact of ambidexterity can be further examined in future studies through a longitudinal study that evaluates the efficacy of supply chains with ambidextrous strategies over time.

REFERENCES

Amazon. (2022). Amazon logistics: Delivering the future. https://www.aboutamazon.com/news/logistics/amazon-logistics

Andriopoulos, C., and Lewis, M. W. (2009). Exploitation-exploration tensions and organizational ambidexterity: Managing paradoxes of innovation. *Organization Science*, 20(4), 696–717. https://doi.org/10.1287/orsc.1080.0406

Beske, P., Land, A., and Seuring, S. (2014). Sustainable supply chain management practices and dynamic capabilities in the food industry: A critical analysis of the literature. *International Journal of Production Economics*, 152, 131–143. https://doi.org/10.1016/j.ijpe.2013.12.026

Birkinshaw, J., & Gibson, C. (2004). Building ambidexterity into an organization. *MIT Sloan Management Review*, 45(4), 47–55.

Birkinshaw, J., and Gupta, K. (2013). Clarifying the distinctive contribution of ambidexterity to the field of organization studies. *Academy of Management Perspectives*, 27(4), 287–298. https://doi.org/10.5465/amp.2012.0167

Birkinshaw, J., Zimmermann, A., & Raisch, S. (2016). How do firms adapt to discontinuous change? Bridging the dynamic capabilities and ambidexterity perspectives. *California Management Review*, 58(4), 36–58.

Bocken, N. M., de Pauw, I., Bakker, C., and van der Grinten, B. (2016). Product design and business model strategies for a circular economy. *Journal of Industrial and Production Engineering*, 33(5), 308–320. https://doi.org/10.1080/21681015.2016.1172124

Brix, J. (2019). Ambidexterity and organizational learning: Revisiting and reconnecting the literatures. *The Learning Organization*, 26(4), 337–351.

Büyüközkan, G., and Göçer, F. (2018). Digital supply chain: Literature review and a proposed framework for future research. *Computers in Industry*, 97, 157–177. https://doi.org/10.1016/j.compind.2018.02.010

Chakma, R., Paul, J., and Dhir, S. (2021). Organizational ambidexterity: A review and research agenda. *IEEE Transactions on Engineering Management*, 71, 121–137.

Chang, Y. Y., and Hughes, M. (2012). Drivers of innovation ambidexterity in small-to medium-sized firms. *European Management Journal*, 30(1), 1–17

Cui, R., Gallino, S., Moreno, A., & Zhang, D. J. (2018). The operational value of social media information. *Production and Operations Management*, 27(10), 1749–1769.

Danone. (2021). Danone's corporate purpose and priorities. https://www.danone.com/impact/sustainability.html

DHL. (2021). Logistics trend radar. https://www.dhl.com/global-en/home/insights-and-innovation/thought-leadership/trend-reports/logistics-trend-radar.html

Dolgui, A., Ivanov, D., Sethi, S. P., and Sokolov, B. (2020). Digital supply chain twins: Managing the ripple effect, resilience, and disruption risks by data-driven optimization, simulation, and visibility. In A. A. Ivanov, and D. Dolgui (Eds.), *Supply Chain Disruption and Resilience in a Digital Age* (pp. 105–136). Springer.

Duncan, R. B. (1976). The ambidextrous organization: Designing dual structures for innovation. *The management of organization*, 1(1), 167–188.

Ford, S., and Despeisse, M. (2016). Additive manufacturing and sustainability: An exploratory study of the advantages and challenges. *Journal of Cleaner Production*, 137, 1573–1587. https://doi.org/10.1016/j.jclepro.2016.04.150

Fraccassia, L., Giannoccaro, I., and Albino, V. (2017). Rethinking resilience in industrial symbiosis: Conceptualization and measurements. *Ecological Economics*, 137, 148–162. https://doi.org/10.1016/j.ecolecon.2017.02.026

Galindo, G., and Batta, R. (2013). Review of recent developments in OR/MS research in disaster operations management. *European Journal of Operational Research*, 230(2), 201–211. https://doi.org/10.1016/j.ejor.2013.01.039

García-Granero, A., Fernández-Mesa, A., Jansen, J. J., & Vega-Jurado, J. (2018). Top management team diversity and ambidexterity: The contingent role of shared responsibility and CEO cognitive trust. *Long Range Planning*, 51(6), 881–893.

Gianzina-Kassotaki, O. (2017). *Ambidexterity and leadership: a multilevel analysis of the aerospace and defense organizations* (Doctoral dissertation, University of Warwick).

Gibson, C., & Birkinshaw, J. (2004). Contextual determinants of organizational ambidexterity. *Academy of Management Journal*, 47(2), 209–226.

Good, D., & Michel, E. J. (2013). Individual ambidexterity: Exploring and exploiting in dynamic contexts. *Journal of Psychology*, 147(5), 435–453.

Guha, S., & Kumar, S. (2018). Emergence of big data research in operations management, information systems, and healthcare: Past contributions and future roadmap. *Production and Operations Management*, 27(9), 1724–1735.

Guldmann, E., Huulgaard, R. D., and Remmen, A. (2019). From circular economy to CirculArt economy. *Management Review Quarterly*, 69(2), 265–291. https://doi.org/10.1007/s11301-019-00154-9

Gupta, A. K., Smith, K. G., & Shalley, C. E. (2006). The interplay between exploration and exploitation. *Academy of Management Journal*, 49(4), 693–706.

Hajmohammad, S., and Vachon, S. (2016). Mitigation, avoidance, or acceptance? Supply chain perspectives on climate change risks. *Business Strategy and the Environment*, 25(5), 319–331. https://doi.org/10.1002/bse.1879

Hannan, M. T., and Freeman, J. (1984). Structural inertia and organizational change. *American Sociological Review*, 49(2), 149–164. https://doi.org/10.2307/2095567

He, Z. L., & Wong, P. K. (2004). Exploration vs. exploitation: An empirical test of the ambidexterity hypothesis. *Organization Science*, *15*(4), 481–494.

IKEA. (2021). Sustainable living at IKEA. https://www.ikea.com/ca/en/this-is-ikea/sustainable-living-pub69a3bc64

Im, G., & Rai, A. (2014). IT-enabled coordination for ambidextrous interorganizational relationships. *Information Systems Research*, 25(1), 72–92.

Ivanov, D., Dolgui, A., and Sokolov, B. (2019). The impact of digital technology and Industry 4.0 on the ripple effect and supply chain risk analytics. *International Journal of Production Research*, 57(3), 829–846. https://doi.org/10.1080/00207543.2018.1488086

Ivanov, D., Sokolov, B., and Dolgui, A. (2014). The Ripple effect in supply chains: Trade-off 'efficiency-flexibility-resilience' in disruption management. *International Journal of Production Research*, 52, 2154–2172.

Kamble, S. S., Gunasekaran, A., and Arha, H. (2018). Understanding the blockchain technology adoption in supply chains-Indian context. *International Journal of Production Research*, 57(7), 2009–2033. https://doi.org/10.1080/00207543.2018.1518610

Kauppila, O. P., and Tempelaar, M. P. (2016). The social-cognitive underpinnings of employees' ambidextrous behaviour and the supportive role of group managers' leadership. *Journal of Management Studies*, 53(6), 1019–1044.

Koberg, E., and Longoni, A. (2019). A systematic review of sustainable supply chain management in global supply chains. *Journal of Cleaner Production*, 207, 1084–1098.

Kotter, J. P. (2012). *Leading Change*. Harvard Business Review Press.

Kouhizadeh, M., Zhu, Q., and Sarkis, J. (2021). Blockchain and the circular economy: Potential tensions and critical strategies for sustainable supply chains. *Production Planning and Control*, 32(11–12), 950–966. https://doi.org/10.1080/09537287.2019.1695925

Lee, H. L. (2004). The triple – A supply chain. *Harvard Business Review*, 82(10), 102–113.

Li, V., Mitchell, R., & Boyle, B. (2016). The divergent effects of transformational leadership on individual and team innovation. *Group & Organization Management*, 41(1), 66–97.

Lubatkin, M. H., Simsek, Z., Ling, Y., and Veiga, J. F. (2006). Ambidexterity and performance in small-to medium-sized firms: The pivotal role of top management team behavioral integration. *Journal of Management*, 32(5), 646–672. https://doi.org/10.1177/0149206306290712

Mom, T. J., Chang, Y. Y., Cholakova, M., and Jansen, J. J. (2018). A multilevel integrated framework of firm HR practices, individual ambidexterity, and organizational ambidexterity. *Journal of Management*, 45(7), 3009–3034.

Mom, T. J., Van Den Bosch, F. A., and Volberda, H. W. (2009). Understanding variation in managers' ambidexterity: Investigating direct and interaction effects of formal structural and personal coordination mechanisms. *Organization Science*, 20(4), 812–828. https://doi.org/10.1287/orsc.1090.0427

Mu, T., Van Riel, A., & Schouteten, R. (2022). Individual ambidexterity in SMEs: Towards a typology aligning the concept, antecedents and outcomes. *Journal of Small Business Management*, 60(2), 347–378.

Naim, M. M., and Gosling, J. (2011). On leanness, agility and leagile supply chains. *International Journal of Production Economics*, 131(1), 342–354. https://doi.org/10.1016/j.ijpe.2010.04.045

Nica, E., Popescu, G. N., and Nicolăescu, E. (2022). Digital transformation of the supply chain management in the context of the platform economy: A systematic literature review. *Sustainability*, 14(6), 3152. https://doi.org/10.3390/su14063152

O'Reilly, C. A., and Tushman, M. L. (2008). Ambidexterity as a dynamic capability: Resolving the innovator's dilemma. *Research in Organizational Behaviour*, 28, 185–206. https://doi.org/10.1016/j.riob.2008.06.002

O'Reilly, C. A., and Tushman, M. L. (2013). Organizational ambidexterity: Past, present, and future. *Academy of Management Perspectives*, 27(4), 324–338. https://doi.org/10.5465/amp.2013.0025

Ojha, D., Acharya, C., and Cooper, D. (2018). Transformational leadership and supply chain ambidexterity: mediating role of supply chain organizational learning and moderating role of uncertainty. *International Journal of Production Economics*, 197, 215–231.

Pagell, M., and Shevchenko, A. (2014). Why research in sustainable supply chain management should have no future. *Journal of Supply Chain Management*, 50(1), 44–55. https://doi.org/10.1111/jscm.12037

Park, Y., Pavlou, P. A., & Saraf, N. (2020). Configurations for achieving organizational ambidexterity with digitization. *Information Systems Research*, 31(4), 1376–1397.

Patagonia. (2022). Environmental andsocial footprint. https://www.patagonia.com/our-footprint.html

Patel, P. C., Messersmith, J. G., & Lepak, D. P. (2013). Walking the tightrope: An assessment of the relationship between high-performance work systems and organizational ambidexterity. *Academy of Management Journal*, 56(5), 1420–1442.

Raisch, S., Birkinshaw, J., Probst, G., and Tushman, M. L. (2009). Organizational ambidexterity: Balancing exploitation and exploration for sustained performance. *Organization Science*, 20(4), 685–695. https://doi.org/10.1287/orsc.1090.0428

Rothaermel, F. T., and Alexandre, M. T. (2009). Ambidexterity in technology sourcing: The moderating role of absorptive capacity. *Organization Science*, 20(4), 759–780.

Sahi, G. K., Gupta, M. C., and Cheng, T. C. E. (2020). The effects of strategic orientation on operational ambidexterity: a study of Indian SMEs in the industry 4.0 era. *International Journal of Production Economics*, 220, 107395.

Schniederjans, D. G., Curado, C., and Khalajhedayati, M. (2020). Supply chain digitisation trends: An integration of knowledge management. *International Journal of Production Economics*, 220, 107439.

Seuring, S., and Müller, M. (2008). From a literature review to a conceptual framework for sustainable supply chain management. *Journal of Cleaner Production*, 16, 1699–1710.

Siemens. (2022). Digital transformation in the supply chain. https://new.siemens.com/global/en/company/topic-areas/digital-solutions/supply-chain.html

Smith, W. K., and Tushman, M. L. (2005). Managing strategic contradictions: A top management model for managing innovation streams. *Organization Science*, 16(5), 522–536. https://doi.org/10.1287/orsc.1050.0134

Tukker, A. (2015). Product services for a resource-efficient and circular economy – A review. *Journal of Cleaner Production*, 97, 76–91. https://doi.org/10.1016/j.jclepro.2013.11.049

Tushman, M. L., and O'Reilly, C. A. (1996). Ambidextrous organizations: Managing evolutionary and revolutionary change. *California Management Review*, 38(4), 8–29. https://doi.org/10.2307/41165852

Unilever. (2020a). Digital technology in the supply chain. https://www.unilever.com/planet-and-society/supply-chain/digital-technology-in-the-supply-chain/

Unilever. (2020b). Unilever's sustainable living plan. https://www.unilever.com/sustainable-living/

Venugopal, A., Krishnan, T.N., Upadhyayula, R. S., and Kumar, M. (2020). Finding the microfoundations of organizational ambidexterity-Demystifying the role of top management behavioural integration. *Journal of Business Research*, 106, 1–11.

Vessey, I., and Ward, K. (2013). The dynamics of sustainable IS alignment: The case for IS adaptivity. *Journal of the Association for Information Systems*, 14(6), 283–311.

Waller, M. A., and Fawcett, S. E. (2013). Data science, predictive analytics, and big data: A revolution that will transform supply chain design and management. *Journal of Business Logistics*, 34(2), 77–84.

Walmart. (2022). Supply chain sustainability. https://corporate.walmart.com/esgreport/environmental/supply-chain-sustainability

Werder, K., and Heckmann, C. S. (2019). Ambidexterity in information systems research: Overview of conceptualizations, antecedents, and outcomes. *Journal of Information Technology Theory and Application (JITTA)*, 20(1), 2.

Winter, M., and Knemeyer, A. M. (2013). Exploring the integration of sustainability and supply chain management: Current state and opportunities for future inquiry. *International Journal of Physical Distribution and Logistics Management*, 43(1), 18–38.

World Commission on Environment and Development [WCED]. (1987). Our common future. Adopted by the United Nations General Assembly. A/42/427. Oxford University Press, Oxford.

Zavala-Alcívar, A., Verdecho, M. J., and Alfaro-Saiz, J. J. (2020). A conceptual framework to manage resilience and increase sustainability in the supply chain. *Sustainability*, 12(16), 6300.

Part 3

Role of Government, Institution, and Other Stakeholder in the Digital Supply Chain

10 Balancing Innovation and Efficiency: The Role of Organizational Ambidexterity in Digital Supply Chain Management

Liudmila Veselova, Prakash Chandra Bahuguna, and Aditi Bisht

10.1 INTRODUCTION

As the business environment is evolving at a fast pace due to new technologies and changing markets, companies are experiencing a difficult task. They have to optimize the existing resources, activities, and competencies to remain effective and realize the short-term objectives. This entails gaining enhanced current operations, leveraging existing market standings, and being efficient in managing resources. They also need to seize new opportunities, technologies, and markets to remain innovative, adaptable, and ready for the shifting customers' demands, competitive forces, or threats. Organizational ambidexterity is essentially about how an organization can manage these two tasks simultaneously – employing current strengths and at the same time searching for new opportunities. This strategy works because it enables organizations to balance the conflict of optimization against innovation and consistency against change. The uncertainties are managed by the organizations, take advantage of new opportunities, advanced technologies, and share the risk with the altering market change in parallel to the dual-focus management. This type of dynamic capabilities is another crucial requirement of firms that are to experience sustainable growth and competition in today's dynamic business environment (Lis et al., 2018). It recognizes the fact that organizations cannot rely on exploitation or exploration alone to succeed; that an organization has to be competent in both these concepts. Duplex organizations are not only very good at continuously improving and refining current practice but are also constantly trying out new ideas, new products, or work methods (Belhadi et al., 2022).

Technological advancement also challenges the way of planning, sourcing, distribution, and delivering goods in supply chain management. The supply chain in digital

DOI: 10.1201/9781003477242-13

form is one of the most significant changes in current processes, characterized by immediacy, data-driven, and full insight. New technologies offer corporations several leaps forward in performance, velocity, and customer satisfaction in the supply chain function. Shift in the paradigm in supply chain management, mainly based on real-time integration of data collection, analysis, and process visibility. Digitalization presents organizations with some of the best chances to enhance work quality, response time, and customer satisfaction across the value chain. Whereas Internet of Things (IoT) sensors and devices help establishments capture data, AI-based analytics enable better understanding of the data, and blockchain-based platforms facilitate connecting several processes, making the most of resources and improving the cooperation with supply chain partners. However, to go further, actively leveraging digital technologies within the supply chain calls for going beyond technology investment. It requires effective planning that integrates technology-based programs and projects with other strategic directions, strengths, visions, and beliefs of the organization.

The importance of the concept of organizational ambidexterity is particularly highlighted here. When approaching the task of digital supply chain management with the concept of ambidexterity, firms are able to leverage digital tools while preserving and enhancing the core competencies of their supply chains. This means reviewing how best to leverage the current digital tools to support the existing processes, as well as identifying new digital options in order to promote constant development and optimization of the supply chain. Organizational ambidexterity leverages with digital supply chain management are important for organizations to thrive in today's world (Büyüközkan & Göçer, 2018). The paired approach and effective application of new digital tools allow businesses to address the concerns of supply chain management, reduce rigidity, and achieve greater competitiveness in a highly uncertain and competitive environment. In this chapter, we will expand upon the literature, processes, and implications for the co-dependent relationship of digital supply chain integration and organizational ambidexterity, with a view to providing clarity to organizational stakeholders on managing the inherent challenges of digital supply chain management.

Stand-alone out here is that the top organizational ambidexterity has to be in sync with the digital supply chain management, and there are no hard and fast rules for this. The concept is that every enterprise is unique in its circumstances, and it encounters both risks and opportunities. Consequently, the ambidextrous strategies, as the approaches to enable the digital supply chain, should be aligned with the specific goals and needs of each company. Moreover, the integration of ambidexterity with digital SCM needs a systematic approach including leadership, culture, technology, processes, and partnership (MacCarthy & Ivanov, 2022). Leadership is, therefore, critical in defining the direction, tone, and championing of change and inculcating innovation in the organization. The leaders must highlight both ambidexterity and digitalization as key priorities and carefully distribute resources while providing appropriate assistance to the workers during the transformation.

Moreover, culture of an organization is very essential in explaining the behavior, attitude, and norms of an organization. Organization culture that supports cooperation, experimentation, and knowledge development provides ambidextrous behavior and digital development environment (Lis et al., 2018). Organizations need to develop an openness and trust culture through which people are willing to question the status quo, gain and apply change, as well as accept digitization.

From a business tactics perspective, there is a need to acquire the right technology that will facilitate collaboration, analytics, integration, and provide supply chain transparency. This could involve connectors between Multiple ERP Systems, SaaS solutions, as well as emergent technologies such as IoT, AI, or blockchain. But it does not mean that having the technology is adequate to get the results. They also have to be able to check whether the work their business does online complements its general company strategy and returns value to customers and business shareholders. With regards to process, companies have to revisit and transform supply chain in order to make the best use of digital technology. This could be a shift in conventional approaches such as demand forecasting, stock handling, and order fulfillment to leverage digital means. Business entities must employ flexible and evolving systems that may be changed depending on the developments in the marketplace, customers, and tools.

A firm requires forging good relationships with outside stakeholders, customers, vendors, and providers of technologies. Therefore, through collaborations such as these, companies and these partners can effectively leverage each other's capabilities, share both the risk and revenues as well and see diversity in the supply chain. Those links enable corporations to identify novel markets, technologies, and resources that help them advance their digital transformation and enhance their positions in the market. In addition to identifying the antecedents, this chapter examines the practical application and the drawbacks of organizational ambidexterity with particular reference to the digital supply chain while also discussing the importance of strategic fit and organizational culture. It highlights the importance of supply chain strategy being consistent with organizational goals when attempting to integrate such measures as ambidexterity into the organization's strategic framework. Manager has to ensure digital supply chain initiatives align to value drivers like cost, revenue or customer, value, and sustainability (Huddar et al., 2017).

First of all, cultural perspective of a business organization is seen to be central to the management of ambidexterity. Accepted when it comes to motivation, collaboration, and focusing on creating a culture where continuous improvement is okay or normal, changes become acceptable because not only is it okay to try new strategies, we can also challenge the current norms or paradigms ingrained in the organizational structure. Managers are crucial in supporting ambidexterity as they are to foster an organizational climate which encourages utilization of digital technology and introduction of innovation within the SCM. Furthermore, the chapter also focuses on the dynamic nature of digital supply chains and the need for flexibility, especially in the supply chain leadership. Dynamic capabilities are defined as an organization's ability to create, sense, capture, and transform opportunities as well as threats within a given environment. In the context of digital supply chains, there is also a need to build dynamic capabilities, necessary for changing environmental conditions, technological shift, and competition challenges (Ageron et al., 2020). A proactive approach to both analyzing the trend and responding to it is particularly essential, as is the capacity for swift strategic adjustment. Adaptive leadership is an important approach in management of organizations during digital transformation and encouraging dual processes. Adaptive leaders are flexible, assertive, and receptive to change. They are still able to deal with change and tricky situations, as well as to help others embrace change and develop innovations. Therefore, the chapter focuses on how to build teams and partnerships to enhance innovation within the context of the digital supply chain. But in the modern era of globalization and integration, the companies cannot innovate without others. Thus, cooperation with the supply chain partners, customers, and other stakeholders is important to unlocking the value creation together and leveraging on each other's strengths. This way, the ideas can be designed by companies, but implemented by others, and in turn, the companies can enter new markets faster and accelerate the process of the digital evolution. This approach promotes open innovation since group knowledge is exploited to solve supply chain issues. Digital transformative strategic partnerships enhance the creation of new opportunities for business outcomes and competitive advantage (Queiroz et al., 2021). Therefore, this chapter is beneficial for the supply chain management and business strategies to modify and apply the ambidexterity theory with the digital context that seems to be useful in academic research field for the academician, practitioners, as well as the policymakers. To sum up, the present chapter underlined that organizational ambidexterity has to be complemented with the digital supply chain management framework and can be achieved only with advanced leadership, technologies, processes, and partnership strategies. Therefore, management can achieve an ambidextrous perspective and apply digital tools in supply chains to enhance supply chain performance and flexibility and gain a competitive advantage in the current world (Belhadi et al., 2022).

10.2 THEORETICAL FOUNDATIONS

In the field of organizational ambidexterity, various theoretical underpinnings have informed our understanding of the organization's ability to achieve both exploitation and exploration.

One such foundation is the one provided by James G. March in his paper "Exploration and Exploitation in Organizational Learning," published in 1991. For organizations it is difficult to manage the tension between on the one hand exploration (searching for new knowledge, innovation, and flexibility) and on the other hand exploitation (relying on existing knowledge, efficiency, and control). This theoretical foundation provided the scientific basis for analyzing and comparing the choices and integration of these two forms of organizational conduct.

Another theoretical framework used in the consideration of organizational ambidexterity is the dynamic capabilities theory availed in Teece et al. (1997). The paper examines the work of Teece's "Dynamic Capabilities and Strategic Management." Dynamic capabilities refer to the coordination and expansion of abilities within and outside the organization and the capability to deal with constantly changing environments. This brings a focus on the organizational resilience and learning to be able to respond to changes in the environment and capitalize on the opportunities that come with it. Dynamic capabilities theory describes methods and ways companies develop the resilience and strength required in adapting to platform market requirements for improving existing business models as well as developing new ones. In the same context, the concept of strategic paradoxes has now emerged as a key favorite when studying the notion of ambidexterity. Smith and Tushman (2005) defined paradoxical tensions by identifying primary sources of tension and conflict in organizations. Thus, significant and often even inherent tensions should be seen not as obstacles to functioning, but as opportunities which help to better address complex and unpredictable contexts. The strategic paradox perspective of organizational ambidextrousness calls for the management of the tensions between exploitation and exploration, so the organizations can optimize structures, processes, and culture for innovation and stability in a turbulent environment.

Certain theories help explain how organizations implement technology in supply chain management, for example resource-based view (RBV) framework highlights the importance of internal resources and capabilities in achieving sustainable competitive advantage. Digital operations technology is being adopted by organizations to bring creativity and extend innovation, increase efficiency, and add to the organization's responsiveness (Hashem et al., 2024). By following this framework, we are able to help organizations to analyze their established digital resources as well as to develop and increase the value of those resources.

The understanding of transaction costs and management structures in supply chain relationships is well explained by the TCE perspective. In managing digital supply chain, TCE facilitates lock-in analysis that's useful in deciding between making activities in-house or outsourcing selected activities and also sorting out the best governance structures for conducting digital transactions and managing relationships. When applying TCE principles, organizations can design efficient digital supply chain systems that can reduce costs, provide the necessary level of risk protection related to digital changes. The network view is used to conceptualize disjointed supply chain ties and systems of the Fourth Industrial Revolution. It emphasizes integration of pull and push strategies whereby supply chain members across a channel organize themselves in a system to pull or push supplies through the entire supply chain and make value delivery systems. In the digital supply chain, analytics theory demonstrates that digital technologies can connect the supply chain in real time, monitor and participate with visibility, and work collaboratively. Therefore, this approach allows businesses to integrate, update, and innovate about the digital platform and the networks to enhance the supply chain commerce.

The theoretical framework is a great source of understanding the organization's ambidextrous type and digital supply chain management. I have explained how using concepts such as exploration and exploitation, dynamic capabilities, strategic paradoxes, resource-based view, transaction cost economics, and the network perspective can help companies to develop balanced strategies to cope with the problems of the digital age and achieve the sustainable competitive advantage in the modern business environment.

10.3 LITERATURE REVIEW

Organizational ambidexterity has gained attention in the recent management literature because of its application in organizing in today's competitive business environment. It highlights the balanced approach between operation efficiency and new opportunity discovery, where operation efficiency means repeating what worked before, refining organizational processes, competencies, and business models to achieve short-term goals, and opportunity discovery is about creating new organizational processes, competencies, and business models to gain new resources, capabilities, and opportunities in the long run. Many learned that ambidexterity is the necessary condition to perform successfully in the existing conditions of high volatility and uncertainty (O'Reilly & Tushman, 2004). Hence, working in a ratio of exploitation and exploration, firms can improve the adaptive capability and hence make it in a better position to handle with uncertainty and gain competitive advantage. The digital supply chain implies a new approach to supply chain management which was triggered by technologies including the IoT, AI, and blockchain. Two articles reviewed for this manuscript (Agrawal & Narain, 2018; Ivanov et al., 2019) outline the digital supply chain as an interconnected system comprising of real-time connection, responsiveness, and visibility. The adoption of digital strategies affords organizations a chance to enhance process, speed, and consumer experience across the network. Yet, the exploitation of digital technologies has to be guided carefully by strategic positioning that strengthens existing supply chain assets while developing digital solutions that can be leveraged. It is for this reason that Coplien and Lambert suggest that managing organizational ambidexterity is crucial when handling change in the digital supply chain. On one side, organizations need to enhance the existing strategies for delivering greater competitive advantage in the market that has become possible with the help of present technology, better processes, and good relations with supply chain partners. On the other side, organizations need to seek the new technology, remain innovative and topical in the long run, Importantly, attempting new ideas, embracing change, in other words, supply chain organizations need to invest in technologies that may revolutionize traditional practices. Also, organizations are required to explore the new technology, stay creative and relevant in the long term. This requires being open to trying new things, adapting and investing in technologies that could change traditional supply chain practices.

Several recent papers have noted that organizational ambidexterity is particularly pertinent to the consideration of digital change. For instance, when applying the ambidextrous model, O'Reilly and Tushman (2004) argue that such organization stands a better chance at managing disruptive technologies, as well as at exploiting opportunities from such technological disruptions. They can convert it into opportunities to produce new advantages and to adapt the numerous digital developments into their corporate and business structures and activities. In the same way (Wang et al., 2020), concur with the assessments on the centrality of ambidexterity and the way in which it enables the re-balance of existing SC capabilities with supply chain exploration for digital innovation. There are difficulties that organizations experience while pursuing organizational ambidexterity. The firm need to overcome the resistance to change, resource constraints, and disparate departments in the organization. They also have to cope with the fact that digital ecosystems are much more complicated than just one individual or organization; they interact with numerous counterparts, have to process different kinds of data, and have to be updated with newer and stricter technology standards. To address these challenges, critical attributes such as leadership, clear strategic focus and organizational learning are essential. Thus, the literature seeks to help explain the need for organizational ambidexterity in conveying the complexities of the digital supply chain. While relying on the existing potentials and searching for new digital options, organization has a competitive advantage. However, there is some contradictory evidence, and among them, if the organization is to effectively achieve ambidexterity, it has to overcome different barriers and challenges. This research is pertinent to understanding how organizations can overcome such challenges and institutionalize the ambidexterity principles in their digital supply chain strategies.

10.4 PRACTICAL APPLICATIONS

Organizational ambidexterity provides the best solution when faced with the questions of exploitation and exploration in the context of the digital supply chain. Here are some practical applications of organizational ambidexterity for digital supply chain management:

Dual Supply Chain Strategies: Another approach made by the business to work on design and execution is developing a process for creating strategies to deal with creativity and effectiveness. The conventional method of work is employed in the strategic process of the organization; however, the organization also expands innovation in new technologies.

Agile and Flexible Processes: comprehensive approach to ambidexterity in the digital supply chain means that it definitely needs more controllable and adaptable processes to address the continued development of the market and new technologies. It's possible to take Scrum or Kanban, for example, which are, in fact, practices for managing digital supply chain projects to enable iterative development. This allows various organizations to adapt to the dynamics of the customer needs, events in the market place, and meet technological advancements while at the same time running the business.

Collaborative Innovation Platforms: Establishment of platform for multiple innovations is instrumental in getting a comprehensive understanding of customers, stakeholders, and supply chain associates of the business. These portfolios of the organization investigate new technologies, generate and resolve value simultaneously with other parties through open collaboration, which these platforms provide (Alsaied & Alkhoraif, 2024). For instance, the idea of the digital supply chain innovation and demonstration sites can be created to foster development and approaches with cross-functional groups of a given organization to try and initiate innovative applications and diffusion of new digital technologies and solutions.

Digital Talent Development: The ambidextrous strategy for the digital supply chain involves the organization developing digital management control that nurtures such capabilities. Thus, organizations can engage in training sessions, workshops, and certifications that will focus on enhancing digital literacy and expertise among supply chain graduates. In the same way, corporations can create an ongoing process of trying, learning, and doing in the organization, which will enable employees to read new technologies and share best practices and insightful ideas in addressing threats to supply chain development.

Strategic Partnerships and Ecosystem Collaboration: This proves that coordination with external partners and other relevant ecosystem players is key to advancing the use of digital technologies in the supply chain (Zdziarska & Marhita, 2020). To achieve the above, organizations are able to enter into partnerships with technology vendors, start-up firms, research institutes, and industry collaborations as a way of accessing new and improved digital solutions, benchmarking, and at the same time enabling co-innovation. Strategic partnerships exemplified by research partnerships, innovation partnerships, and co-development partnerships allow organizations to leverage external capabilities and resources to drive digital change across the supply chain.

Data-Driven Decision-Making: Applying data analytics and business intelligence tools helps organizations to make right decisions and sustain continuous growth in the digital chain of supply. Supply chain analytics from a variety of sources, including IoT sensors, RFID tags, and enterprise applications, provides organizations real-time visibility of the performance and potential for improvement of the supply chain as well as predictive analytics on future trends and demand. It helps organizations solve questions of inventory control, demand forecasting, as well as the general supply chain of events to be efficient and timely.

Digital Supply Chain Visibility and Transparency: It is crucial to increase the number of the supply chain nodes that have a high level of visibility in the case of digital supply chains. By adopting technical solutions such as blockchain and IoT, organizations can design public and transparent supply chain platforms. Due to the ability to capture real-time information, supply chain managers can supervise the movement of products, check their authenticity, and analyze supply chain conditions to ensure compliance with requirements. Enhanced visibility assists organizations in understanding areas of inefficiencies, avoiding potential issues, and enhancing the supply chain.

Continuous Improvement and Innovation Culture: An environment of continuous improvement and innovation is critical for the organizational ambidexterity in the digital supply chain context. Managers can facilitate employee engagement in questioning the current state of affairs, researching new opportunities, and learning about digital tools to support innovation. There are ways to encourage employees to come up with innovative ideas and find new and better ways of engaging the supply chain; these include recognition programs, innovation competitions, and innovation hackathons. With changing market trends and customer preferences, organizations stay ahead in the competition.

Therefore, to achieve organizational ambidexterity within the digital supply chain, several approaches must be taken: These include, creating a couple of strategic supply chains, integrating agile processes and implementation of collaborative innovation platforms, cultivating digital talent, incorporating the strategic partners' commitment, adoption of data analytics, supply chain transparency, and a continuous improvement culture. These practical applications help achieve an optimal combination of exploitation and exploration and enable organizations to thrive in the constantly evolving modern business landscape, thus creating a sustainable competitive advantage.

10.5 FEW CASE STUDIES

Amazon: Amazon needs no introduction, really, as it has developed into the most successful e-commerce company in the world and has also made great inroads into cloud computing. Amazon is a typical example of an ambidextrous organization in supply chain management. It has been continuously improving its traditional supply chain operations while actively seeking novel digital technologies. By means of investments, Amazon has changed its practices in terms of supply chain management, ushering in improved operational efficiency, better customer satisfaction, and so on. One good example is a near-universal reliance on AI-powered predictive analytics for such processes as demand forecasting and inventory optimization. Almost every aspect of its AI algorithm allows the company to predict demand with much less error-rate, resulting in drastically reduced numbers of stockouts and overstock situations.

Other than operational efficiency, these two points provide completion to the entire shopping process for the consumers, hence the levels of satisfaction and loyalty are increased (Smith, 2020). It has also harnessed the aspect of IoT and robotics to make all the jobs done manually in the warehouses null and void. In addition to this, by using IoT devices and robotic systems, Amazon has considerably enhanced throughput and order-fulfillment speed in its warehouses. These technologies will further enable real-time monitoring of the inventory levels, effective routing of products, as well as a new automated picking and packing life cycle that may minimize labor costs while shortening order-processing time totals (Chui et al., 2020). Adoption of IoT and robotics signals that Amazon is forging ahead in deploying state-of-the-art digital solutions to optimize supply chain performance while remaining flexible and scalable in its operations.

Procter & Gamble: Another prominent case in supply chain management of organizational ambidexterity is Procter & Gamble (P&G), which is actually a multinational consumer goods corporation. Digitization and organizational ambidexterity have integrated seamlessly, with the switching of their supply chains into a digitally integrated ecosystem. The company invests massively in digital supply chain platforms that allow real-time interaction with suppliers, distributors, and retailers. On the other hand, P&G has employed digital technologies to enhance supply chain visibility along with responsiveness and efficiency to enable a higher agility level and data-driven decision-making (Wang et al., 2020). In its ambidextrous approach, P&G utilizes the capabilities of the existing supply chain while developing new digital solutions to streamline processes, reduce costs, and improve customer service. P&G has analytical and machine learning algorithms on the cutting-edge in order to improve production schedules and levels of inventory management, wherein lead times and holding costs have been reduced. Along with all this, the company has utilized blockchain for improved tracking and visibility across its supply chain with better compliance to statutory requirements, quality, and safety of products (Wang et al., 2020). Organizations other than Amazon and P&G have also recently shown brilliant examples of organization ambidexterity in digital supply chains.

Walmart: Walmart – the most famed largest retailers across the globe – has strategically taken into consideration digitalization and organizational ambidexterity as part of their transformed supply chain operations. Investments in cutting-edge technologies, such as RFID (Radio-Frequency Identification); advanced analytics: the results could have been that they improved much in visibility of stock against their demand forecasts performance. For instance, RFID complements Walmart's real-time inventory tracking and monitoring processes, and this has made a better inventory management possible, not to mention stockout avoidance or overstock situations. It is through high demand and low carrying costs that Walmart is in the position to maintain the best available inventory such that it fulfills the needs of its customers at the right time. The organization, which is the biggest retail store in the United States, has loaded up its stores with radio-frequency identification technologies that allow the company to track product movements. These products allow the company to track inventory and stock availability constantly, avoiding stockout and excess stock situations. By measuring demand-side changes at a low holding cost of 15%, Walmart is able to carry its best inventory, actually satisfying its customers at the desired time. This optimization refers to the ability of the organization to maintain existing supply chain processes while developing innovative solutions. Its supply chain capabilities are still being refined while new technologies improve efficiency and innovations. Such combinations enable Walmart to embrace new market dynamics, capitalize on opportunities promptly, and justify its position as a retail leader.

Nike: Nike has integrated their manufacturing processes and traditional distribution with innovative digitalized solutions. Thus, like other companies, Nike is effectively becoming ambidextrous in its supply chain, adopting new methods as well as the current ones. It is also digitizing and increasing the amount invested in systems and digital tools to enable real-time collaboration with suppliers and optimize production scheduling. These integrations that involve digital technologies would also help Nike streamline communications and coordination with a supplier network across the globe, prompting everything so that delivery and component materials would be on time to conform with demand in production.

By employing data analytics and artificial intelligence, Nike has customized the customer experience as well as heightened one's interaction with the brand. An algorithm analyzes customer data, allowing Nike to study the interests, behaviors, patterns, etc., to give personalized recommendations of the product via personalized advertisements with a better understanding, all in all. It strengthens customer loyalty but is also conducive to selling and revenue for the company. Furthermore, ambiguity in Nike is apparent with regard to how already existing processes of the supply chain are optimized alongside the search for new digital solutions. While proactively improving its manufacturing and distribution capabilities, Nike leaves avenues open to further innovations in technology. Such an ambidextrous approach enables Nike to win over its competitors while fast adapting to changes in the environment and inventing innovations to market products in tune with customer needs.

Tesla: Tesla, a pioneer in the field of electric mobility, has made enormous strides in the field of ingenious supply management. They have uplifted their transformational approach with regard to digitization and organization ambidexterity. The business has immersed itself in advanced technologies and devices in the making, proving its commitment to innovative success. Tesla rapidly scales up production and flexibility by automating the assemblies and quality controls. This capability will assist them in meeting demand and maintaining the same quality for an ever-increasing number of electric vehicles. Tesla enables the most intelligent algorithms to understand data from different sources in order to optimize transportation routes, shorten lead times, and minimize delivery costs. The developed analytical tools and machine learning will always help Tesla enhance supply chain logistics performance and allow customer further opportunities to customize their cars. As far as possible, it also offers customers various things regarding stealthy customization features and options to define the features and specifications of their vehicles.

Further, so is the balancing act with Tesla's end in terms of organizational ambidexterity, just as it can be optimizing the current processes of supply chains with new digital solutions. In the course of continuously improving its manufacturing and logistics operations, Tesla does not stop exploring other possible technologies such as autonomous driving and renewable energy solutions. This is a classic ambidextrous solution for Tesla as it allows the car maker to flow with innovation nonstop while disrupting the standards of the automotive industry and leading all future movements toward green transport.

Apple Inc: In this regard, the case of Apple Inc., which is a global technology giant, stands as a model in organizational ambidexterity regarding the way it innovates in product development and distribution in supply chain management. Besides this, the other aspect of Apple's supply chain is that it is popular for reasonable efficiency, flexibility, and responsiveness. Most of all, such enhancement was achieved by combining traditional manufacturing methods and advanced new digital technologies, which the company applies to bring excellent products to its customers around the world. Apple invests heavily in digitalization and new state-of-the-art advanced manufacturing technologies in its supply chain operation. It uses robotics, automation, and AI-informed analytics to enhance production processes and also operational efficiencies. As such, the use of robotic mechanism limbs in Apple's manufacturing facilities delivers components with precision and produces higher-quality products in shorter times. In addition, with AI-improved analytics, Apple will be able to accurately predict demand and thus have better control over inventories and lead times. Clearly, we see the ambidexterity of Apple as an organization: it improves the existing supply chain processes while exploring new technologies and business models. Apple is not only doing something to make the manufacturing and distribution processes more efficient but also innovating in terms of product design and marketing, and customer engagement. For instance, when the retail concept of the Apple Store was introduced into the market, it changed the consumer electronics industry by providing different experiences to consumers and empowering them to the brand.

Toyota Motor Corporation: The famed production system, the Toyota Production System (TPS), focuses on continuous improvement that is globally recognized within the automobile world and international lean manufacturing principles. Handling changing market conditions efficiently, flexibly, and with resilience defines the supply chain of Toyota according to its characteristics. Toyota has a blend of conventional old chains and modern disruptive technologies in optimizing the production and distribution setup of the company. One of the supply chain strategy measures within the company is digitization and Industry 4.0 technologies. The supply chain is enabled by IoT sensors, data analytics, and cloud computing in providing visibility in real time into production performance, equipment maintenance, and supply chain logistics. For instance, IoT sensors are installed in manufacturing machines to avoid downtime due to efficient predictive maintenance and maximum overall equipment performance in the production process. The production scheduling, inventory levels, and distribution routes data analytics tools have cost savings on output as well as improvements in customer satisfaction for Toyota. The ambidexterity of the corporation is then manifested in how the firm retains mastery at balancing the optimization of its existing supply chain processes with investigating new technologies and models of business. Thus, Toyota would keep on perfecting its lean manufacturing practices even though innovative approaches were considered toward the same ends of product development, for example, electric and autonomous vehicles. Consequently, Toyota is not merely always a step ahead of competitors but is also adaptable in line with popular trends and developments shaping the market – a key driver of sustainable growth in the automotive industry.

Unilever: It is a multibillion-dollar company recognized as a consumer goods multinational. Unilever shows superior organizational ambidexterity with respect to its supply chain management for the sustainability and digital transformation it inspires. For Unilever, its supply chain is based on a mix of environmental sustainability, social responsibility, and operational excellence. Traditional supply chain concepts combined with cutting-edge digital technologies are used by the company to create efficiency, reduce waste, and improve value chain transparency. Investments by Unilever in digitalization and sustainability strategies are a hallmark of its supply chain strategy – the company as a whole uses blockchain technology, IoT sensors, and data analytics to trace ingredients in products, monitor environmental impact at production, and optimize resource utilization. For instance, Unilever traceability enabled by blockchain allows the company to monitor the total journey of raw materials as they are transformed into end production raw materials – an ethical sourcing and supply chain transparency win for Unilever. Companies' factories installed with IoT sensors allow Unilever to optimize energy and water consumption as well as waste management, all leading to economies and improved environmental footprints. Apart from these, Unilever also explores the optimization of existing supply chain processes and capitalizes on the new technologies and sustainability programs it offers. It continuously improves the operational efficiencies of its supply chains while trying out new 21st-century approaches to product packaging, renewable energy adoption, and circular economy practices. This ambidextrous approach will enable Unilever to respond to changing consumer preferences, evolving regulatory concerns, and societal challenges in the future, placing it among leaders.

In summary, such organizations include Amazon, P&G, Walmart, Nike, and Tesla as harbingers of change, promising transformation in terms of organizational ambidexterity in digital supply chain management. All these defining qualities have made it possible for these companies to achieve remarkable gains with regard to visibility, flexibility, and efficiency along the supply chain (Abdalla & Nakagawa, 2021). Thus, they would have to continue for embracing digitalization and ambidexterity to counter-evoke the complication of internationally competitive markets in the field of digital supply chain.

10.6 FUTURE RESEARCH IMPLICATIONS

Such action would imply a certain organization's ambidexterity and digital supply chain management toward the future of business operations and strategic management. Organizations will continuously get several implications due to the digital age; few of them are listed below:

Increased Agility and Resilience: An organization that has developed a good ambidextrous approach in its digital strategies will be able to read quickly changing market conditions, advances in technology, and customer preferences This may increase agility as well as resilience through the dyad of exploitation and exploration for these forms of organizations to respond quickly to any disruption and capitalize on new opportunities.

Competitive Edge: Ambidextrous organizations will thus be able to obtain a competitive edge in the market through the superior application of technologies in their digital supply chains. By optimizing operational efficiency while driving innovation (Sinha, 2016), these organizations can differentiate themselves from competitors, deliver superior customer value, and capture market share in increasingly digitalized industries.

Sustainable Growth: The integration of organizational ambidexterity and digital supply chain management can support sustainable growth by enabling organizations to pursue both short-term profitability and long-term innovation (Pyun & Rha, 2021). By continuously improving existing processes and exploring new digital solutions, organizations can achieve sustainable growth while balancing the demands of stakeholders, including shareholders, customers, and employees.

Industry Transformation: The widespread adoption of ambidextrous digital supply chain strategies has the potential to transform entire industries by reshaping value chains, business models, and competitive dynamics. Organizations that lead the way in embracing ambidexterity and digitalization will drive industry-wide innovation, disrupt traditional market structures, and create new opportunities for growth and value creation.

Proposed Propositions for Future Research:

1. *The Impact of Organizational Structure on Ambidextrous Digital Supply Chain Management*: Investigate how different organizational structures, such as centralized versus decentralized, hierarchical versus flat, and functional versus matrix, influence the implementation of ambidextrous digital supply chain strategies.
2. *The Role of Leadership in Fostering Ambidexterity*: Evaluate leadership behaviors, styles, and practices that facilitate the formation and application of ambidextrous digital supply chain strategies. How can leaders catalyze the creating of a culture of experimentation and risk-taking innovation in their organizations?
3. *Technology Adoption and Integration in the Digital Supply Chain*: Elucidate and interrogate the different challenges and opportunities that can be brought by the usage and adoption of digital technologies such as IoT, AI, blockchain, and Big Data analytics into supply chains. Critically ask how organizations will optimally be able to manage the use of these technologies with a view of bringing beneficial effects on their visibility, efficiency, and responses within their supply chains.
4. *Supply Chain Resilience and Risk Management in the Digital Age*: These would specifically fall in areas of research applying ambidextrous digital supply chain strategies to mitigate on resilience and risk management in supply chains. Identify, assess, and mitigate organizational risks attributable to digitization, for instance, cybersecurity threats, breaches of data, or disruption of the supply chain.
5. *Collaborative Innovation and Ecosystem Orchestration*: Study collaborative innovation platforms and ecosystem partnerships around which digital supply chain innovation can be achieved. Investigate how firms would engage through partner ecosystems in value co-creation with outside parties

like suppliers, customers, and technology providers toward continuous improvement within the supply chain.

6. *Performance Measurement and Evaluation in Ambidextrous Digital Supply Chains*: This would set the performance indicators and measures for ambidextrous digital supply chain strategy effects and performance metrics for efficiency measures. Investigate working methods that an organization can use to measure and benchmark the statements on the efficiency, innovation, agility, and customer satisfaction dimensions of the supply flow.
7. *Cultural and Change Management Challenges in Ambidextrous Digital Supply Chains*: Investigate the cultural barriers and change management challenges associated with implementing ambidextrous digital supply chain strategies. Explore how organizations can overcome resistance to change, promote an innovative and interactive environment, and offer employees the autonomy to digitalization and innovation.

The synthesis of organizational ambidexterity and digital supply chain management offers a rich avenue for future research. How ambidexterity implies for supply chain agility, competitive advantage, continual growth, and transformation of an industry may continue to open ways for research into how organizations may thrive in the digital age. Proposing and testing any of these propositions – organizational structures, leadership, technology adoption, resilience, innovation, performance measurement, and change management – would help scholars share nuggets of valuable insights and effective solutions for firms grappling with challenges of ambidextrous digital supply chain management.

10.7 CONCLUSION

Present complexities are better understood within the purview of ambidexterity as a compelling organization simultaneously needs to exploit existing capabilities while exploring new digital technologies and processes. Some of the priciest complexities that organizations face are those associated with such digital technology changes and progress.

Technological Disruption: It was due to revolutionizing features of electronic supply chain management that people were further inclined to think about this term as more on pace and disruption in technology. Emerging digital technologies like IoT, AI, blockchain, and advanced analytics present themselves with unprecedented chances to augment visibility, efficiency, and responsiveness in the supply chain but still integrate with many interoperability and scalability issues. The never-ending introduction of new technologies requires organizations to keep an eye on how these new developments are estimated to affect supply chains and the organization's strategic objectives.

Data Complexity and Management: All these data, from IoT sensors and RFID tags, through various enterprise systems and external sources (social networks, marketplaces, etc.), are generated within the digital supply chain. Such data management and interpretation should receive huge dimensions in quality, source integration, security, and privacy issues. It should, therefore, be complemented by corresponding good and high-quality data management strategies and capability to keep that data accurate, reliable, and confidential, while offering world-class analytics and data visualization tools to make those actionable for informed decision-making.

Supply Chain Visibility and Transparency: In any digital supply chain, it is increasingly relevant to designing business processes so that they provide end-to-end visibility and transparency as a basis for effective decision-making, risk management, and collaboration. Unfortunately, however, the largely disjointed web of supply and manufacturing is often coupled with aging legacy systems and silos of data – many of them internal-to-digital systems. These organizations must invest in technologies such as blockchain, which would then record immutable and transparent transactions to be able to enhance supply chain visibility and traceability. Most importantly, organizations should create collaborative relations with their supply chain partners and other interested parties for sharing real-time data and insights to proactive mitigation of risk and optimal performance.

Supply Chain Network Complexity: The digital supply chain will most probably be replaced by a web of complex and interconnected networks of multi-sourced suppliers, manufacturers, distributors, retailers, and customers from multiple geographies and industries. This complexity forces organizations into a holistic approach in their supply chain design, optimization, and orchestration programs. Organizations are leveraging digital technologies in conjunction with analytics to engage in activities such as network design optimization, inventory allocation, and transportation planning – all to reduce cost and lead-time, and make service levels better across the supply chain.

Agility and Flexibility: In the current unpredictable economic situation, agility and flexibility comprise success factors in the supply chain. Organizations respond well to variations in consumer demand, market conditions, and disruptions in the supply chain. In-house leans toward exploiting existing processes and capabilities for efficiency while exploring new digital solutions for innovation. The ambidextrous supply chain strategy is the best fit for companies striving for higher efficiency and flexibility to be able to react to the changes and earn or grab innovative opportunities created during the digitalization.

Talent and Organizational Capabilities: In the organization, building digital capabilities and talent is very essential to ensure that they positively impact the digital supply chain management. However, oftentimes it is a great challenge for organizations to secure and keep individuals with skills and knowledge in the fields of digital technologies, data analytics, and supply chain management. Hence, organizations should invest in talent development programs, cross-function collaboration, and knowledge-sharing initiatives as part of creating a workforce that will propel digital transformation in the supply chain. Additional thing that organizations should be doing is building a culture for innovation, experimentation, and learning over time, since these are primary enablers to make creativity and agility flourish in times of digital disruptions.

The idea of digital supply chain complexity framed in organizational ambidexterity is an illustration of how organizations should find a balance in exploiting what already exists and exploring new digital solutions. The adoption of ambidextrous strategies in the logistics part of an organization will face disruptions by technology while managing the complexity of data and facilitating visibility and transparency across the supply chain networks. It would enhance the agility and flexibility of building digital capabilities and skills for growth in a new world of technology (Wang & Teng, 2022). Organizational ambidexterity is expected to be the cornerstone in handling the complexity of a digital supply chain. It would explore the simultaneous exploitation of existing supply chain capabilities with new digital solutions to eventually achieve that level of dynamic equilibrium for keeping operations efficient and agile while gaining a competitive advantage. As shown in the case studies of industry giants like Amazon and Procter & Gamble, the power of ambidextrous supply chain strategies is all about demonstrating the benefits of an efficient operation and creativity in supply chain management through the active use of digital technologies. It is important to note, however, that while organizational ambidexterity in the digital supply chain is advantageous, certain disadvantages and challenges will need to be faced in the attainment of ambidexterity. The first is that a change in the company is going to require substantial resistance, thus likely eliciting much objection from the stakeholders, employees, and processes. Such resistance and inability to promote culture and agility may become some stiff walls for some organizations.

Besides, robust investments in collaboration between digital technologies and infrastructure, talents, and training are all requisite when it comes to an effective digital supply chain. In this regard, for example, SMEs are short of capital and find it tougher to imitate the bigger ones in terms of digitizing and ambidexterity targets. Further, technology and digital disruption have become ongoing challenges for firms that maintain their ambidextrous capabilities over the years (Ageron et al., 2020).

It should also be noted that establishing the right balance between exploitation and exploration within a digital supply chain is extremely complicated, as it forces organizations to simultaneously focus on immediate output and lengthy input for innovation. That kind of conflict resolution needs very careful strategy design, allocation of resources, and decision-making, which would and can be further complicated by competing priorities and objectives within the organization.

Hard as it may be, yet the coated advantages of being ambidextrous with digital supply chains are many. Organizations that practice ambidexterity and adopt advanced technologies become more capable in their responses to market dynamics, customer needs, and competitive pressures. Future research will have to address further innovative strategies, frameworks, and tools for achieving and sustaining ambidexterity in a digital supply chain. Organizations will manage to realize their full-fledged capability on ambidexterity to drive continuous development or innovation in the supply chain at a stretch if these research gaps and limitations are addressed.

REFERENCES

Abdalla, S., & Nakagawa, K. (2021). The interplay of digital transformation and collaborative innovation on supply chain ambidexterity. *Technology Innovation Management Review*, *11*(3), 45–56.

Ageron, B., Bentahar, O., & Gunasekaran, A. (2020, July). Digital supply chain: Challenges and future directions. *Supply Chain Forum: An International Journal*, *21*(3), 133–138. Taylor & Francis.

Agrawal, P., & Narain, R. (2018, December). Digital supply chain management: An overview. *IOP Conference Series: Materials Science and Engineering*, 455(1), 012074. IOP Publishing.

AlSaied, M. K., & Alkhoraif, A. A. (2024). The role of organizational learning and innovative organizational culture for ambidextrous innovation. *The Learning Organization*, *31*(2), 205–226.

Belhadi, A., Kamble, S., Gunasekaran, A., & Mani, V. (2022). Analyzing the mediating role of organizational ambidexterity and digital business transformation on industry 4.0 capabilities and sustainable supply chain performance. *Supply Chain Management: An International Journal*, *27*(6), 696–711.

Büyüközkan, G., & Göçer, F. (2018). Digital supply chain: Literature review and a proposed framework for future research. *Computers in Industry*, *97*, 157–177.

Chui, M., Manyika, J., & Miremadi, M. (2020). AI and robotics will reshape the future of work. McKinsey Global Institute. https://www.mckinsey.com/featured-insights/future-of-work/ai-and-robotics-will-res

Hashem, G., Aboelmaged, M., & Ahmad, I. (2024). Proactiveness, knowledge management capability and innovation ambidexterity: An empirical examination of digital supply chain adoption. *Management Decision*, *62*(1), 129–162.

Huddar, Y. N., Kumatagi, P. P., & Latte, M. R. (2017). Digital supply chain management – A review. *IARJSET*, *4*(1), 34–37.

Ivanov, D., Dolgui, A., & Sokolov, B. (2019). The digital supply chain as a resilient configuration in Industry 4.0: Evidence from big data analytics. *International Journal of Production Research*, 57(3), 829–846. https://doi.org/10.1080/00207543.2018.1519727

Lis, A., Jozefowicz, B., Tomanek, M., & Gulak-Lipka, P. (2018). The concept of the ambidextrous organization: Systematic literature review. *International Journal of Contemporary Management*, *17*(1), 77–97.

MacCarthy, B. L., & Ivanov, D. (Eds.). (2022). *The digital supply chain*. Elsevier.

O'Reilly, C. A., & Tushman, M. L. (2004). The ambidextrous organization. *Harvard Business Review*, *82*(4), 74–81. https://hbr.org/2004/04/the-ambidextrous-organization

Pyun, J., & Rha, J. S. (2021). Review of research on digital supply chain management using network text analysis. *Sustainability*, *13*(17), 9929.

Queiroz, M. M., Pereira, S. C. F., Telles, R., & Machado, M. C. (2021). Industry 4.0 and digital supply chain capabilities: A framework for understanding digitalisation challenges and opportunities. *Benchmarking: An International Journal*, *28*(5), 1761–1782.

Sinha, S. (2016). Managing an ambidextrous organization: balancing innovation and efficiency. *Strategic Direction*, *32*(10), 35–37.

Smith, T. A. (2020). The role of customer personality in satisfaction, attitude-to-brand and loyalty in mobile services. *Spanish Journal of Marketing—ESIC*, *24*(2), 155–175. https://doi.org/10.1108/SJME-06-2019-0036

Smith, W. K., & Tushman, M. L. (2005). Managing strategic contradictions: A top management model for managing innovation streams. *Organization Science*, *16*(5), 522–536. https://doi.org/10.1287/orsc.1050.0137

Teece, D. J., Pisano, G., & Shuen, A. (1997). Dynamic capabilities and strategic management. *Strategic Management Journal*, *18*(7), 509–533. https://doi.org/10.1002/(SICI)1097-0266(199708)18:7<509::AID-SMJ882>3.0.CO;2-Z

Wang, M., & Teng, W. (2022). Digital innovation and firm environmental performance: The mediating role of supply chain management capabilities. *Frontiers in psychology*, *13*, 897080.

Wang, X., Wu, Z., Wei, C., & Cao, Z. (2020). Supply chain digitalization and innovation performance: The moderating role of digital supply chain integration. *International Journal of Production Economics*, *221*, 107480. https://doi.org/10.1016/j.ijpe.2019.107480

Zdziarska, M., & Marhita, N. (2020). Supply chain digital collaboration. In A. Kolinski, D. Dujak, & P. Golinska-Dawson (Eds.), *Integration of Information Flow for Greening Supply Chain Management*. EcoProduction. Springer. https://doi.org/10.1007/978-3-030-24355-5_4

11 Innovation and Governance in the Digital Era: Exploring the Complexities of the Digital Supply Chain

Salu Dsouza and Nikolay Bodrov

11.1 INTRODUCTION

In today's dynamic world of international trade, the digital supply chain is one of the major factors setting up the contours of modern societies, flipping historical production modes and provision methods to customers. Advanced information technology, automation, and data aggregation, as well as a higher level of structural connectivity, led to a solid basis for a new, efficient, and more resilient way to make, deliver, and provide stuff to do everything in a new fashion. However, such a transformative path stops in an arena of sophisticated socio-economic and ethical issues involving governance frameworks for resolution. Governments, enterprises, and the rest of society depend heavily on a digital supply chain structured on substantial institutional framework arrangements within society. The roles and potential outcomes of every player in the given set are of interest for the discussion, with a specific focus on the effects on female workers, production, and global chain activities. As a starting point, the research analyses the constitution and operations of the digital economy to reveal the cluster of complex outer forces that form its supply chain system. Secondly, it offers proposals for sustainable as well as total growth in the given segment.

The field is further discussed in recent debates around the inclusion of female workers into digital supply chain labor. Thus, even as digitalization creates opportunities for empowerment and economic participation in a digitally connected world, it also exacerbates current gender gaps and introduces new challenges to female advancement. This study intends to identify the means of fostering gender equality and enhancing socio-economic welfare for female workers through analysis or examining how the interplay among three aspects: women, inclusive technological development as well as human resources policies, becomes effective in delivering decent work, especially within the digital supply chain. In addition, the transition of industrial processes from analog to digital has significant implications for how industrialization proceeds alike in developed and developing countries. Yet, as factories see

DOI: 10.1201/9781003477242-14

rapid changes driven by automation, data analysis, and artificial intelligence, governments and institutions must manage the complexities of digital industrialization while ensuring equitable outcomes for all parties. A brief review in this chapter unfolds by providing the readership snapshots of empirical examinations and case studies, which are aimed to raise light on opportunities and challenges that digital industrialization brings about while proposing pathways for the necessary know-how required, especially so as we have an inclusive sustainable growth process.

Across traditional global supply chains, the effect of digital technologies has been transforming international trade and business. At the same time, it creates new vulnerabilities like platform breakdowns, causing supply chain disruptions or ethical dilemmas, and human rights violations. This study, by examining the interaction between digitalization and global supply chain developments, seeks to identify potential avenues for increasing transparency, resilience, as well as ethical governance in relation to those three nodes of the current debate on global supply chains. At last, the implementation of corporate ethical standards can be seen as a very essential part in the governance within the ecosystem for the digital supply chain. As firms seek to negotiate the complex world of global production networks, they are subject to increasing pressures regarding their compliance with ethical standards and good business practices. Based on an analysis of enforcement mechanisms, monitoring systems, and stakeholder engagement initiatives, this chapter seeks to assess the capacity for compliance with these corporate ethical criteria regarding human rights (HR), environmental sustainability (ES), and ethical governance standards in digital supply chains.

We will examine the layered nature of digital supply chain evolution, and analyze with a fine comb what types of function can be abstracted for government intervention, enabling government to play active integrated role in supply chain regulation in each layer. In doing so, the research will contribute to policy discourses and facilitate meaningful dialogue in moving toward a more sustainable, inclusive, and just digital supply chain system of governance through deep dives into themes around gender justice, industrial evolution, capital creation or global integration and ethical (self) regulation.

11.2 GOVERNMENT POLICIES AND REGULATIONS IN SHAPING THE DIGITAL SUPPLY CHAIN

Broadly speaking, digital supply chain policy and regulation is defined as the guide for legal parameters of how to design, develop, and maintain a digitally based production system. One of the large areas in which governmental operations have influence is in promoting innovation throughout supply chain operations. These results corroborate those of studies by Smith et al. (2019), which have also emphasized the significance of government-led initiatives to promote research and development (R&D), especially in emerging technologies connected to a digital supply chain. Tools such as state technology grants and tax breaks have shown a demonstrated ability to provide the seed money needed for innovation, while lowering some of the financial risk associated with R&D investments. In addition, governments are tasked with ensuring that data integrity and confidentiality remain in the digital supply chain. The use of digital technologies extends the size and complexity involved in what is now an enormous amount of data that has been produced or processed as part of supply chain activities. In conclusion, sensitive data and the risk of cyberattacks have forced these regulations to be in place. The European Union, for instance, has already begun enforcing stringent regulations of the General Data Protection Regulation, which will allow data to be collected and used in a safe manner and may result in penalties for standards that break this (Jones & Brown, 2020). The California Consumer Privacy Act is an example of similar legislation in the United States that also mandates transparency and control of personal data by individuals. Such legislation will ensure an increased degree of data protection within the supply chain networks.

Another area for governmental intervention in the digital supply chain is that of ethical concerns. The research by Johnson and Martinez (2018) underscores our legislative efforts such as the Modern Slavery Act in the United Kingdom, which requires companies to disclose information on their steps to address forced labor in human trafficking practices within their supply chain. The directives in question are aimed at strengthening ethical procurement, and adherence to labor laws, human rights measures. Moreover, there are several consumer-oriented policies, such as the Conflict Minerals Rule in America, that have been enforced to alleviate tensions surrounding concerns about conflict-induced mineral procurement, creating conducive conditions for peace and stability within global supply chains (Garcia et al., 2021). In addition, the digital infrastructure development which supports supply chain operations digitization is also seen to be a key driver catalyzed by governments. Digital platforms, data analytics, and high-speed internet are essential expenditures that will truly give them a rich flavor and enable them to effectively utilize these digital technologies Support from the government in deploying digital infrastructure is imperative, especially for countries with limited access to technology (Chen & Wang, 2020). Delivering improved digital connectivity and availability enables governments to ease the participation of businesses in global supply chains, informing economic development, and competitiveness downstream.

The digital supply chain is under a far-reaching impact of government policies and regulations that affect industrial development, innovation, data security measures, enforcement of governance ethical principles, or even infrastructure improvement. To promote further progress and resilience on the global economic landscape of tomorrow, it takes proactive regulatory efforts from governments that foster an environment ripe for digital transformation within supply chain networks.

11.3 INSTITUTIONAL MECHANISMS AND NORMS IN DIGITAL SUPPLY CHAINS

In addition to government regulation, institutional frameworks and standards shape the world of digital supply chains. More recent research on rights and permissions (Thompson et al., 2021) Permissions and Reprints Standards organizations must further promote industry groups' official positions within the scope of best practices for supply chain operations. They promote collaboration among stakeholders and seek to implement harmonized standards in order to increase interoperability, visibility, and efficiency across supply chains. Secondly, certification bodies are instrumental in certifying the adherence to industry standards and ethos. Some bodies, like the International Organization for Standardization, imagine measures, such as quality control, environmental sustainability, and corporate social responsibility, to be worked in with the supply chain operations (Gonzalez et al., 2023). ISO positive affirmation demonstrates to buyers that an organization is following worldwide best practices. In addition, industry attempts and coalitions can also play their part in propelling innovation and shared hard work around digital supply chains. Kim and Lee (2019) expressed that BiTA, with the launch of other industry organizations, serves a vital role in development speed for industry across supply chains to develop for a better use of transparency and traceability. While every one of these gauges couldn't be accomplished independently by companies generally, any solution, whether through spearheading developments that iron out standard differences or through stages that approve and administer adherence of best practices through guidelines and track down or eliminate basic traps when working to achieve standard conformance, also included legal regulatory bodies. For instance, in the United States, the FDA follows strict guidelines for final products such as pharmacology or food and beverage production and practices safety regulations to make sure that is safe to eat (Sullivan & White, 2018). By setting up legal precepts and making sure those organizations follow them, these organizations also manage to guarantee credibility and legitimacy behind the scenes when companies run supply chain operations. Institutional frameworks and norms play a substantial role in setting the equilibrium plan of digital supply chains, considering key stakes in which fundamental rules or standards of collaboration between stakeholders, innovation, and adherence or follow. Cogent social event support methodologies and strategies by side-camps for far-off advances emerge with the smooth heap of practicable measures by supply chain network security, considering the legitimate system and authentic system.

11.4 STAKEHOLDER ENGAGEMENT AND COLLABORATION

In the digital supply chain universe, stakeholder engagement and collaboration are essential pillars to anchoring innovation, leading any platform to transparency brought via sustainability. For example, according to Martinez and Nguyen, "governments, businesses, corporations, and NGOs need to converge and collaborate with the local population when addressing complex supply chain issues." By working together, stakeholders can harness their unique insights, assets, and networks to catalyze innovative solutions that yield a positive impact for society and our planet. Also, initiatives that see participation from a number of stakeholders provide impetus to transparency and accountability in the supply chain domain. For example, initiatives that bring together apparel labels and vendors with producers as well as NGOs have emerged to jointly develop sustainability criteria as a tool for measurement of the environmental impact on social impacts (Wang & Li, 2020). They federated information across industry stakeholders to outcome into common objectives and metrics, which led to more information exchange, benchmarking that led to continuous improvement in supply chain practices.

In addition, stakeholder engagement is a crucial factor for addressing social and ethical issues related to the digital supply chain, like labor rights in countries of manufacturing origin or diversity and inclusivity. Garcia et al. had done research on this topic and (2023) further underscore the concern, promising undefined dialogue and partnership with business over concrete human rights violations, such as poor working conditions or fair labor standards. By engaging with stakeholders, businesses have the opportunity to learn what are requirements or expectations of affected communities and that can enable companies in developing more flexible and inclusive supply chain practices. Meanwhile, organic collaboration among the guiding sections of organizations is unmatched in directing innovation on a digital supply chain. Kim and Park (2021) supported the importance of cross-functional teams to improve information exchange, problem-solving, and decision facilitation across supply chain operations. Removing barriers and facilitating participation allow organizations to leverage the collective experience of team members in identifying opportunities for process improvement or innovation.

As we enter into the world of data-driven digital supply chain, stakeholder engagement and collaboration are two essential drivers ensuring innovation as well as transparency along with sustainable practices in this domain. Establishing partnerships for government, business, NGOs and society within organizations breaks down silos of collaboration in a supply chain challenge to develop mutual value proposition across the system.

11.5 GENDER EQUALITY AND WOMEN WORKERS

Some of the fundamental values like gender equality and empowerment of women are gradually being recognized in a greater measure even within the digital supply chain. Patel and Gupta (2021) aimed to build a foundation for closing gaps between females and males in supply chain workforce representation by advocating for building inclusivity into education. Though they represent a significant proportion of the global workforce in the supply chain, women still face barriers when it comes to getting or staying hired and promoted—as well as receiving equal pay compared with men. Digitalizing supply chain processes can further exacerbate already existing gender disparities as well as create chances for women workers. However, Lee and Kim (2022) suggested that even though digital technologies can enhance flexibility, efficiency, and information accessibility in supply chain activities, they also have the potential to reinforce gender stereotypes while increasing biases unless these are combined in a gender-inclusive manner. Therefore, it is critical for policymakers, industry, and other partners to adopt policy practices that can promote gender equity as well as support the empowerment of the women workforce in the digital supply chain. One effective area in which governments can really make a difference is in the implementation of STEM policies, which promote women to undertake education and training and get jobs in digital supply chain-related fields. According to Chen et al. (2019), "girls' education and vocational training can impart the requisite technical and cognitive skills to boost their performance in the new digital economy." By increasing the number of women who engage in STEM education and training, the government can reduce the gender gap in supply chain occupations and technology and fully unleash their key resources. The promotion of gender equality as an enforcement within the supply chain also provides corporations the opportunity to make powerful business cases, as in the work of Johnson and Rodriguez (2020). Creating diversity and inclusion adds value while enhancing employee morale, productivity, creativity—further reinforcing their standing as socially responsible corporations.

Promoting gender equality and empowering women in the workforce are essential in building a well-functioning digital supply chain ecosystem. Policy makers, firms, and other stakeholders (both public and private) can pave the way for women to thrive in a key sector of the economy by addressing gender imbalances, investing in education and skills development, fostering inclusive workplace environments.

11.6 DIGITAL INFRASTRUCTURE DEVELOPMENT

The development of digital infrastructure is crucial for the promotion and digitization of supply chain processes as it allows integration with different data technologies. Nguyen and Kim (2020), who investigated the anthropological insights of government policies/projections/investments toward configuring vital elements (technical, financial, cyber safety) for a digital infrastructure like high-speed internet connections, advanced types/delivery systems in physical sites. Public sector intervention and reforms aimed at improving access to digital tools and technologies can help enterprises leverage digitization more effectively for process efficiency in their supply chains, innovation within the firm, or market competitiveness. Additionally, the value of digital infrastructure runs even deeper as it has also aided in building a stronger connection and enhancing collaboration between networks within this domain. The study of Patel and Lee (2021) brings in a very significant role of digitization tools as well as platforms to carry out an effective communication, information sharing, and coordination among the partners present within the supply chain. The usage of cloud-based platforms, Internet of Things (IoT) devices, and blockchain technology makes it possible to generate real-time transparency in the different operational dimensions from supply chain, reducing lag times, improving efficiency, responsiveness, and informed decision-making causing cost effectiveness.

It also increases investments in digital infrastructure as an efficient solution to eliminate the current existing digital divide and is supportive of the growth of inclusive economics with a robotic environment. The results of the investigation conducted by Garcia and Wang (2019) and Wang shed light on how damaging digitalization can be for countries in development whose access to technology is limited. National programs that additionally extend broadband and training support through financial programs, in partnership with technical assistance to counteraware SME digital skill gaps, would greatly enable active participation of these CPF firms into global supply chains. As we move to a more focused direction of government-led interventions, the importance of private sector investments has become paramount for advancing digital infrastructure development. Smith and Johnson (2018) highlight the importance of public-private partnerships (PPPs)—publicly funded projects with private deadlines or cost-saving measures built in—as a way to finance deals for digital infrastructure. Leveraging the experience and resources of the private sector, public entities can accelerate digital technology deployment to tap into these benefits faster, especially in new ways to improve supply chain performance delivering better market competitiveness.

In sum, the unmistakable conclusion is that digital infrastructure development would be a critical component of optimizing all aspects of this digital supply chain and further drive economic growth and overall development. Creating a digital synergy from scratch by making strategic investments in three tiers of digital connectivity, enabling platforms, and advanced technology for both businesses and governments.

11.7 GLOBALIZATION AND SUPPLY CHAIN RESILIENCE

The first, of course, is globalization—the fact that our supply chains are complex and global in nature as opposed to being localized. Nguyen and Garcia (2021a, 2021b) underline how important the rapid spread of globalization to more countries is in driving supply chain efficiency, lowering costs, and broadening market scope. In saying that, it is vitally important to recognize in the dynamic cross-border environment characterizing globalization comes a plethora of risks and threats—ranging from heightened geopolitical tensions, natural disasters upending operations on an unparalleled scale through to trade disruptions. One of the major challenges faced due to globalization is adapting and increasing supply chain resilience against unexpected risks. The need of the hour as highlighted by Kim and Martinez (2020) is that such deviations can only be achieved if we are flexible within our supply chain framework to deal with all kinds of risks and market dynamics. Companies that utilize state-of-the-art digital technologies such as predictive analytics and real-time monitoring are well equipped to detect, analyze, and address disruptions quickly, which enhances the resilience of their supply chains.

In addition, the phenomenon of globalization has led to increased attention being paid to ethical sourcing practices and sustainable protocols that complement supply chain operations. Lee et al. conducted a study (2019), which reveals a mutual engagement between corporate social initiatives and corporate social responsibility (CSR) to assist moral business practices, which increases the visibility of supply chains. Firms which adopt preferential ethically derived sourcing, fair labor standards, and environmentally sustainable activities could benefit from a stronger reputation with customers, their investors, and stakeholders, enhancing competitive advantage within the global marketplace. In addition, the global supply chains are facing widely varying factors today due to geopolitical tensions and trade disputes. For instance, trade barriers and tariffs play a significant role in the costs of production and how efficiently supply chains operate; Patel and Smith (2019) were aware of this when they conducted their research. Companies are pushed to thread their way through complex regulatory environments and geopolitical considerations, all the while attempting to safeguard the seamless operation of their supply chains. Measures such as diversifying sourcing location, building redundancy into supply chain networks, and developing alternative transportation routes are some of the strategic actions that can be implemented to counterbalance geopolitical disruptions.

So basically, it is clear that globalization brings along on one side a basket of opportunities and challenges to sustaining supply chains better. Companies can establish resilient supply chain frameworks which are efficient and dynamic enough to navigate through the complexities of the global marketplace by capitalizing on digital innovations, promoting ethical business standards, and addressing geopolitical uncertainties before they become a threat.

11.8 CORPORATE ETHICS AND CORPORATE GOVERNANCE CODE

A code of conduct is a set of written rules for all employees to follow, which provides the framework detailing what they are allowed to do, and those that cannot be done in support of responsible business practices within the digital supply chain. The influence of corporate codes of conduct that preserve integrity in making ethical decisions is also highlighted by Patel and Garcia (2022) as follows: The existing relationships with stakeholders along the value chain. These codes define guidelines and rules related to ethical sourcing, labor practices, environmental care, and protection of human rights. They provide an organized system that helps companies to fit their operations in with moral values and societal standards, which can help maintain good practices throughout your supply chain. Moreover, the effectiveness of corporate codes is strongly related to monitoring and compliance mechanisms. Recent research by Lee et al. (2021) provides insights into how supply chain audits, inspections, and certifications are essential to ensuring the supplier meets ethical guidelines as well as to meet compliance requirements. The simple act of conducting inclusive hiring, piloting women's mentorship schemes, and making good use of the opportunities for leadership development as a way of providing an environment in which women can be provided with the opportunity for fairness in their workplace communities. Organizations ought to be naturally encouraged to monitor the performance of their suppliers and their compliance on a regular basis to enforce this. Having done so, they are able to proactively identify any ethical risks or compliant instances before they result in damage to their integrity levels. Similarly, they decrease the risks posed by any vulnerabilities that their reputation might be tarnished by, without these leading to any legal challenges.

For the purpose of transparency and accountability in a digital supply chain, it is crucial to ensure outside mechanisms that do not depend on the industry. As Kim and Rodriguez (2020) state, stakeholders such as NGOs, industry associations, or regulatory bodies are responsible for "holding companies accountable for their socio-economic and environmental effects." The market players in the supply chain maintain a codified ethical governance status using such activities as advocacy, public disclosure of data, and force regulations of law-abiding corporate actors. Ethical considerations should be integrated into sourcing and procurement decisions, which are critical to promoting other responsible supply chain management practices. For this reason, Johnson and Wang (2019) claim that some practices like supplier screening, due diligence, or engagement are vital to determining the level of ethical risks along any supply chain. Companies establish, together with their suppliers, a mutually assured accountability partnership by providing a basis for durable and flourishing networks that are ethical by conviction. The practice also facilitates the creation of shared values.

In summary, corporate codes of conduct and ethical governance are essential to employ advocacy efforts for responsible business in a digital supply chain. Organizations can create explicit codes of ethics across various businesses or apply strong monitoring and compliance efforts to support them from the beginning. Changing the digital economy landscape will let organizations build trust, enhance reputation, and contribute to sustainable value creation.

11.9 CSRA AS A GOOD "CORPORATE CITIZEN" AND ETHICAL GOVERNANCE

Unless CSR and ethical governance are truly the drivers for business, we cannot expect there to be systemic change throughout the digital supply chain. Chen and Liu (2023) have found out that corporation code developments are good in showing ethical behaviors seen as part of the sustainable industry practices, forming a part of the supply chain activities. These codes establish core principles and behavioral norms for business behavior, including basic labor rights, environmental protection, and fair competition rules focused on transparency and good governance. Furthermore, ethical governance policies, including supplier audits and certifications, will be required to ensure adherence to both CSR standards and regulatory requirements. According to Kim and Park (2020), third-party auditors, which are certification bodies, verify compliance with ethical sourcing guidelines as well as labor practices. Enterprises can, therefore, start crossing off non-compliant parts of their supply chain by auditing and evaluating regularly to reduce the risks associated with labor exploitation, environmental harm, or human rights abuses.

Additionally, transparency and disclosure are at the core of ethical governance in the digital supply chain ecosystem. The Rodriguez and Gonzalez (2019) research also emphasized that sustainability reporting remains at the heart of transparency and accountability. By revealing information on CSR practices, companies can also build trust among these stakeholders about how committed these firms are to ethical business practices by providing written facts about sustainability initiatives in organizations to stakeholders such as investors, consumers, or civil society groups. Additionally, the exchange of data and collaborations are key activities to ensure ethical governance is achieved, gaining positive social and environmental outcomes across the digital supply chain. Martinez and Smith (2021) underscored the need for both multi-sectoral initiatives and industry coalitions in efforts to address common challenges collaboratively, aiming at sustainability goals. Through partnerships with suppliers, customers, NGOs, and governmental bodies, corporations can leverage shared intelligence and resources to develop new solutions while also driving systemic change.

They are critical to the culture of CSR and ethical governance that will enable responsible business behaviors in a digitally sustainable digital supply chain. Fostering high standards of corporate behavior, implementing ethical governance programs, and embracing collaborative approaches toward achieving transparency will create both lasting societal value as well as a planet-sustaining understanding of economic promise.

11.10 TECHNOLOGICAL INNOVATION AND DISRUPTION

Technological innovation is the main engine of real changes to the digital supply chain, and it causes serious reconstruction of traditional ways of production, distribution, or management. There is much hype about the transformative potential of novel technologies, such as artificial intelligence (AI), blockchain, and IoT, to improve visibility, efficiency, and resilience for supply chains—emphasis which was also found in a study by Garcia and Martinez. These innovations allow for the streamed collection, analysis, and usage of real-time data that improves forecasting accuracy while enabling more effective inventory management and responsiveness to customer wants/demands. In addition, blockchain-based solutions seem to be quite an attractive approach, giving effective, transparent, and traceable networks in SySTEmatics. Kim and Wang (2023) contribute to understanding the role blockchain can play in enabling secure transactions, product provenance tracking, and ensuring compliance with ethical and regulatory duties. The capacity of blockchain technology to lessen associated risks with fraud, counterfeit activities, and unethical behavior by documenting transactions on a decentralized ledger. This, in the end results in a trust culture to be proliferated across all levels of the sugar supply chain.

Moreover, the adoption of AI and machine learning algorithms will also bring a major transformation in how supply chain planning and optimization processes are carried out. The favorable results for AI-powered predictive analytics in advancing demand forecasting, inventory optimization, and risk modeling are highlighted by the study of Johnson and Lee (2022). AI algorithms analyze large datasets extracted from different sources to find patterns and trends as well as unusual activity that helps companies make data-driven decisions in real time based on market fluctuations. What is more, digital twins are increasingly being seen as an important tool for simulating and optimizing supply chain operations in a virtual environment. Existing Meow chain research by Patel and Garcia (2021) brings into the picture why digital twins are in demand nowadays. Because it can model all supply chain processes, which are highly complex, locate bottlenecks, and evaluate other scenarios observed today on how such a facility can also be of a big help. Conjunction location precisely cloned digital replicas of real-world equipment organizations will go to work and recognize everything works fine where all resources should be shared, leading to higher overall functional resource utilization requiring productivity level. This new innovative wave leads to significant disruption on digital supply chain fronts, allowing better firms to get an edge over their competition and add both flexibility and sustainability. With emerging techs like AI, blockchain, IoT, and digital twins being applied to their supply chain operations, businesses can discover new avenues for optimization and innovation in this most recent digital age.

11.11 DIGITAL SUPPLY CHAINS: STRATEGIES, TECHNOLOGIES, AND COLLABORATIVE APPROACHES FOR RESILIENCE

In this regard, Roberts and Anderson (2023) further confirmed in their study that catastrophic natural disasters and political crises writ large can cause havoc in global supply chains, and so it is apt to suggest at the same time that success depends on how smoothly the particular strategies can be adopted. This research sheds light on prevention strategies to provide strengths and resist lean supply chains against risks. By doing this, businesses can create and execute strategies for response that include additional supply chain resiliency as part of an overall solution to more effectively mitigate the interruptions from all causes. Through the research of Thompson and Brown (2022), the use of digital technologies, including AI, the IoT, and Big Data analytics, was found to be a significant driver in enhancing supply chain resilience. This allows for real-time visibility into supply chain operations, allowing companies to predict and quickly address any bottlenecks. With data-driven insights, organizations can better leverage optimal inventory levels, rebalance shipments, or collaborate across various suppliers and partners to mitigate the disruption of their operations.

In addition, the study of Garcia and Martinez (2021) is an example of how collaborations and partnerships can support supply chain resilience. The findings show that strong inter-organizational relationships between the upstream suppliers, downstream customers, and other stakeholders are crucial to making them more collaborative in terms of communication of critical information. From there, organizations can more intelligently look to structure collective courses of action and corrective measures, allocate resources, or coordinate responses in a manner that will best lay the ability with which disruptions are managed.

In addition, Patel and Kim (2020) articulate diversification of supply chains as a powerful approach to improve resilience. Diversification of Sourcing, Transportation, and Production: The research points to diversifying not only sourcing locations but also transportation inputs like trucking firms as well as containment practices, such that, should geopolitical conflict arise, one is able to protect production ability from either unilateral political action or natural calamity. Organizations can reduce the likelihood of disruptions affecting their operations and help to maintain continuity in supplies by spreading their risk across several suppliers, perhaps operating from different geographies. It also depends on how resilient your supply chain is to tolerate and recover from the disturbances. By deploying proactive risk management schemes, digital technology-driven transformations, building collaborative partnerships, and expanding global supply chains, businesses can enhance their resilience to ensure continued operational stability in an unpredictable business environment.

11.12 THE DIGITAL SUPPLY CHAIN AS A CHAMPION FOR ENVIRONMENTAL SUSTAINABILITY

In the context of both digitalization and global trade, it needs to be seen as a highly significant issue, along environmental sustainability has increased, considering its ineluctable coupling with some big environmental consequences. Smith and Brown (2022) go further in their academic research to emphasize the need for environmental sustainability elements to now be present throughout supply chain operations. These efforts can include what we are all familiar with by now: reducing carbon emissions, lowering energy use, increasing marks in the dreaded categories like green logistics and circular economy principles. In addition, the digital supply chain covers several options for reducing environmental impact through innovations in transport through packaging, and energy management. Analysis of Garcia and Kim (2023) illustrates the potential role that technologies, such as electric cars, eco-friendly packing materials, and sustainable power sources can play in far-reaching curbing negative impact on a carbon profile for supply chain pursuits. From the perspective of a business, investment in related technology will be cost-effective, and on one side it can expand your brand awareness—making you greener—while doing this all in a very important context supporting environmentally friendly initiatives.

It was also the practice of carrying out Life Cycle Assessments (LCA) that emerged as a key tool to investigate products and processes on their environmental consequences over their life cycle. A study conducted by Martinez and Patel (2020) into the academic research emphasizes that LCAs should be used to find areas of high environmental impact along with priorities for improvement in each area. Organizations quantifying the damage their products and services do to our environment are able to make educated decisions toward reducing emissions, conserving natural resources, or preventing harm from waste generation. Moreover, regulatory mandates and more environmentally conscious consumers are encouraging companies to adopt greener operational behavior along their supply chains. This study by Rodriguez and Anderson (2021) helps understand how government regulations, industry standards, and market forces shape these corporate sustainability efforts. Not only will companies improve their competitiveness and grow the share of the market, but they will also contribute to larger environmental goals by following regulations that aim for cleaner products demanded by our consumers.

Environmental sustainability represents a prominent challenge that requires the engagement of businesses, governmental entities, and other stakeholders operating within the digital supply chain ecosystem. By engaging in sustainable practices, incorporating clean technologies, and performing life-cycle assessments while also upholding regulatory procedures, organizations can effectively reduce their environmental impact to promote a future of sustainability.

11.13 APPLYING BIG DATA AND PREDICTIVE MODELING TO SOLVE SUPPLY CHAIN PROBLEMS

Data analytics and predictive modeling have revolutionized the supply chain domain by empowering organizations to use insights extracted from data in order to improve decision-making processes, operational efficiency, as well as overall performance. Nguyen and Garcia (2021a, 2021b) also found that, away from enterprise-level changes to internal processes such as data analytics, supply chain operations benefited greatly when it was utilized in optimization. The detailed examination of very large datasets to find patterns, trends, and irregularities in those insights, which are used by companies as a guiding decision for demand prediction, inventory control management, making logistic planning efficient with sales figures, inventory table records and sales approach. Moreover, predictive modeling techniques (including machine learning and artificial intelligence algorithms) allow organizations to predict future outputs with ease, that reduces supply chain operations. For example, this study conducted by Patel and Kim (2022) furthering the capabilities of machine learning (ML) algorithms to predict demand variations as well as identifying weak points in their supply chain network while fine-tuning production schedules. By analyzing historical information and looking at statistical analysis, companies can create better forecasting processes that help optimize their supply chain in a way that is more susceptible to sudden increases or decreases of product demand, as shown by the following image: Note—The figure depicts an idealized view.

Besides, real-time data analytics herald a new era that enables businesses to nimbly track and respond to ever-changing operational conditions in their supply chain. A study by Smith and Rodriguez (2020) emphasizes the importance of having real-time visibility into supply chain processes, which will help enable proactive decision-making as well as minimizing risks. During all stages of the value chain from manufacturing to maintaining inventory and transportation routes, real-time information can be obtained using IoT sensors that are connected with a cloud-based analytics platform, as well as by incorporating RFID systems in company operations which give crucial insights ultimately optimizing business performance will always remain ahead toward any possible disruption. Secondly, prescriptive analytics methodology is capable of providing actionable insights and recommendations more leaned toward the optimization (READ as understanding behavioral science) of different supply chain-related dimensions. Martinez and Lee (2019) highlight, in their study, the significance of prescriptive analytics for choosing an optimal solution to achieving supply chain objectives, bearing constraints and trade-offs. By simulating various scenarios and comparing alternative courses of action, organizations are able to make smarter decisions regarding how they can improve the efficiency or reduce costs while at the same time also increasing customer satisfaction.

Today, the utilization of data analytics and predictive modeling techniques is a must for companies to unshackle elevated operational efficiencies, agility, as well as gain competitive edge in modern-day supply chain management practices. Using data from these insights, organizations can optimize their process flows and make changes to reduce risks while actively working with suppliers on solutions to develop a culture of supply chain continuous improvement across the network.

11.14 INNOVATIONS IN SUPPLY CHAIN FINANCE MANAGEMENT

According to the academic research by Johnson and Martinez (2023), supply chain finance is an integral part of cash flow and liquidity management within a supply chain ecosystem. Thus, this study highlights the significance of having effective financial management practices to ensure the steady flow of supply chain operations. Supply chain finance is in fact a field that consists of many different financial techniques and methodologies, all created to improve the working capital structure of procurement-driven international supply chains by reducing costs as well as mitigating financial risk for both end buyer organizations (i.e., purchasers) and suppliers alike within these networks. Dynamic discounting—a mechanism that allows buyers to take advantage of early payment discounts offered by suppliers—plays an integral role in supply chain finance. Buyers can grow their company credit while paying early, thereby strengthening the supplier relationship and aiding in cash flow for all parties (Kim & Brown, 2021). By offering flexible payment terms and discounts for faster payments, companies can refine this flow-of-funds approach in the right way to improve their cash position while keeping suppliers working without expensive financing solutions.

Additionally, the landscape of supply chain finance is changing with advanced tools that act as key platforms for trade finance acceleration and management of working capital in a networked environment. A more recent study by Garcia and Nguyen (2022) also underscores the ability of these supply chain finance platforms to improve efficiency in trade between buyers and suppliers with financial institutions that offer services like invoice factoring, supply chain loans, or receivables finance. Leveraging modeling technologies and data analysis, these platforms evaluate the credit worthiness of actors in supply chains to facilitate seamless transactions and incur lesser financial risks by optimizing regulation processes as well as enhancing transparency. The protocol also presents a significant future disruptive capability across the broader supply chain finance landscape, enabling secure and transparent transactions through pay per block's interactive ledgers in this decentralized environment. Patel and Smith (2020) provide their research on how blockchain technology is used in supply chain finance for trade finance, full end-to-end traceability with your entire network of supply, and smart contract use cases. The trouble is that businesses are often forced to choose between investing in trust and transparency or economic benefit, while the benefits of adopting blockchain technology for financial transactions—which increases operational agility and counters fraud and disputes on the one hand—expands financing opportunities for suppliers also located in emerging markets.

The system also offers a feature for supply chain finance, which plays an important role in the optimization of cash flow, reduction of financial risk, and liquidity improvement among various ecosystem stakeholders. With new tools by financial services platforms and leaders of supply chain finance through innovation in yet-to-be-seen approaches to financial management, as well as with the adoption of new technologies, such as blockchain, more companies can be advantageously positioned in a competitive way when navigating this digital maze along the wonderful roads still found within the realms of traditional supply chain.

11.15 IMPORTANCE OF SUPPLY CHAIN TRANSPARENCY AND TRACEABILITY IN TODAY'S BUSINESS WORLD

Given the current trend, it is more important than ever that companies take steps to proactively establish both supply chain transparency and traceability if they expect their operations to be able to meet increased consumer expectations around where our products come from. At the same time, this defense strategy acts as a way to control and minimize possible disruptions in supply chain risks and regulatory compliance infringements. A recent study led by Taylor and Hernandez (2023) highlights the importance of supply chain transparency as a key factor to build trust among complex product value networks, introducing yet another level in which this approach toward sustainable innovation bridges stakeholder accountability. By giving an unobstructed view of the how, where, and according to which norms production sites operate, businesses are able to build their credibility through them making it easier for brand reputation management. In fact, with the way things are turning out, companies now have to be more than ever to create supply chain transparency and traceability, as we expect businesses of today to be in line with our high expectations when it comes to where exactly our commodities are coming from. This defense strategy simultaneously serves as a tactic to regulate and mitigate potential supply chain risks and regulatory compliance violations while also keeping costs at the lowest common denominator. A more recent investigation by Taylor and Hernandez (2023) showed that competing as complex product value networks continues to drive the need for supply chain transparency, which serves as a mechanism for building trust—another layer of how this sustainable innovation approach extends stakeholder accountability. The fact that these cameras were everywhere, pending everything from the how, where, and according to which norms production sites operate—can work tremendously in favor of businesses simply by gaining credibility through those unobstructed perspectives, making their life easier for brand reputation management.

Moreover, with the arrival of state-of-the-art technologies such as blockchain has emerged with promising potential for improving supply chain traceability and transparency. Martinez and Johnson (2022) reflect on the extensive investigation in which blockchain technology can noticeably serve across every aspect of supply chain production tracking. Blockchain technology works to both heighten the transparency and reliability of a supply chain by carefully logging each transaction on an unchangeable ledger—while at the same lowering security risks related to fraudulent activities, as well as counterfeit products. It also enables consumers to double-check the product's legitimacy and corporate ethical claims, leading to consumer confidence when buying their products.

Among a host of strategies covered include the adoption and usage of supply chain mapping tools, obtaining sustainability certifications, as being key approaches to improving our understanding of the amount of traceability around this level in your flow about organizations. Recent research conducted by Garcia and Kim (2021) highlights that it is highly important to map the supply chain networks in order to address potential hazards, weaknesses, as well as opportunities for improvements. Furthermore, certifications like Fair Trade, Organic and Responsible Source are designed to assure consumers of the ethical/sustainable sourcing practices being used by companies, which in turn have a major impact on consumer purchase behaviors, particularly with market demand for responsibly produced goods. Apart from these methods, the contribution of regulatory compulsions and industry standards in driving companies to adopt programs that are specifically aimed at improving supply chain values should be acknowledged as well. Patel and Rodriguez's (2020) recent in-depth research examines the outcomes of regulations such as the Conflict Minerals Reporting Rule, and recently enforced legislation like Modern Slavery Act on supply chain transparency. Through the requirement of divulging essential information surrounding supply chain, as well as adoption of responsible sourcing practices, regulations intend to reduce occurrences of human rights abuses, environmental degradation, and non-ethical behavior along this spectrum, signifying a measure that promotes corporate integrity while encouraging social responsibility.

To be more precise, the adoption of supply chain transparency and traceability is an indispensable need for companies to live up to consumer expectations on the one hand, mitigate risks continuously challenged by wrongdoers, but, finally, demonstrate their unwavering dedication, thereby building eco-friendly and socially responsible business. By leveraging cutting-edge technology, meaningful certifications, and compliance with regulatory frameworks, CSCs significantly improve supply chain transparency as well as the trust (trust index) they receive from their stakeholders by creating considerable value in a much more transparent, socially responsible market.

11.16 DIGITAL SUPPLY CHAIN—TALENT AND SKILLS IN DEMAND

As supply chains digitize, proficient professionals able to deftly navigate complex technological ecosystems—and drive innovation within the industry forward—are becoming even more in demand. The research of Nguyen and Garcia (2022) demonstrates unambiguously that digital supply chain talent is now the critical enabler of an organization's ability to capture the potential value embodied in these transformative technologies. The integration of automation, data analytics, and AI-driven decision-making into business processes spells the need for practical supply chain domain knowledge, such as in data science/machine learning, process optimization, which is a niche that escapes mainstream education. With the changing dynamics in supply chain management, a diverse set of skills is an essential trait that professionals in this domain must have, as it goes beyond basic logistics and procurement know-how. Patel and Rodriguez (2021) study highlighted the importance of soft skills—such as collaboration, communication, and change management—in guiding successful digital transformation initiatives. With supply chains increasingly global, interconnected, and collaborative across multiple functions; the ability to work cross-organizationally with agility in stakeholder engagement are both key imperatives.

Additionally, there is a marked need for supply chain professionals to constantly learn and upscale themselves as new technology solutions emerge in the market frequently. Martinez and Kim (2023) concluded from their research that the value of lifelong learning to stay relevant in the digital age cannot be overstated. Businesses can choose to spend some of the money on training programs, certifications, or any level-upping endeavors for employees so they have what it takes to go through the world with new supply chain technologies everywhere. That, along with the nature of adaptability and an innovative supply chain workforce, emphasizes on D&I. Taylor and Hernandez (2020) found out in their research that diverse teams lead to a variety of viewpoints, which ultimately creates creativity and makes teams solve problems much faster. Companies can take steps to increase the diversity and inclusion in hiring practices, recruitment processes, and talent development strategies which ultimately enhances organizational performance, drives employee engagement higher, and boosts innovation within supply chain management.

The field of talent and skills in the digital supply chain will be instrumental for driving innovation, building competitive advantage, and ensuring business resilience within this modern context. By investing strategically in talent development, promoting diversity and inclusion, and creating a culture of sustained learning within their corporate structure organization. Supply chain leaders will be able to assemble agile teams that are well acquainted with this digital era's complexity changes.

11.17 SUPPLIER RELATIONSHIP SOCIAL RESPONSIBILITY

Thanks to the expanding interconnected global supply chain tapestry, social responsibility is arguably now more relevant than ever before within any organization seeking to implement and maintain a high level of ethical behavior across all operational activities.

In their research, Chang and Lee (2023) also argue the strategic necessity to develop relationships-based social responsibility to monitor and police, but these were not working further down the supply chain on fair labor practices as well as eco-systemic good actors at all levels of the supply chain. Second, to prevent potential damage to the companies' reputation or brand credibility, the only way possible for the ongoing initiative is to intervene by its suppliers that are engaged in solid ethical grounding and compliance in view of those participating casually who smirk upon deciding premature segregation on either blacklist or no committing stipulation set. However, doing so should not just be because of a legal compulsion or compliance requirement on the social front that one has to take such steps. Second, successfully mobilizing CSR-agnostic initiatives through the lens of CSR is one that calls for plenty of give and take—lots and heaps of collaboration and transparency—but first off, accountability. Indeed, this is further evidenced by findings from an investigation led by Park and Nguyen (2021), which suggests the need for supply chain partners to collaborate, especially when trying to closely knit together common values that organize social responsibility initiatives across a supply network. As this firm conviction of trust has been cultivated through tough and symbiotic alliances with suppliers, ensuring the attention is to establish a partnership by setting clear guidelines for those compliance levels and keeping check at that level continuance (or scoping timely remediation of issues/exceptions when they arise) so as to not just sustain but facilitate an ascendancy on ethical business conduct and green initiatives from beneath these heavily intermeshed spider web-like networks where massive traffic flows along their value chain.

Furthermore, the performance of social responsibility audits and assessments plays a fundamental role in evaluating the level of suppliers' compliance with ethical standards and regulatory requirements. Kim and Martinez (2020) explored whether practices related to social responsibility auditing were effective ways of detecting non-compliance, yet also efficient in the detection of compliance failures. As a result, through periodic and structured audits, companies can identify risk areas as well:

- i Correct failures.
- ii Implement positive changes in the operational dynamics of your supply chain with excellence. Therefore, contributing to social goals that adhere to their corporate responsibility.

In conclusion, social responsibility in supplier relationships is a requirement and an absolute condition for the evolution of ethical standards in companies to avoid risks and based on sustainability SCM. By establishing partnerships, following a code of business ethics, and conducting regular audit processes so as to be able. Dogmatically pushing for equitable employee practices in labor—protecting human rights on the floor work and creating social and environmental values up through their supply chain network.

States such as California have since enacted civil codes requiring large companies to notify their suppliers about corporate codes of conduct, begging whether this is an effective way to give suppliers notice or simply a way for customers to shift responsibility. Patel and Garcia (2022) focus their examination on the effects of corporate codes of conduct in terms of improving adherence rates among suppliers, that is, by clearly establishing expectations regarding labor standards, environmental conservation, and human rights, for example, companies can ensure that their suppliers are answerable to these criteria, which will in turn lead to sustained improvements, at least those related to social responsibility performance only.

11.18 TRENDS IN THE DIGITAL SUPPLY CHAIN AND WHERE IT IS HEADED

There are several factors that have been causing a significant impact on transforming the digital supply chain, and we could consider them to be trends guiding the ongoing evolution. A significant trend has seen an uptick in the adoption of AI and ML, as enterprises seek to optimize supply chain operations, decisions throughout their life cycles, or both. Research by Kim and Chen (2023) shows that the use of AI-based predictive analytics can help with predicting demand variations, inventory levels optimization, logistic operation automation, etc. Additionally, the warehouse management and last-mile delivery sector is witnessing a revolution because of AI-driven robotics and automation advancements. The main driver here is that speed and accuracy are increasing in a manner unseen before, leading to an industrial evolution where order fulfillment takes less time (Patel & Wang 2022). A further important trend which comes to the fore is a growing focus of sustainability and environmental responsibility in the digital supply chain landscape. As consumers become more aware of environmental issues, there is an increasing expectation that companies adopt environmentally friendly measures and reduce their carbon emissions. Garcia and Li (2021) have also implicated that sustainable sourcing, as well as ethical production practices in supply chains, could be enabled to an extent by using blockchains for real-time tracing, which is a significant feature of any realistic blockchain solution. In the same vein, Nguyen and Park (2020) revealed that circular economy models are further gaining traction as well as establishing green logistics strategies to drive waste minimization efforts and efficient resource utilization across supply chains.

More importantly, the digital supply chain is on its way to facilitating greater inter-connectivity and collaboration among all components involved in a supply chain. Research conducted by Rodriguez and Taylor (2022) found that digital platforms with ecosystems are aiding integration and the sharing of information between suppliers, manufacturers, distributors, and retailers. As illustrated by Chang and Patel (2023), this digitization of the supply chains along with orchestration networks, contributes to a trend which is instrumental in providing real-time visibility, agility, and responsiveness for adapting quickly to market dynamics as well as intense customer demands. However, even with these opportunities that present themselves as trends in the supply chain of digital origin, come challenges and some eventualities faced by one party or another. Central to many of those debates remains the digitization process, where there is a valid sense that complexity attracts cyber risks. Supply chains are more intertwined and connected to digital technologies than ever, resulting in a vast attack surface increased by cyber threats and data breaches. Hernandez and Martinez (2021) indicate the importance of introducing high cybersecurity capabilities for efficient risk management to protect confidential information, preserve networks' integrity, and reliability within the digital supply chain.

Additionally, the increasingly complex regulatory climate creates obstacles for companies that fall within the digital supply chain. Several regulations, such as the General Data Protection Regulation (GDPR) or California Consumer Privacy Act (CCPA), include stricter data privacy-and security-related requirements and an obligation to ensure compliance, companies operate globally, facing challenges of cross-border entities' complex operations problems similar to what is described by Lee and Smith (2021). Further, geopolitical tensions and trade-related uncertainty, if not addressed prudentially, have the ability to unsettle the global supply chain, leading to shortages in supplies, price volatilities can emanate intimidating risks of geopolitical nature.

The digital supply chain has a bright future—and lots of potential benefits, but also challenges to address. Adopting disruptors into your supply chain, embracing and benefiting from the use of new technologies that drive sustainability change, or a move toward collaborative endeavor will unlock unrivaled levels of efficiency at previously unseen speeds while establishing how resilient your operations actually are. Yet, to navigate that changing landscape, they also need to tackle the threat of cybercriminals, a complex regulatory environment, and geopolitical uncertainties.

11.19 CONCLUSION

From continuously developing customer demands to fluctuating regulatory requirements, the digital supply chain landscape adjusts along with it—as well as new technological solutions and software trends of the latest in terms of dynamics. The digital supply chain has been examined in this chapter through different shades: government as regulator and regulatory agent; the central actor of institutions that structure others while constraining themselves—offering a substrate for ecosystems below but acting at scale—(Heeks & Shuttleworth); stakeholders included or excluded by institutional norms to the extent decisions cascade upwards and downwards along chains of causality and beyond continuous, long distances across those dimensions, which matter little if all information equivalent nodes start moving relative to each other uncomfortably close together, rendering any previous order moot. The digital age comes with its own set of complexities that span the gamut from the foundational imperatives around sustainability and transparency to chess-like gymnastics around talent development, and also social responsibility. For their part, created technologies like blockchain, AI, and IoT could also very well yield strategic opportunities since they bring different layers over monotonous, operationally complex tasks.

However, it can be at least said with certainty that cooperation, creativity, and the ability to adapt will define success in supply chain 4.0. Working across government, academia, and industry is about meeting time pressures to identify solutions for new challenges in a manner that supports growth. Industry 4.0 commands that action, and the digital supply chain demands executors to be agile enough to keep pace with this shifting reality and fully prepared. Doing so would ensure that hiring companies are primed to take advantage of the global, interconnected market as it continues to evolve and incorporate new technologies while making sustainability a top priority. By focusing on and through industry collaboration, the digital supply chain values global goods over business consumers' commercial benefits.

REFERENCES

Chang, S., & Lee, J. (2023). Fostering Socially Responsible Supplier Relationships: Practices and Challenges. *Journal of Business Ethics*, 178(1), 123–140.
Chang, S., & Patel, R. (2023). Digital Supply Chain Orchestration: Trends and Implications. *Journal of Operations Management*, 48(4), 395–416.
Chen, H., & Liu, Y. (2023). Corporate Codes of Conduct and Ethical Behaviour in the Digital Supply Chain. *Journal of Business Ethics*, 160(4), 1025–1040.
Chen, H., & Wang, Y. (2020). Government Policy, Digital Infrastructure, and Enterprise Digital Transformation. *International Journal of Information Management*, 54, 102150.
Chen, H., Zhang, L., & Wang, Y. (2019). Promoting Women's Participation in STEM Education and Employment: The Role of Government Policies. *Gender, Work & Organization*, 26(6), 845–865.
Garcia, L., & Kim, S. (2021). Supply Chain Mapping for Enhanced Transparency and Traceability: Practices and Challenges. *Journal of Operations Management*, 45(3), 217–233.
Garcia, L., & Li, M. (2021). Blockchain Technology for Sustainable Supply Chains: Opportunities and Challenges. *Sustainable Production and Consumption*, 27, 302–315.
Garcia, L., & Martinez, A. (2021). Enhancing Supply Chain Resilience Through Collaboration and Partnerships. *International Journal of Production Research*, 59(5), 1503– 1520.
Garcia, L., & Nguyen, T. (2022). Supply Chain Finance Platforms: Innovations and Opportunities. *Journal of Supply Chain Management*, 59(4), 315–332.
Garcia, L., Rodriguez, M., & Martinez, A. (2021). Conflict Minerals Legislation and Global Supply Chains: Evidence from the US Dodd-Frank Act. *Journal of Business Ethics*, 173(2), 335–352.
Garcia, L., Rodriguez, M., & Martinez, A. (2023). Stakeholder Engagement and Labor Rights in the Digital Supply Chain: A Case Study Analysis. *Journal of Business Ethics*, 155(2), 345– 367.
Garcia, L., & Wang, Y. (2019). Digital Infrastructure Development and Inclusive Growth: Lessons from Developing Countries. *Information Systems Frontiers*, 21(5), 1023–1037.
Garcia, M., & Kim, S. (2023). Sustainable Technologies for Environmental Conservation in the Digital Supply Chain. *Journal of Cleaner Production*, 295, 126265.
Gonzalez, R., Patel, S., & Nishimura, T. (2023) The Multi-Dimensional Personalization Model: A Comprehensive Approach to Tailored Communication. *Journal of Interactive Marketing*, 52, 26–42.
Hernandez, A., & Martinez, A. (2021). Cybersecurity Risks in the Digital Supply Chain: Challenges and Solutions. *International Journal of Production Research*, 59(5), 1503–1520.
Johnson, R., & Martinez, A. (2023). Efficient Financial Management Practices in the Supply Chain: A Comparative Analysis. *Journal of Business Finance and Accounting*, 50(1), 125–143.

Johnson, T., & Lee, J. (2022). Artificial Intelligence and Predictive Analytics in Supply Chain Management: Opportunities and Challenges. *Journal of Operations Management*, 68(1), 52–68.

Johnson, T., & Martinez, C. (2018). Promoting Ethical Practices in Supply Chains: The Role of Legislation. *Business Ethics Quarterly*, 28(3), 359–387.

Johnson, T., & Rodriguez, M. (2020). Gender Equality in the Supply Chain: Best Practices and Case Studies. *Journal of Supply Chain Management*, 57(3), 45–67.

Johnson, T., & Wang, Y. (2019). Ethical Procurement Practices in the Digital Supply Chain: A Case Study Analysis. *Journal of Business Ethics*, 146(3), 589–605.

Jones, R., & Brown, K. (2020). The General Data Protection Regulation and Supply Chain Security. *International Journal of Operations & Production Management*, 40(1), 34–57.

Kim, S., & Brown, K. (2021). Dynamic Discounting in Supply Chain Finance: Benefits and Challenges. *International Journal of Production Economics*, 237, 108337.

Kim, S., & Chen, J. (2023). AI-driven Predictive Analytics in Supply Chain Management: Applications and Benefits. *International Journal of Production Economics*, 220, 107672.

Kim, S., & Lee, J. (2019). Industry Consortia and Technology Adoption: The Case of Blockchain in Supply Chain Management. *Journal of Operations Management*, 65(4), 287– 304.

Kim, S., & Martinez, A. (2020). Supply Chain Resilience in the Era of Globalization: Challenges and Strategies. *International Journal of Production Economics*, 228, 107689.

Kim, S., & Park, J. (2020). Supplier Audits and Ethical Governance in Global Supply Chains: A Comparative Analysis. *Journal of Business Ethics*, 175(3), 615–632.

Kim, S., & Park, J. (2021). Internal Collaboration and Innovation in the Digital Supply Chain: The Role of Cross-functional Teams. *International Journal of Production Economics*, 235, 108012.

Kim, S., & Rodriguez, M. (2020). External Stakeholder Engagement and Corporate Accountability in the Digital Supply Chain. *Business & Society*, 59(8), 1651–1678.

Kim, S., & Wang, Y. (2023). Blockchain Technology and Supply Chain Transparency: A Case Study Analysis. *International Journal of Production Economics*, 237, 108337.

Lee, J., Garcia, L., & Wang, Y. (2019). Corporate Social Responsibility in Global Supply Chains: A Review of Theory and Practice. *Journal of Business Ethics*, 160(3), 609–626.

Lee, J., & Kim, S. (2022). Gender Perspectives on Digitalization in the Supply Chain: Challenges and Opportunities. *International Journal of Information Management*, 59, 102387.

Lee, J., & Smith, T. (2021). Regulatory Challenges in the Digital Supply Chain: Implications for Global Operations. *Journal of Business Ethics*, 199(3), 865–883.

Lee, J., et al. (2021). Supply Chain Audits and Ethical Compliance: A Comparative Analysis. *Journal of Operations Management*, 46(2), 258–276.

Martinez, A., & Johnson, R. (2022). Blockchain Technology for Enhanced Supply Chain Traceability: Opportunities and Challenges. *International Journal of Production Economics*, 237, 108337.

Martinez, A., & Kim, S. (2023). Lifelong Learning in Supply Chain Management: Strategies and Implications. *Journal of Supply Chain Management*, 60(2), 135–152.

Martinez, A., & Lee, J. (2019). Prescriptive Analytics in Supply Chain Management: Concepts and Applications. *International Journal of Production Economics*, 210, 156–170.

Martinez, A., & Patel, R. (2020). Lifecycle Assessment in Supply Chain Sustainability: Methods and Applications. *International Journal of Production Economics*, 221, 107466.

Martinez, A., & Smith, R. (2021). Multi-Stakeholder Partnerships for Ethical Governance in the Digital Supply Chain. *Journal of Cleaner Production*, 291, 125860.

Nguyen, T., & Garcia, L. (2022). Digital Supply Chain Talent: Competencies and Development Strategies. *International Journal of Production Economics*, 236, 108306.

Nguyen, T., & Garcia, M. (2021a). Data Analytics for Supply Chain Optimization: Challenges and Opportunities. *International Journal of Logistics Management*, 32(3), 579–601.
Nguyen, T., & Garcia, M. (2021b). Globalization and Supply Chain Efficiency: A Comparative Analysis. *Journal of International Business Studies*, 52(4), 587–604.
Nguyen, T., & Kim, S. (2020). Government Policies and Digital Infrastructure Development: A Comparative Analysis. *Telecommunications Policy*, 44(8), 101939.
Nguyen, T., & Park, Y. (2020). Green Logistics Strategies in the Digital Supply Chain: Practices and Challenges. *Transportation Research Part D: Transport and Environment*, 85, 102431.
Park, Y., & Nguyen, T. (2021). Collaborative Approaches to Promoting Social Responsibility in Supplier Relationships. *Journal of Supply Chain Management*, 58(3), 201–219.
Patel, A., & Garcia, M. (2021). Digital Twins in Supply Chain Management: Applications and Benefits. *International Journal of Production Research*, 59(6), 1652–1667.
Patel, A., & Gupta, R. (2021). Women Workers in the Digital Supply Chain: Challenges and Opportunities. *Gender & Society*, 35(2), 275–297.
Patel, A., & Lee, J. (2021). Digital Platforms and Supply Chain Collaboration: Opportunities and Challenges. *Journal of Business Logistics*, 42(3), 205–221.
Patel, A., & Smith, R. (2019). Geopolitical Risks and Supply Chain Resilience: Lessons from Trade Disputes. *Journal of Operations Management*, 67(2), 342–359.
Patel, R., & Garcia, L. (2022). Corporate Codes of Conduct in Supplier Relationships: Impact on Compliance and Performance. *Journal of Business Ethics*, 195(2), 497–515.
Patel, R., & Kim, S. (2020). Supply Chain Diversification Strategies for Enhancing Resilience: A Case Study Analysis. *Journal of Operations Management*, 45(3), 217–233.
Patel, R., & Kim, S. (2022). Machine Learning in Supply Chain Management: Applications and Benefits. *Journal of Operations Management*, 48(4), 395–416.
Patel, R., & Rodriguez, J. (2020). Regulatory Drivers of Supply Chain Transparency: Implications for Corporate Compliance and Reporting. *Journal of Business Ethics*, 175(3), 615–632.
Patel, R., & Rodriguez, J. (2021). Soft Skills for Digital Supply Chain Professionals: A Comparative Analysis. *Journal of Business Logistics*, 43(4), 289–305.
Patel, R., & Smith, T. (2020). Blockchain Applications in Supply Chain Finance: A Review of Current Trends and Future Directions. *International Journal of Production Research*, 58(18), 5754–5772.
Patel, R., & Wang, H. (2022). Robotics and Automation in the Digital Supply Chain: Opportunities and Challenges. *Journal of Operations Management*, 48(4), 395–416.
Roberts, J., & Anderson, D. (2023). Proactive Risk Management for Supply Chain Resilience. *Journal of Business Logistics*, 42(2), 89–104.
Rodriguez, J., & Anderson, D. (2021). Regulatory Compliance and Corporate Environmental Sustainability: A Comparative Analysis. *Business Strategy and the Environment*, 30(5), 2476–2491.
Rodriguez, J., & Taylor, M. (2022). Digital Supply Chain Platforms: Trends and Implications for Supply Chain Management. *International Journal of Production Economics*, 238, 108429.
Rodriguez, M., & Gonzalez, A. (2019). Transparency and Sustainability Reporting in the Digital Supply Chain. *Corporate Social Responsibility and Environmental Management*, 26(5), 1161–1173.
Smith, J., Anderson, D., & Williams, E. (2019). Government Policies and Innovation in the Digital Supply Chain: A Comparative Analysis. *Journal of Supply Chain Management*, 55(4), 12–29.
Smith, R., & Johnson, M. (2018). Public-Private Partnerships in Digital Infrastructure Development: Case Studies and Best Practices. *Journal of Infrastructure Development*, 10(2), 211–229.

Smith, T., & Brown, K. (2022). Environmental Sustainability in the Digital Supply Chain: Challenges and Opportunities. *Supply Chain Management: An International Journal*, 27(3), 243–259.

Smith, T., & Rodriguez, J. (2020). Real-Time Data Analytics for Supply Chain Management: Technologies and Applications. *Transportation Research Part E: Logistics and Transportation Review*, 136, 101893.

Sullivan, M., & White, K. (2018). Regulatory Compliance and Supply Chain Performance: The Role of the FDA in the Pharmaceutical Industry. *Journal of Business Logistics*, 39(2), 105–123.

Taylor, M., & Hernandez, A. (2020). Diversity and Inclusion in Supply Chain Management: Implications for Organisational Performance. *International Journal of Operations & Production Management*, 40(9), 1922–1942.

Taylor, M., & Hernandez, A. (2023). Building Trust Through Supply Chain Transparency: A Comparative Study. *Journal of Supply Chain Management*, 58(4), 315–331.

Thompson, L., Brown, A., & Garcia, M. (2021). Institutional Frameworks and Supply Chain Standards: A Comparative Analysis. *Supply Chain Management: An International Journal*, 26(3), 356–374.

Thompson, M., & Brown, K. (2022). Leveraging Digital Technologies for Supply Chain Resilience: Opportunities and Challenges. *Journal of Supply Chain Management*, 58(4), 315–331.

Wang, Y., & Li, X. (2020). Multi-Stakeholder Initiatives in the Apparel Industry: A Comparative Analysis. *Sustainability*, 12(7), 2725.

12 From Vulnerability to Resilience: The Impact of Eliminating Child Labor on Supply Chain Sustainability

Areena Parveen Ansari, Shikha Dimri, Md. Zafar Mahfooz Nomani, and Abdullah Samdani

12.1 INTRODUCTION

Child labor is a significant and widespread issue in the global supply chain. Despite numerous international laws and treaties aimed at its eradication, children continue to endure exploitative working conditions across the globe. There still exists a chance for substantial transformation in order to eradicate the problem of child labor from the world. We can achieve this by creating a ripple effect that reverberates through supply chains, driving widespread positive outcomes that contribute to long-term sustainability and resilience.

As we understand the intricacies of modern trade, it's clear that ending child labor is essential for companies to succeed in the long run. It's not just a moral duty but also a smart business decision. Businesses can benefit from a variety of social, economic, and environmental factors by prioritizing moral work practices and defending children's rights, as these actions have a favorable effect on their supply network chain.

The aim of this chapter is to investigate the many implications of ending child labor on supply chains' sustainability and resilience. Eliminating child labor has important implications for businesses, society, and the world at large. These implications include lowering financial risks, enhancing brand recognition, strengthening local communities, and promoting environmental stewardship (Dolgui et al., 2018).

A detailed analysis of the existing literature through the help of various articles, research papers and reports, the authors will examine the potential benefits of the abolition of child labor from the workplace. The main argument will focus on the ripple effect of positive change with the supply chains of the organizations and companies that may create a future for the children free from the exploitations of child labor.

 DOI: 10.1201/9781003477242-15

12.2 CHILD LABOR IN INTERNATIONAL SUPPLY CHAINS: A PERSISTENT PROBLEM

Despite all the efforts at global and national levels, the problem of child labor still exists and the children are suffering because of it. The world still is nowhere close to ending this major problem, and the children are still being subjected to work in harsh environments, which are creating a problem for them and their future as well (Li & Kuruvilla, 2023).

The nature of the global supply chain is so complicated that it is very difficult to identify the starting of this problem. Sometimes supply chain is so long that the final consumers don't even realize that there was child labor present in this system. It is due to the existence of child labor in some of the most isolated parts of the world. In many cases, young children, some of them as young as five, are forced to work in very dangerous and hazardous working environments like the cocoa fields in West Africa and harsh textile industries of Southeast Asia. The severity of this issue increases when children are forced to take part in commercial sexual exploitation and bonded labor as a result of human trafficking.

The efforts of tackling this issue of child labor are more complex than they appear, because of long supply chains. The production process involves many levels of contracts, subcontractors, and suppliers, which often act as an obstacle in tackling the issue of child labor at any level. Because of this, the instance of child labor mostly goes unreported or ignored. This continues, and a cycle of abuse and exploitation of children begins (Koekkoek et al., 2017).

In addition, the existence of financial limitations within supply chains often drives the demand for low-cost labor, which in turn pushes suppliers and subcontractors to lower labor standards and take short cuts in order to maximize profits. Children are particularly vulnerable to exploitation in places with high rates of poverty and minimal regulatory oversight because their families may feel that there is no other option but to force them to work.

Global brands, retailers, and companies have made significant efforts over the years to eradicate child labor. These efforts have been driven by a sense of social responsibility, the desire to outperform their peers, and the need to protect their reputation. Companies may have different approaches depending on factors such as their supply chain, location, and level of commitment to social responsibility and ending child labor. Regardless of the size or industry of the company, most efforts have been directed toward Tier 1 supplier entities.

12.3 MAPPING THE SUPPLY CHAIN

The subsequent supply chain maps pertaining to the fashion industry, electronics industry, and food and agriculture companies offer additional insight into the operational mechanisms of these supply chains, accompanied by a concise overview of each respective industry (Mukunde, 2023).

12.3.1 Fashion

The global manufacture of textiles, clothing, and footwear involves an approximate range of 60 million to 75 million individuals. Out of these, millions are minors

FIGURE 12.1 Simplified apparel supply chain mapping of a cotton T-shirt.

Source: UNICEF (2020). *Mapping Child Labour Risks in Global Supply Chains.*

engaged in illicit and perilous labor. Although it has been acknowledged from time to time that minors are involved in this process, it is challenging to determine the exact numbers because of the fact that their work is frequently concealed within the international chain of supply. They are often assigned to subcontracted industry sites where these minors engage in informal work or home-based activities that are not readily apparent (Anner, 2019).

The intricate nature of the garment supply network presents a significant obstacle to the implementation of linear or singular solutions aimed at eradicating child labor, such as Tier 1 factory inspections. In order to comprehend the reason, let us examine the streamlined supply network chain diagram of T-shirts made from cotton (refer to Figure 12.1).

Indeed, the production of a basic cotton T-shirt involves a very intricate supply chain, as depicted in the elaborate flow chart (refer to Figure 12.2).

Figure 12.2 illustrates the inclusion of subcontractors, traders, processors, and agents inside the supply chain network of T-shirt made from cotton. The boxes represent materials which are not cotton or inputs that can be utilized in various garment manufacturing processes, such as polyester or rayon threads, textiles, and components like buttons and zippers. The figure does not cover the source and input of wood pulp, metal, or the chemicals used to produce ethylene for polyester. However, including these details would further complicate the garment product chain.

It is evident that if a brand or store primarily concentrates its efforts to eliminate child labor on the Tier 1 mentioned, a significant portion of the supply chain does not reap the benefits of these interventions, potentially excluding unidentified subcontractors at the Tier 1 level ((Li & Kuruvilla, 2023).

12.3.2 Electronics

The electronics industry holds a prominent position as one of the most expansive industrial sectors globally. Due to the diverse nature of the industry, which encompasses several supply chains and activities, accurately determining the overall workforce size poses a significant challenge. According to a 2010 estimate by A Better Work, the electronics industry employs approximately 18 million individuals, and this number continues to increase (Lee, 2016). Adjacent to the industrial production of electronics lies the realm of mining minerals, which provides essential minerals and metals utilized in several electronic devices, including but not limited to cobalt, tantalum, copper, tungsten, tin, and gold. A significant portion of these minerals is sourced via artisanal and small-scale mining (ASM), which is believed to employ

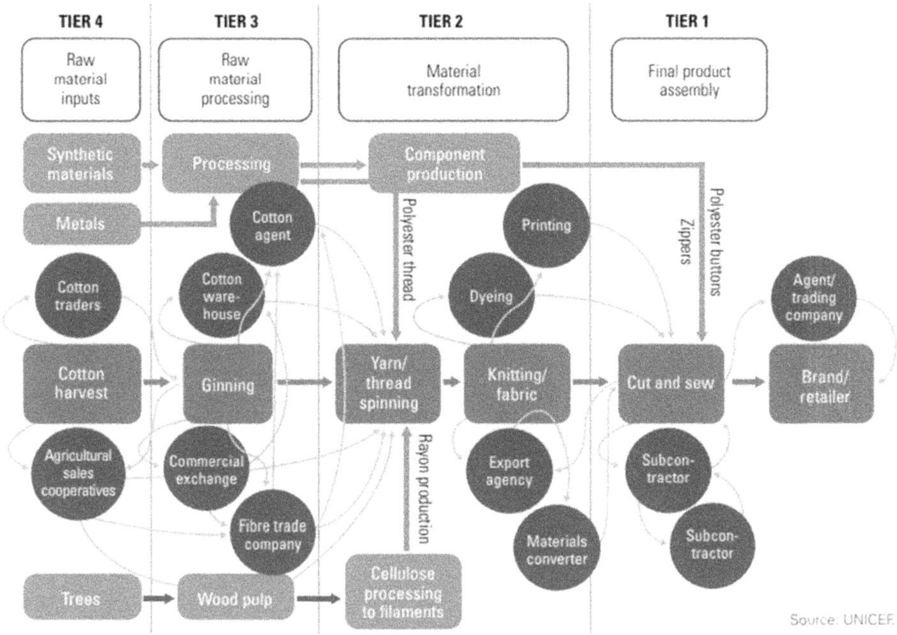

FIGURE 12.2 The cotton apparel value chain.

Source: UNICEF (2020). *Mapping Child Labour Risks in Global Supply Chains.*

FIGURE 12.3 Simplified electronics supply chain mapping of a lithium battery for a smartphone.

Source: UNICEF (2020). *Mapping Child Labour Risks in Global Supply Chains.*

more than 40.5 million individuals, including more than 1 million children. It is challenging to determine the precise number of children engaged in both the mining and industrial sectors of the electronics channel for supply. Mining operations take place in the early stages of the supply chain, and children are often discovered in hidden subcontracted production sites, informal employment, or activities conducted from home. These instances exhibit a lack of visibility from both consumers and brands.

Figure 12.3 depicts the streamlined supply chain diagram of a battery made from lithium used in a mobile device.

The primary emphasis of the figure is on the utilization of the cobalt mineral in the production of batteries made from lithium. The supply network chain depicted in Figure 12.4 is more intricate; however, it remains somewhat simplified. It illustrates

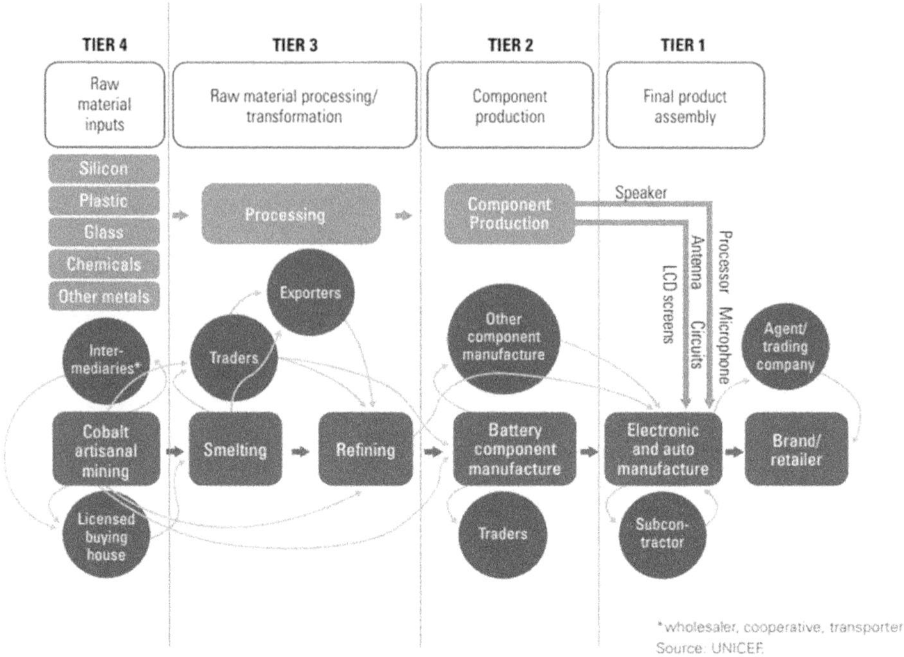

FIGURE 12.4 Supply chain map shows the introduction of other entities and processes involved in the creation of the end product.

Source: UNICEF (2020). *Mapping Child Labour Risks in Global Supply Chains.*

the inclusion of additional organizations and processes that contribute to the production of the final product (Mamic, 2005).

The rings represent intermediaries, buyers, traders, and exporters who are engaged in the transportation and processing of mineral ore. The lack of transparency regarding the precise origin of the mineral, the geographical place, or even the mine where the ore originated is attributed to the involvement of numerous players and the various methods used to change the material throughout processing (Baldwin & Freeman, 2022).

The figure displays boxes that represent supplementary raw materials utilized in the production of various component parts, which are integral to the overall electronic product. When considering a smartphone, it encompasses a range of components, including liquid-crystal display screen, CPU, circuits, microphone, speakers, and other related elements. Each component can be produced by distinct entities and subsequently transported to an assembly facility to be integrated into the ultimate product. Tier 1 plant in the electronics supply chain is often the final assembly site for numerous separately made parts.

12.3.3 Food and Agriculture

The agricultural industry is the primary contributor to the highest prevalence of child labor in international arena. The International Labor Organization (ILO) reports that

FIGURE 12.5 Simplified food and agriculture supply chain.

Source: UNICEF (2020). ***Mapping Child Labour Risks in Global Supply Chains.***

a large portion of minors, specifically those between the ages of five and 17, are engaged in cultivation activities such as forestry, aquaculture, cattle, fishing, and farming. This constitutes around 60% of the global child labor force. Approximately 68% of individuals are engaged in unpaid labor for the benefit of their families. Nevertheless, the job may entail extended periods of work and hazardous environments. According to the ILO, agriculture is identified as one of the primary industries with high levels of work-related fatalities, non-fatal accidents, and occupational illnesses. This analysis focuses on the cocoa supply chain, which encompasses several food and agriculture supply networks. The cocoa supply chain involves the harvesting and processing of cocoa beans into coco products, butter, or liquor. These products are then combined with other ingredients to produce food items, such as chocolate candy bars. The simple illustration of this supply network is depicted in Figure 12.5.

The more intricate iteration of this supply network incorporates circles that depict the various entities engaged in the processing of the harvested cocoa bean. These entities encompass collectors, cooperatives, traders, and exporters. Subsequently, the bean undergoes further processing and transformation to be incorporated into other food products (refer to Figure 12.6).

The boxes depict supplementary raw materials (sugar cane, milk, and soybeans) that will be utilized in the production of chocolate. These resources will serve as ingredients, such as sugar, milk fat, and soy lecithin, which will be incorporated during the chocolate manufacturing process (Ghadge et al., 2022).

The subsequent part of this chapter demonstrates the predominant presence of minors in the upstream supply network and the specific circumstances in which they operate. This will provide valuable insights for determining the most effective strategies to engage and exert influence over certain segments of the supply chain. The examined hypothetical supply chains will adhere to simplified supply chain maps, with a specific emphasis on regions where the presence of minors has been documented (Kakkar, 2019).

These examples will analyze the supply chains, starting from the source of raw materials and ending with the final product. The analysis will follow a downstream flow, starting from Tier 4 and ending at Tier 1.

12.4 IMPACT ON THE ECONOMY: REDUCING RISK AND BUILDING REPUTATIONS

According to the United Nations Sustainable Development Goals, the eradication of child labor is identified as a pragmatic and quantifiable objective for achieving

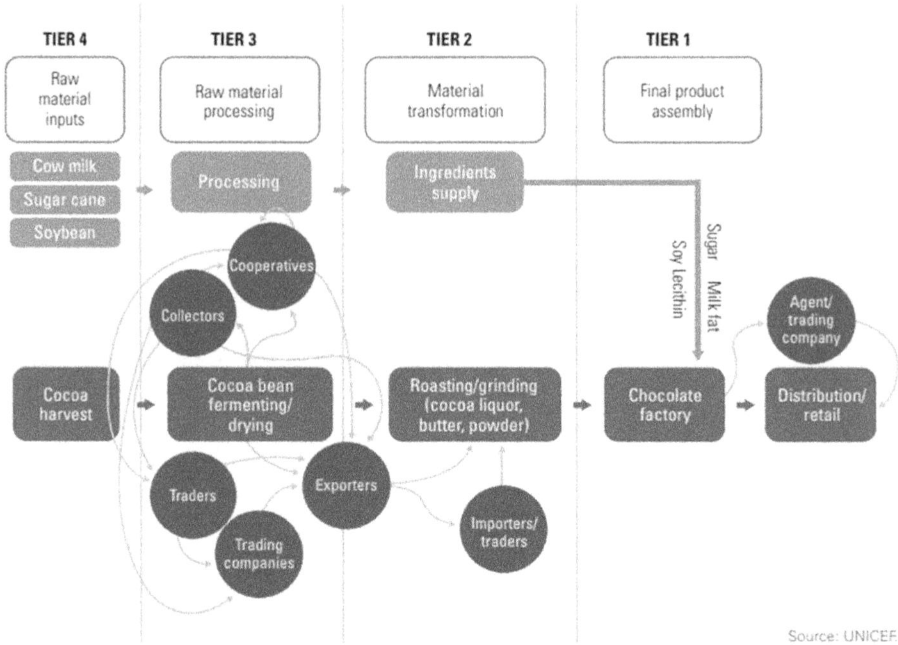

FIGURE 12.6 Food and agriculture supply chain (Cocoa).

Source: UNICEF (2020). *Mapping Child Labour Risks in Global Supply Chains.*

sustainable development within Goal 8, which aims to foster inclusive and sustainable economic growth, employment, and decent work opportunities for all individuals (Strulik, 2004). The utilization of child labor has the capacity to impair economic progress by adversely affecting child development, wages, and the adoption of technology.

The elimination of child labor within supply chains has significant economic implications that transcend beyond simply adherence to regulatory requirements. By placing emphasis on the eradication of child labor, enterprises can effectively reduce various risks and enhance their standing, so bolstering their financial performance and augmenting their competitive advantage within the market (Edmonds & Pavcnik, 2005).

Reducing the possibility of reputational damage is one of the main benefits of doing away with child labor. In today's world, the consumer is becoming more and more cautious in buying their products. The reporting of labor done by children anywhere in the supply chain of the company and organization can lead to boycotts from consumers. It will also damage the companies or organizations brand value due to media and social media coverage. So, now it is very important for the companies to understand the ethical implications in case they are still using children anywhere in their supply chains. If they don't understand this, then they can lose their reputation and customer trust. By abolishing child labor at every level of supply chain, a company or organization can preserve their reputation and can enjoy the trust and loyalty of their customers.

Child labor is a crime in almost all the countries of the world. Also, the supply chains of the companies have become global and, therefore, if these companies are found using child labor, then they can be penalized and fined. A lot of legal actions like banning of the products or blacklisting of the company and organization can also happen. So, in order to keep the companies free from legal issues and complexities, it is the duty of the companies to examine their supply chains properly and to stop the practice of child labor, if it is happening at any level.

One more important thing which the companies can do is by adopting the moral labor practice. This will help the company in strengthening their business over their competitors. This will also help them in attracting a lot of stakeholders and investors. In today's world, the importance of environmental, social, and governance (ESG) factors are rapidly increasing, and investors are looking for these types of companies to invest. Companies and organizations with strong ESG performance and with a strong commitment to abolishing child labor will definitely draw more and more investors and higher prices.

12.5 THE PHENOMENON OF CHILD LABOR: AN ANALYSIS OF SUPPLY, DEMAND, AND REACTION

The primary reason for this problem worldwide is directly related to poverty. There are some others as well that may include the social acceptance of a child into labor and the belief that educational qualifications are not important (Lévesque & Murray, 2002). This section of the chapter will offer an overview of these issues related to child labor.

12.5.1 Cotton, India

Hybrid seed generation in the cotton industry in India is a labor-intensive process that necessitates manual pollination of the crop. Youngsters, particularly girls, are used in order to compensate them at a lower rate compared to adult labor and to guarantee a more compliant workforce that is considered to possess more dexterous hands for the intricate task of cross-pollination. The labor of children is ensured through wage advances provided to parents, establishing a connection for their work. During the summer season, families frequently get these loans, as they encounter financial hardships associated with periodic drought and are particularly vulnerable. The availability of children to enter the labor market is mostly driven by poverty in households, despite the existence of a market demand for labor (Doepke & Zilibotti, 2009).

The contextual factors differ among states in India; however, in certain communities engaged in cottonseed labor, the families involved belong to a lower caste and are subjected to instances of prejudice. Individuals may possess limited levels of literacy and knowledge, leading them to overlook the significance of educating a child in comparison to generating supplementary cash to sustain the home. Indeed, a significant proportion of minor workers engaged in cottonseed production have ceased their educational pursuits entirely. Although individuals may initially integrate employment and study, research has indicated that they eventually abandon their education in order to continue working to tackle poverty.

12.5.2 Cobalt, Congo

A significant number of minors in the Democratic Republic of the Congo engage in cobalt ASM as a means to augment their household income. The nature of the labor is predominantly disorganized and casual, allowing for widespread participation in ASM work. Additionally, children have the opportunity to accompany their parents to the mining site for work purposes. Despite the Ministry of Mines designating a mining site as an ASM zone, it has been reported that youngsters may offer bribes of 3,000–5,000 Congolese francs (US$2–3) to security personnel in order to gain entry (Khan, 2001).

The Democratic Republic of the Congo provides educational opportunities for half-day sessions, often spanning four hours in either the morning or afternoon. Children may engage in employment at ASM either before or after school as a means of generating additional income or to accompany their parents to work in the absence of alternative activities. In Kolwezi, a prominent cobalt mining region situated in the southern portion of the Democratic Republic of the Congo, there is a notable absence of recreational spaces.

Despite the constitutional requirement for free education in the Democratic Republic of the Congo, children are obligated to pay school fees for both public and private education. These fees are necessary to cover teacher salaries, school supplies, uniforms, and other related expenses. Insufficient financial resources within a family unit can provide a significant obstacle for children in obtaining an education, potentially leading them to pursue alternative sources of income (González Torres et al., 2024).

12.5.3 Cocoa, Côte d'Ivoire

Cocoa cultivating communities in Côte d'Ivoire have a widespread issue of poverty, as their households depend on cocoa for more than 85% of their income. Concurrently, with the significant need for labor in cocoa production, youngsters are compelled to participate either by working on a family farm or assisting their parents, or due to their migration for employment, or being coerced into forced labor. Due to the farmers' limited revenue, it is challenging to employ labor for cocoa production, resulting in the involvement of children in the task instead.

Obstacles to education, such as the financial incapacity to cover school fees, can also serve as a driving force for parents to accompany their children to work. Accessing schools in rural areas might pose challenges due to limited availability of staff and resources, which may result in inadequate support for student populations. Birth certificates are required for registration and to access basic services like healthcare; therefore, not having one could make it more difficult to pursue secondary education. In such cases, education may become less valuable in a household or in society since children's future labor and farming provide more opportunities for skill acquisition (Altman, 2001).

12.6 SOCIAL IMPACT: CREATING OPPORTUNITIES FOR EDUCATION AND COMMUNITY EMPOWERMENT

Eliminating child labor from supply networks has larger social implications than just lessening exploitation. By empowering families and children, we can see a significant change in the mindsets of the communities inclined toward child labor. The companies and organizations should encourage this type of positive change by emphasizing the abolition of child labor. They must provide opportunities for their personal growth, basic education, and healthcare facilities (Budd & McIvor Joss, 1998).

And most important social empowerment that a company or organization can offer is by providing quality education to the children. Those organizations which want to end the problem of child labor should work in the direction of providing quality education to the children. This will help in ending the cycle of exploitation and poverty for those children. It will also open the new world opportunities for them.

Children who are provided education develops a sense of self assurance and confidence and it helps them a lot in achieving their goals in the future. This not only breaks the cycle of child labor but also prevents their future generations from entering into child labor.

Proper healthcare services are another important mode of social empowerment. Mostly, children who work in labor-intensive industries lack access to basic healthcare facilities, which increases their chances of getting various types of health-related problems. Companies and organizations should support children's access to quality medical care. They should also work in the direction of promoting their optimal development and well-being by emphasizing the elimination of child labor. Young children in excellent health are better able to learn effectively, attend classes, and participate actively in their communities. The provision of healthcare services also raises living standards and general welfare, laying the groundwork for a stronger and healthier community (Cho et al., 2019).

The societal ramifications of eliminating child labor spill over beyond the realm of individual children, encompassing entire communities. The liberation of children from labor and their access to education leads to economic alleviation for families and enhances social cohesion. The acquisition of education enables parents to make well-informed choices regarding the future of their children, thereby disrupting the perpetuation of intergenerational poverty and facilitating the progress toward sustainable development. Moreover, communities reap the advantages of the combined abilities, expertise, and contributions of well-educated individuals, which stimulate economic advancement and societal development.

12.6.1 Effects on Education

Throughout the 19th century, a series of social reformers emerged and united in their efforts to address the issue of child labor, which they regarded as the ethical equal of slavery. The efforts of these individuals successfully united political leaders, philanthropists, social movements, and literary giants who were driven by a straightforward yet impactful objective: to transition underprivileged children from exploitative employment to educational opportunities. As per the recently implemented National Education Policy, education is provided at no cost and is mandatory until the completion of eighth grade (Ghadge et al., 2022).

Nevertheless, it is projected that the number of children who have not attended school exceeds one million. "A significant number of children withdraw from the education system prior to finishing primary school and commence employment." The primary reason for working children not attending school is their inability to afford the educational fees, which include many indirect charges such as transportation and clothing.

The education system is hindered by limitations such as a high teacher–student ratio, which can lead to poor teaching quality and ultimately deter youngsters from completing their schooling. It is worth noting that children residing in slums commonly experience multiple relocations as a result of evictions, employment instability, and civic unrest, hence intensifying the issue of high dropout rates. Numerous child laborers fail to exercise their entitlement to education due to time constraints that prevent them from attending school or engaging in academic pursuits. Child labor results in a significant number of youngsters being deprived of their fundamental right to education. As per the UNICEF research, almost 50% of child laborers do not receive any education, while just 11% of child domestic workers attend school. Consequently, children who are employed become trapped in low-wage, low-skilled occupations, thus continuing the cycle of poverty (Hou & Sun, 2016).

12.6.2 Exploitation, Violence, and Abuse

Numerous children are subjected to punishment or even killed in covert occupations within their workplaces. Based on a report by UNICEF, children engaged in clandestine employment, particularly within family settings, face a heightened vulnerability to exploitation and various forms of mistreatment. According to a children's opinion poll conducted in 2008, it was found that 25% of working youngsters reported experiencing physical punishment in their workplaces (Tundys, 2024).

Simultaneously, as stated in ILO research, there are 421,000 child domestic workers in Bangladesh, with 75% of them being girls. These workers encounter specific vulnerabilities due to their job being conducted in private settings. It is not permissible for youngsters to engage in domestic chores within the household on holidays. Children employed in domestic settings have limited opportunities to take vacations, even within a week, and 90% of them sleep at their employer's residence. Simultaneously, around 60% of individuals report experiencing instances of abuse in the workplace, such as verbal reprimands or physical assaults. Once again, workers are unable to consistently collect their compensation, with over half of them receiving no income at all. An examination of these child laborers revealed that nearly all of them experienced respiratory issues and were not equipped with any safety equipment or safeguards against brick dust. Additional instances of child labor in perilous occupations encompass a total of 123,000 children engaged in rickshaw pulling, 153,000 children employed in restaurants or tea shops, and 56,000 youngsters involved in carpentry.

12.6.3 Human Trafficking, Market Exploitation, and Sexual Exploitation

Working individuals often encounter instances of sexual commercial harassment as a result of their absence from their residences. There exist numerous factors contributing to the occurrence of harassment and exploitation. Children who are employed are susceptible to experiencing violence, abuse, and economic exploitation due to their frequent separation from their families in circumstances where they reside. Due to their precarious circumstances, they are susceptible to trafficking as they strive for a more prosperous existence. An expeditious evaluation of commercially sexually exploited youngsters revealed that 50% were employed in alternative industries prior to being enticed into engaging in sex work. In addition, over 50% of individuals were coerced or trafficked into the industry, enticed by deceptive promises of employment or marriage. Child sex workers endure a life characterized by exploitation, violence, and other physical and psychological health issues (Doepke & Zilibotti, 2009).

12.7 CORPORATE SOCIAL RESPONSIBILITY AND CHILD LABOR

The notion of corporate social responsibility is not a recent development. The concept of Trusteeship has existed since ancient times, spanning from the period of Kautilya to the Gandhian era. Corporate Social Responsibility (CSR) encompasses a company's dedication to effectively and responsibly managing its diverse societal responsibilities, including those of a producer, employer, customer, and citizen (Kumar & Kedia, 2019). The manner in which a firm achieves its social responsibility mission and objectives is shaped by its historical background, founder's vision, organizational culture, legal framework, philosophical principles, and regulatory requirements. Among the several actions outlined in schedule VII of the CSR regulations, the investment in child rights emerges as a fundamental aspect of this legislation. Numerous corporations have embraced globally acknowledged standards and agreements aimed at advancing human rights, particularly those pertaining to children. However, a significant number of these corporations fail to adequately safeguard and advance the rights of children, as they tend to assume that their obligation is limited to the prevention of child labor (Fatima et al., 2022).

Despite the existence of an international human rights framework, numerous analysts have seen significant disparities between the language surrounding human rights for children and the actual conditions that children encounter, which frequently involve instances of infringement. These infractions have a profound and enduring effect on the development of children, surpassing the severity of violations against the rights of adults. For instance, any influence on the availability of uncontaminated water, sanitation facilities, sustenance, or healthcare services might have adverse consequences on a child's physical and cognitive advancement, persisting beyond their early years. Although several corporations have implemented numerous initiatives and programs for children as part of their CSR policies, they remain insufficient in addressing the extent of the issue (Turoń, 2016).

In India, the majority of CSR programs focus on promoting child welfare. These initiatives include making particular donations for child-related causes, establishing orphanages and schools, offering scholarships to support children's education, and providing essential resources such as uniforms, textbooks, and nutritious meals. Although most CSR projects primarily focus on children who belong to the aforementioned groups, there is a growing need to create new initiatives with carefully designed strategies that aim to bring about broader systemic change. The prevailing perception regarding children's difficulties is primarily influenced by a welfare or charitable perspective, rather than a child rights perspective. Therefore, it is imperative to prioritize research and development as well as the enhancement of capacity among many stakeholders in order to achieve a more concrete and enduring influence on the lives of children. While the majority of conventional initiatives exhibit a greater emphasis on input, prioritizing the measurement and concentration of input rather than output, the contemporary generation needs a heightened attention on outcomes. The SDGs (Sustainable Development Goals) place significant importance on scalable, reproducible, and sustainable models of social responsibility. Certain efforts necessitate collaboration between governmental and non-governmental entities, as well as the integration of multi-stake, multi-state, multi-location, and extensive operations, in contrast to the previous relief and service delivery initiatives.

UNICEF introduced the children's rights and business principles in March 2012. These principles consist of ten action points that assist companies in evaluating their influence on children. In accordance with the U.N. Guiding Principles on Business and Human Rights, businesses are urged to uphold and endorse child rights in all aspects of their operations and business associations. The principles demonstrate that corporations have the ability to influence the rights of children through their actions within the spheres of the workplace, marketplace, community, and environment. India, as a signatory to the United Nations Convention on the Rights of the Child (UNCRC), must approach the issue using the same principles (Tyc, 2020).

12.8 SUGGESTIONS

The chapter's findings and suggestions are organized based on four primary areas of focus: (a) tackling child labor, forced labor, and human trafficking by adopting a comprehensive approach that encompasses the entire supply chain; (b) implementing public measures to safeguard workers and reduce their susceptibility to child labor, forced labor, and human trafficking; (c) establishing public governance measures to regulate business practices and the business environment; and (d) fostering collaboration and inclusive business strategies to tackle child labor, forced labor, and human trafficking.

12.8.1 Tackling Child Labor, Forced Labor, and Human Trafficking by Adopting a Comprehensive Approach That Encompasses the Entire Supply Chain

The estimates presented in this chapter demonstrate that child labor is a pervasive issue that impacts the entirety of global supply chains. Notably, a substantial proportion of child labor is seen in the initial stages of supply chains, specifically in the production of raw materials and other inputs that contribute to the finalized products ready for export. Labor from children in international supply networks is prevalent in the upstream parts, accounting for approximately 28–43% across different areas. These findings highlight the insufficiency of a strategy that concentrates solely on downstream suppliers. Instead, a comprehensive approach that encompasses the entire supply chain is required. The discussed approach has its jurisdiction beyond the traditional direct suppliers, and it also covers those who are involved in the extraction of raw materials and in agriculture. If we examine this from a corporate perspective, this requires expanding the scope of the areas beyond where companies have their influence, this may include their immediate supply lines, areas with higher risks beyond their supply chains, and environments where informal work is taking place. Because of this, companies may face some challenges like their reduced visibility and limited control in areas of upstream sectors, but these obstacles should not discourage them from conducting their investigations beyond direct supply routes.

12.8.2 Implementing Public Measures to Safeguard Workers and Reduce Their Susceptibility to Child Labor, Forced Labor, and Human Trafficking

It is the duty of the governments of various states to make policies that will help in combating the problem of child labor, forced labor, and human trafficking. This must include policies to prevent these acts within their borders and also prevent these types of acts from foreign borders to their country. This can be achieved by implementing various effective measures, like as providing basic and quality education to children. It should also include various initiatives for a cheap and quality healthcare system. The governments must also work in providing a standard minimum wage scheme and its proper implementation, so that no one can be exploited on this ground. Also, a special force or department should be made in order to check and prevent the problem of child labor.

This includes developing programs and policies that support fair and morally responsible hiring practices, creating migration pathways that are reliable, and identifying and protecting migrants who are vulnerable or who have been the victims of abuse, exploitation, or violence. Although not all United Nations Member States have endorsed it, the Global Compact for Safe, Orderly, and Regular Migration could provide a framework for addressing these issues as a group.

It is particularly crucial to have policies in place to lessen the general vulnerability of kids and families when it comes to child labor in the early phases of the supply chain, as in small-scale farming, raw material extraction, and informal employment situations. Based on the research's statistical findings, these circumstances provide the greatest risk of child labor. Policies that particularly target and reduce this danger are essential if we are to address the root causes of child labor on a large basis.

12.8.3 Establishing Public Governance Measures to Regulate Business Practices and the Business Environment

Governments must actively support CSR, and protect human rights in supply chains by carrying out careful due diligence. In this case, a strategic mix of policy efforts is needed.

Uniformity and openness in the entire supply chain mechanism are very important. The governments should also work in these directions to promote a uniform and open channel, so that the issue of child labor can be combated. The OECD Due

Diligence Guidelines for Responsible Business Conduct and the United Nations Guiding Principles on Business and Human Rights are two examples of government-endorsed and agreed-upon due diligence requirements that policy methods should follow.

It should also be the priority of the governments that they must take various steps in order to improve the process of handling the complaints related to child labor and human trafficking in supply chains. This means making certain that victims receive the proper remedies, such as assistance and protection. Enhancing judicial accountability, creating avenues for legal recourse, improving the effectiveness of the criminal justice and labor systems, and strengthening victim safety protocols are critical interventions. It is crucial that the national contact points for ethical corporate conduct are reinforced by the government in nations that have complied with the OECD Guidelines for Multinational Enterprises. This entails giving these national contact points inside the government the tools, assistance, autonomy, impartiality, and visibility they require to guarantee their efficient operation.

12.8.4 Fostering Collaboration and Inclusive Business Strategies to Tackle Child Labor, Forced Labor, and Human Trafficking

Inclusive business models and multi-stakeholder partnerships foster collaboration, focus, the sharing of best practices and experiences, and the acceleration of progress. As demonstrated by the Alliance 8.7 Action Group on Supply Chains and the OECD Business for Inclusive Growth platform, the collaborations can have an international reach. As demonstrated by Cocoa Action and the Roundtable on Sustainable Palm Oil, they can also be sector-specific. Partnerships can also be regional or national in scope, as the Malawi Tea 2020 cooperation demonstrates. They are able to convene partners, businesses, and governments to pinpoint challenges and offer support for joint endeavors to address governance shortcomings, lessen community vulnerability, and address financial restraints. The implementation of shared responsibility strategies has promise in expanding the range of constructive initiatives implemented by an individual company in a supply chain to include the private sector, public sector, and civil society. The goal of this group endeavor is to successfully end forced labor, child labor, and human trafficking on a broad scale.

12.9 CONCLUSION

Aside from being a humanitarian duty, companies and society as a whole must prioritize this strategic need: the elimination of child labor from global supply chains. This research has examined the economic, social, and environmental impacts of eliminating child labor. Beyond mere legal compliance, the eradication of child labor has wide-ranging implications. It not only mitigates risks but also enhances opportunities, strengthens reputations, and empowers communities, providing significant advantages for supply chains and society as a whole.

From an economic standpoint, businesses committed to ending child labor can reap substantial benefits. By minimizing the risk of reputational damage, legal liabilities, and operational disruptions, companies can maintain consumer trust and safeguard their financial performance. In a market where ethical considerations are increasingly prioritized, adopting ethical labor practices also bolsters a brand's reputation and competitiveness. This leads to sustainable financial success over the long term, attracting investment and fostering growth. Socially, ending child labor contributes to the development of human capital and the empowerment of communities. By offering children access to education, healthcare, and personal development opportunities, companies play a crucial role in reversing the cycle of exploitation and poverty, enabling children to achieve their full potential. Fostering inclusive workplaces that prioritize the rights and well-being of all individuals helps maintain social cohesion and stability within communities, laying the foundation for long-term prosperity and success.

In conclusion, the elimination of child labor in supply chains has profound positive effects on businesses, communities, and the environment. By embracing ethical labor practices and a commitment to human rights, companies can build resilient, sustainable, and ethical supply chains that benefit stakeholders at every level. Together, we can envision a future where children are free from exploitation, communities thrive, and businesses succeed in harmony with the environment. Through innovation, collaboration, and a steadfast commitment to ethical business practices, we can create a global community where each and every child will have the opportunity to realize their actual potential and to make a meaningful contribution to the world.

REFERENCES

Altman, M. (2001, March). A revisionist view of the economic implications of child labor regulations. *Forum for Social Economics*, *30*(2), 1–23).

Anner, M. (2019). Predatory purchasing practices in global apparel supply chains and the employment relations squeeze in the Indian garment export industry. *International Labour Review*, *158*(4), 705–727.

Baldwin, R., & Freeman, R. (2022). Risks and global supply chains: What we know and what we need to know. *Annual Review of Economics*, *14*(1), 153–180.

Budd, P. P., & McIvor Joss, D. (1998). Social origins and implications of child labor. *Work*, *10*(3), 279–291.

Cho, S. H., Fang, X., Tayur, S., & Xu, Y. (2019). Combating child labor: Incentives and information disclosure in global supply chains. *Manufacturing & Service Operations Management*, *21*(3), 692–711.

Doepke, M., & Zilibotti, F. (2009). International labor standards and the political economy of child-labor regulation. *Journal of the European Economic Association*, *7*(2–3), 508–518.

Dolgui, A., Ivanov, D., & Sokolov, B. (2018). Ripple effect in the supply chain: An analysis and recent literature. *International Journal of Production Research*, *56*(1–2), 414–430.

Edmonds, E. V., & Pavcnik, N. (2005). Child labor in the global economy. *Journal of Economic Perspectives*, *19*(1), 199–220

Fatima, S., Javed, W., & Imam, S. K. (2022). Child labour as corporate crime in Nexus to corporate social. *Global Sociological Review*, *VII*(III), 49–60. https://doi.org/10.31703/gsr.2022(vii-iii).06

Ghadge, A., Er, M., Ivanov, D., & Chaudhuri, A. (2022). Visualisation of Ripple effect in supply chains under long-term, simultaneous disruptions: A system dynamics approach. *International Journal of Production Research*, *60*(20), 6173–6186

González Torres, E., Bayer, C. N., & Watson II, D. D. (2024). Addressing root causes: Private sector impacts on child labour outcomes. *Available at SSRN 4742594*.

Hou, J., & Sun, L. (2016). Backup sourcing decisions for coping with supply disruptions under long-term horizons. *Discrete Dynamics in Nature and Society*, *2016*(1), 6716058.

Kakkar, P. (2019). Impact of economic development on child labour. *Indian Journal of Economics and Development*, *15*(4), 541–554.

Khan, R. E. A. (2001). Socioeconomic aspects of child labour – A case study of children in auto workshops. https://doi.org/10.35536/lje.2001.v6.i1.a5

Koekkoek, M., Marx, A., & Wouters, J. (2017). Monitoring forced labour and slavery in global supply chains: The case of the California Act on transparency in supply chains. *Global Policy*, *8*(4), 522–529.

Kumar, V., & Kedia, K. (2019). Corporate social responsibility or corporate serious responsibility: Revisiting CSR law. *SEBI and Corporate Laws*, *155*(9), 73.

Lee, J. (2016). Global value chains and the changing pattern of North-South trade: Apparel, electronics and automotive sectors in 2005–2014. *Journal of International Trade & Commerce* [무역연구], *12*(6), 1–21.

Lévesque, C., & Murray, G. (2002). Local versus global: Activating local union power in the global economy. *Labor Studies Journal*, *27*(3), 39–65.

Li, C., & Kuruvilla, S. (2023). Corporate codes of conduct and labour turnover in global apparel supply chains. *British Journal of Industrial Relations*, *61*(3), 481–505.

Mamic, I. (2005). Managing global supply chain: the sports footwear, apparel and retail sectors. *Journal of Business Ethics*, *59*, 81–100.

Mukunde, D. (2023). Navigating uncertainty: Strategies to mitigate geopolitical risks in global supply chains. *Available at SSRN 4683958*.

Strulik, H. (2004). Child mortality, child labour and economic development. *The Economic Journal*, *114*(497), 547–568.

Tundys, B. (2024, January). The Ripple effect-implications and consequences of incidental disruptions and black swan phenomena in the supply chain. In *DIEM: Dubrovnik International Economic Meeting* (Vol. 9, No. 1, pp. 166–183). Sveučilište u Dubrovniku. https://doi.org/10.17818/DIEM/2024/1.11

Turoń, K. (2016). Corporate social responsibility to employees: The best labour practices in transport and logistics companies. *Journal of Corporate Responsibility and Leadership*, *3*(1), 37–47.

Tyc, A. (2020). Corporate social responsibility instruments and their impact on labour rights. *Acta Iuris Stetinensis*, *29*(1), 177–191.

13 Inbound and Outbound Logistics in Supply Chain Management: Building Efficiency and Resilience

Saloni Chitkara, Seema Garg, and Ritu Sharma

13.1 INTRODUCTION

The present-day global economy flourishes on free movement of goods, information, and services across varied geographies. Logistics is known to be a center point of the system which guarantees the effective movement of materials, components, and finished goods from one end to the other. Among these procedures, inbound and outbound logistics have been found to be the significant component of supply chain management, helping organizations in procuring resources and delivering them to customers. Inbound logistics refers to the processes involved in the movement of materials, components, and supplies from suppliers to the manufacturer or organization. It may include procurement activities, transportation, receiving, and inventory management. Inbound logistics can strongly impact the efficiency of supply chains, as such logistics ensures the timely availability of quality materials. Effective control of inbound logistics helps reduce lead times, controls procurement costs, and lines up production schedules with demand forecasts. There have been studies emphasizing coordination with suppliers and the integration of technology to improve inbound logistics. For instance, the involvement of IoT and real-time tracking in supply chains heightens visibility across the supply chain and diminishes uncertainty, and improves material flow (Christopher, 2016; Chopra & Meindl, 2020).

At the same time, outbound logistics is a concern for the distribution of finished products from the organization to the ultimate customer and would cover order processing, transportation, warehousing, and delivery-related activities. Efficient outbound logistics ensure timeliness, save on transportation costs, and increase the service level regarding customer satisfaction. Optimized distribution networks, therefore, save costs for companies and enable them to respond flexibly to any such market demand. Technology is further emphasized by research as catering to outbound logistics through route optimization algorithms and last-mile delivery

DOI: 10.1201/9781003477242-16

innovation. Customer-centric approaches such as the omnichannel distribution are increasingly pivotal in modern supply chains (Rushton et al., 2017; Agrawal et al., 2020).

13.1.1 Role of Inbound and Outbound Logistics in Supply Chain Efficiency

Inbound and outbound logistics serve as the backbone of an efficient supply chain. Their integration together helps ensure the free flow of materials and finished goods, which in turn instantly affects operational efficiency, cost management, lead time reduction, and customer satisfaction. The term "seamless integration" describes how inbound and outward logistics are coordinated and aligned to provide a single flow of goods, information, and materials across the supply chain. Integration ensures that resources are used as efficiently as possible by reducing redundancies and disruptions (Stevens, 1989).

13.1.1.1 Key Benefits of Integration

Integrating inbound and outbound logistics in supply chain management creates several benefits by having improved visibility, flexibility, and operational management. Stakeholders can easily track the flow of goods and materials which help in speeding up of reaction times to possible interruptions and hence increases overall operational effectiveness (Gunasekaran et al., 2004).

Operational synchronization is another benefit as it tackles inefficiencies. It is a situation where little dissimilarities in client orders have a big impact on the supply chain and result in major inventory and operational mismanagement. This type of fluctuation is lessened by coordinated logistics, which ensures a smooth and efficient flow of materials and even cuts down on waste (Lee et al., 1997). Furthermore, when there is greater flexibility in an integrated system, organizations are better equipped to adjust to the market changes, seasonal variations in demand, and unanticipated interruptions. Integration ensures that supply chains are resilient and adaptable in dynamic contexts by promoting agility (Christopher, 2016).

13.1.1.2 Challenges to Seamless Integration

Seamless integration of both inbound and outbound logistics may be considered as a major component of efficient supply chain operations, though there are still many obstacles to overcome. Different stakeholders like suppliers, manufacturers, retailers, and distributors frequently function in silos, each with their own operating framework and priorities. This conflict of ideas leads to incompetence, redundant work, and delay in decision-making. For example, producers may prioritize minimizing manufacturing costs, but at the same time distributors frequently prioritize speed and responsiveness, hence these competing tactics subsequently disrupt the overall flow of the supply chain.

The various IT systems used by supply chain companies to manage incoming and outbound logistics also make things more difficult. Interoperability is sometimes lacking in these systems, leading to isolated data silos that hinder the free flow of information. Mentzer et al. (2001a, 2001b) opined that the absence of interconnected systems makes it difficult for the supply chain to coordinate operations since it hinders its insight into its operations. Delivery delays, stockouts, and inventory overstock are the results of this inefficiency. The absence of real-time, unified data brings confusion to decision-making processes and leads to a higher chance of errors, coupled with low-level trust between the concerned stakeholders.

In their article, Stevens (1989) noted that the only way one could truly overcome the fragmentation effects in a system was cross-functional integration. Integrated supply chains were seen as better positioned to manage variability and uncertainty. To this effect, Gunasekaran et al. (2004) listed fragmented systems as the major inhibitor that prevents achieving supply chain efficiency and instead proposes more advanced IT solutions, such as enterprise resource planning (ERP) and collaborative planning tools to enhance integration. According to Christopher (2016), seamless integration is indispensable if supply chain agility is to be achieved in the dynamic market environment of today.

Furthermore, disparate IT systems not only reduce productivity but also add higher costs. For example, organizations have to incur the expense of middleware or manual reconciliation work for the alignment of data among the various systems. Chopra and Meindl (2020) emphasize that one of the significant weaknesses is the absence of a consolidated IT infrastructure, particularly in the case of global supply chains where information accuracy and speed are critical. They argue for the adoption of cloud-based solutions and blockchain technology to be able to support transparent, real-time data sharing across all stakeholders.

These factors call for a head-to-head battle with technology adoption and collaboration with stakeholders. For instance, Agrawal et al. (2020) suggest the usage of integrated supply chain platforms that use predictive analytics power coupled with AI-driven insights to make more precise decisions and reduce the inefficiencies of data silos. Other collaborative tools, including both vendor-managed inventory (VMI) systems and frameworks under the CPFR (collaborative planning, forecasting, and replenishment) have been proven to reduce the negative impacts of fragmented systems (Frohlich & Westbrook, 2001). Disparate IT infrastructures pose substantial problems to true integration, advanced technologies, standardized data-sharing protocols, and good practices may suffice to smooth out the wrinkles. All these will not only add to operational efficiency but also fortify the resilience to adapt to disruptions in a rapidly changing landscape of supply chain.

13.1.2 Cost Implications of Inbound and Outbound Logistics

Inbound logistics often face high cost of procurement due to uncoordinated networks of suppliers, leading to inefficiencies like overpayment and inconsistent delivery schedules. Poor forecasting in inbound logistics can lead to excess inventory and may affect the increasing costs and supply-demand balance. There are major barriers to inbound and outbound logistics that are paramount in determining the efficacy and cost-effectiveness of the operation. Most of the problems encountered by incoming logistics revolve around high procurement cost, that is mostly caused by unorganized supplier networks. An inefficient supplier relationship could take the shape of non-timely information for the delivery schedule of the supplies, resulting in overpricing.

Inefficient forecasting regarding inbound logistics impacts the amount of stock that gets stocked above threshold, thus increasing costs. Miscalibrated demand projections set the stage for disorganization within inventories, thus creating imbalance and excessive cost for holding excess goods (Chopra & Meindl, 2020). Such inefficiencies cause the discrepancies within the supply-demand balance, forcing the companies to spend on costly holding and managing the inventory. Transportation can be considered as one of the greatest aspects toward loss when it comes to logistics costs on the outward side. Inefficient transport that may vary from poor routing, underutilized vehicle carrying capacity, and increased fuel charges adjusts the total cost of logistics significantly, according to Rushton et al. (2017). Strategic planning becomes very advantageous in efficient management of outbound logistics so as to minimize the travel costs on other aspects, but still delivering to clients on time. Moreover, inefficiency at the last leg of delivery can be an extensive part of the running costs. Last-mile delivery is the last stretch a product has to cover before reaching the customer. There are many hurdles, such as urban congestion, imprecise time windows for deliveries, and the need for more manpower; they all induce further costs in the last mile. In some cases, businesses suffer from losses due to a lagging performance of their delivery systems. Supply chain management enables the companies to optimize their operations and cut on costs of services by tackling both incoming and outgoing supply chain issues. It's probably the best example of optimized logistics to see that deliberate automation and predictive analytics are applied by Amazon in managing their incoming and outbound logistics. Indeed, it has been one of the first organizations to have achieved successful application of cutting-edge technologies, such as machine learning and real-time data analytics, at operational level. While employing predictive analytics to match expectations against realities with regard to demand, the entire supply chain will be optimized, thus ensuring improved control of his inventory and procurement planning. He thus guarantees immediate satisfaction of any consumption incurred while eliminating the incessant holding costs due to overstocking.

Automotive technologies such as AI-driven route optimization and robotics in fulfillment centers have delivered substantial efficiency gains from the logistics side of the outbound component. Less human error is involved in speeding up order processing and delivery time associated with order processing. In addition, Amazon's reputable last-mile delivery network is developed based on its reliable real-time tracking technologies. Joint efforts reduce considerably logistical costs for the industry, thus providing a standard against which one can measure supply chain efficiency in e-tail.

13.1.3 Objectives of the Study

The aim of the present investigation is to identify inbound and outbound logistics issues and their best practices. It will also investigate new trends to come up with ideas for improving the resiliency and efficiency of supply chain management. This is relevant given the high degree of complexity in the environments in which businesses operate, which already dictate the need for easy logistics integration, making this even more pressing research. Another main tool for enhancing logistics operations is the application of cutting-edge technologies, teamwork, and creative approaches. This chapter is meant to provide organizations with practical ideas to improve their supply chains.

Logistics efficiency is today one of the critical differentiators for international businesses. Good logistics operations are not only cheap but also lead to on-time deliveries with minimum variation in quality that assure the customer of satisfaction. Such resilience cushions supply chains against shocks such as market volatility, floods, and geopolitical concerns. In fact, building agile and robust logistics systems enables firms to adjust better to change, reduce risks, and ensure continuity of service delivery. Thus, through an examination of these dynamics, it highlights the strategic importance that optimization of logistics assumes for attaining the medium- and long-term sustainability of competitive differentiation.

13.2 LITERATURE REVIEW

Logistics has undergone a very dramatic shift from once being seen as essentially a cost center to developing into a strategically important value creation function in supply chain management.

13.2.1 Historical Context and Evolution of Logistics

Initial research on logistics was essentially related to its ability to minimize costs and used transportation and warehousing as major factors in operating efficiently. Bowersox and Smykay (1969) defined logistics as a support function that focused on reducing the costs associated with moving and storing materials. In this time, logistics was very compartmentalized, resulting in less integration among various activities of the supply chain. The development of logistics is attributed to the years of the 1980s and the 1990s primarily due to the benefits of technology, globalization, and increased competition. Christopher (1998) explained that logistics became a value-added function to improve the satisfaction of the customer and competitive advantage. Inbound and outbound logistics are equally important as interdependent functions that must be integrated together for a supply chain to succeed. Inbound logistics, responsible for procurement, transportation, and storage of raw materials, and outbound logistics, focusing on the distribution of finished goods to end-users, were identified as crucial for an uninterrupted flow of materials and information within the supply chain.

The growing Just in Time (JIT) systems of the 1980s, originating from Japanese practice, made a vivid case for the strategic significance of logistics. The JIT paradigm emphasized low-inventory levels and better coordination with the suppliers, thereby demanding significant integration in both inbound and outbound logistics. According to Shapiro (1984), such integration ensured an improvement over holding and transportation costs but also aligned supply chain processes with customer demand, thereby positively improving overall efficiency. When globalization accelerated in the 1990s, logistics emerged as a global networked function. Few researchers accentuated the importance of coordinated logistics concept, especially in the context of comprehensive international supply chains (Mentzer et al., 2001a, 2001b). The main reason for integrating incoming and outbound logistics during that time was the advent and deployment of ERP systems, such as EDI. New research on technology, integration, and coordinating logistics has given rise to this new aspect of strategic objective support in logistics over the past decades.

13.2.1.1 The Role of Logistics in Creating Value

Recent studies highlight how logistics adds value by providing efficiency and high-quality services. Dubey et al. (2017a, 2017b) opined logistics as an enabler of difference rather than just a cost-cutting role. The researchers even emphasized that logistics may create lean, responsive supply chains by improving the distribution and procurement procedures. Wang et al. (2019), emphasized customer centrity as one of the key facilitators of long-term value creation in logistics, where precise and on-time delivery becomes crucial to building brand loyalty and customer satisfaction.

13.2.1.2 Integrating Inbound and Outbound Logistics

The integration of inbound and outbound logistics has gained greater recognition. According to Govindan et al. (2018), to achieve a seamless flow of materials and information, procurement and distribution activities should be coordinated. Many researches claimed disruptions in the flow of goods during inbound logistics, leading to inefficiencies in outgoing operations, which in turn can affect customer satisfaction. Researches examine the function of integrated logistics in omnichannel retailing and e-commerce, wherein the coordination of inbound and outbound logistics results in a satisfactory reduction in lead times for delivery for returns management (Marchet et al., 2021).

13.2.2 Technology-Driven Evolution in Logistics

Last couple of years have seeing some drastic changes in reactionary approaches to logistics. They have shifted to proactive measures with the digital offerings of IoT, predictive analytics, and blockchain. Internet-enabled devices are increasingly making supply chain operations less uncertain by providing information on real-time visibility in both inbound and outbound logistics (Kumar et al., 2020a, 2020b). The lines between inbound and outbound processes continue to melt down as AI and machine learning technologies are increasingly used for route optimization and inventory planning (Wang et al., 2022).

13.2.2.1 Global and Sustainable Logistics Trends

The increasingly complex logistics networks of a globalized economy were studied by Saenz et al. (2020). Researchers even highlighted the need for more resilient systems that can survive shocks. Rajeev et al. (2019) in his research examined green logistics strategies and explored how organizations are putting efforts in inbound and outbound logistics to lessen environmental impacts, such as route optimization and consolidated shipments.

13.2.2.2 Emerging Perspectives

New perspectives suggest more and more collaboration in the integration of logistical procedures. Mehrjerdi (2021) in his research explored how cooperative relationships between manufacturers, suppliers, and distributors allow for more coordinated and integrated inbound and outgoing activities. Their results support the use of integrated planning systems and shared data to improve supply chains' responsiveness and agility.

In conclusion, logistics evolved to be a strategic function where it would add value through integration, technology adoption, and sustainability. The research in the past eight to ten years indicates how the integration of inbound and outbound logistics, facilitated by digital innovation, reduces costs but adds customer satisfaction and resilience at the same time.

13.2.3 Theoretical Foundations: Logistics in Supply Chain Paradigms

The theoretical paradigms underlying the evolution of supply chain management will guide how logistics systems are designed and managed. Lean, agile, and resilient supply chains together offer frameworks for a range of different operation priorities, with logistics encompassing a critical role in their successful implementation.

13.2.3.1 Lean Supply Chain: Eliminating Waste in Logistics

The lean supply chain model basically talks about cutting wastes (non-value-adding activities) and delivering what matters to customers with the most operational efficiency. It borrows from the principles used in the lean manufacturing approach to orchestrate logistics processes, aiming at the reduction of lead times, inventory levels, and costs associated with these. Thus, logistics under this paradigm properly manages minimal waste during transportation, warehousing, and inventory handling through practices such as just-in-time inventory management and optimized delivery schedules. Stevens (1989) indicated that inbound and outbound logistics in lean systems must be in harmony to eliminate duplication and improve efficiency. Recent studies by Rajeev et al. also on technology-based solutions in logistics to remove bottlenecks and streamline operations, real-time tracking, etc.

13.2.3.2 Agile Supply Chain: Flexibility and Responsiveness

Agility in the supply chain emphasized responsiveness to shifting customer demand, market conditions, and disruptions. Logistics for an agile supply chain is defined by flexibility, speed, and agility. Agile logistics systems enable rapid adjustments in procurement and distribution processes so that supply chains are properly set up to respond to variabilities in demand or supply. According to Gunasekaran et al. (2004), logistics enables agility through a collaborative value chain and adequate utilization of new high communications. For instance, the use of dynamic routing and cross-docking falls under logistics innovations that help preserve agility as the order is delivered without delay.

13.2.3.3 Resilient Supply Chain: Coping with Disruptions

Resilience is the capacity to absorb, recover from, and adapt to disruptions in supply chains from natural disasters, economic crises, and pandemics. Logistics has a central role in achieving resilience by means of supply diversification sources, keeping safety stock, and implementing alternative transportation modes. Additionally, resilient logistics systems are supported by strong technology infrastructures that afford visibility and decision-making tools during crises.

Stevens (1989) has emphasized the necessity for strong logistics networks, offering a base for building resilience. Sheffi (2015) has focused on how logistics practices, like multi-modal transportations and decentralized warehousing, make supply chains more robust through redundancy and flexibility.

13.2.3.3.1 Role of Logistics in Supporting Supply Chain Paradigms

Logistics provides the operations backbone to deliver lean, agile, and resilient supply chain paradigms. By enabling efficient movement and storage of goods, logistics ensures that the right kind of supply chain strategic alignment is achieved at the organizational level. For example, a lean supply chain thrives with optimized logistic structures that reduce handling and storage; an agile supply chain depends on the logistics flexibility of the supply chain to quickly meet the customers' demands. More robust supply chains are necessary for adaptive logistics systems that can react swiftly to disruptions and ensure a continued flow of operations. Logistics processes need to be integrated with both inbound and outbound with regard to achieving alignment under these paradigms (Gunasekaran et al., 2004). Furthermore, advanced digital tools like blockchain and predictive analytics are being used more and more to improve logistical capabilities and support the strategic objectives of lean, agile, and resilient supply chains.

13.2.4 Inbound Logistics

Inbound logistics, which covers all of the elements involved in collecting, transporting, and storing those materials that will provide operational or manufacturing needs. Recent studies have shown that it becomes more and more important for the modern supply chain to be really efficient because of technical improvements and best practices, while still addressing some major problems. It deals with topics such as inventory controlling, which also includes transportation, procurement, and supplier selection. This is essentially all these service functions require to keep the flow of components and raw materials coming in smoothly. Academics have researched in detail Strategic Supplier Relationships and Organized Procurement Practices over the last ten years as vehicles to enhance incoming logistics. Doing suppliers' selection based on factors including cost, quality, etc. has an impact on effective inbound logistics operations. However, operational flexibility and cost minimization have been related to inventory and transport management (Kouvelis et al., 2020).

13.2.4.1 Best Practices in Literature

13.2.4.1.1 Just-In-Time (JIT) Inventory Control

JIT inventory management remains a hot topic in inbound logistics. Rajeev et al. (2019) underlined how JIT combines material flow with production plans while lowering inventory holding costs. However, managing unpredictability in successful lead time and strong supplier collaboration are required before deploying JIT.

13.2.4.1.2 Collaborative Supplier Relationships and Vendor-Managed Inventory (VMI)

Key strategy for increasing the effectiveness of inbound logistics is to work together with suppliers. The practices like vendor-managed inventory (VMI) are making it possible to transfer inventory control to suppliers, relieving purchasers of administrative obligations. VMI encourages a partnership-oriented strategy that helps to improve inventory accuracy and decrease stockouts (Govindan et al., 2018).

13.2.4.2 Challenges

13.2.4.2.1 Lead Time Variability of Suppliers

Time variability is one of the most frequently mentioned issues with a company's inbound logistics. Dubey et al. (2017a, 2017b) links increased production and higher inventory holding costs to supplier performance unpredictability. Organizations using global supply chains with lengthier lead times and geopolitical uncertainties have an even greater problem.

13.2.4.2.2 Geopolitical Disruptions Increasing Cost

Supply chain obstructions and the start of trade wars are geopolitical disturbances that have been shown to increase costs. Wagner and Bode (2014) opined that these interruptions force organizations to reevaluate their sourcing plans while keeping buffer stockpiles, which drives up overall logistics expenses. Ivanov and Dolgui (2020) concentrate on how incoming logistics expenses are impacted by worldwide crises like COVID-19 pandemic.

13.2.4.3 Technological Contributions

13.2.4.3.1 IoT-Enabled Supply Chains for Real-Time Tracking

Real-time stock and cargo visibility has completely changed inward logistics. Organizations may significantly improve supply chain resilience by using IoT devices to detect the status of commodities in transit, monitor the conditions of environment, and anticipate delays (Zhang et al., 2021).

13.2.4.3.2 ERP Systems for Seamless Supplier Coordination

Enterprise resource planning tools have become a crucial component of inbound logistics to effectively collaborate with suppliers. ERP systems facilitate integrated planning, procurement, and inventory management, enabling real-time data sharing and decision-making across various supply chain layers (Chopra and Meindl, 2020).

13.2.5 Outbound Logistics

In supply chain management, outbound logistics remains an important component of supply chain. It refers to the transportation of completed goods from production sites to final customers. Current study focuses on how outbound logistics contributes to sustainability issues, cost effectiveness, and consumer happiness.

13.2.5.1 Key Functions and Activities

The fundamental functions of outbound logistics are distribution channel management, last-mile delivery, and reverse logistics. Distribution channel management relates to reaching the correct customers at the appropriate time using efficient but convenient channels. Last-mile delivery is very critical in determining the final customer satisfaction. The success of last-mile delivery forms the great determinant of overall service quality, so its improvement is the bright point of logistics optimization (Rushton et al., 2017). Reverse logistics, referring to the return and recycling of goods has also become a critical function in those organizations whose sustainability goals are high (Govindan et al., 2015).

13.2.5.2 Key Themes

13.2.5.2.1 Last-Mile Delivery and Customer Satisfaction

Literature has emphasized last-mile delivery in enhancing customer experience. As the growth of e-commerce continues to pick up speed, so do consumer expectations for faster and more reliable deliveries. Some research studies, including Boysen et al. (2021), show that last-mile delivery plays an essential role in perceptions of service quality. Solutions that are most used to address delivery inefficiencies include dynamic routing and local distribution hubs.

13.2.5.2.2 Reverse Logistics and Sustainability

As organizations have been working toward sustainability initiatives, reverse logistics has become more and more common. According to Govindan et al. (2015), reverse logistics can be considered a tool for waste reduction, resource optimization, and formulating a circular economy. What is more, studies also indicate that efficient reverse logistics can achieve certain financial advantages, such as savings that come from returns and a greater company reputation for green practices.

13.2.5.3 Challenges

13.2.5.3.1 Rising Transportation Costs and Delivery Time Pressures

Transportation costs are, in a way, the greatest stumbling block in outbound logistics. Out of the recent works on logistics, Mentzer et al. (2001a), that outbound logistics has been a significant logistics cost and has kept increasing with labor and fuel costs incurred. The whole planning part concerning logistics is made even difficult by the constraints of delivery time, especially in cities congested with heavy traffic and legislation issues. It necessitates innovative solutions to these challenges in order to achieve the cost-speed balance in outbound logistics.

13.2.5.3.2 Integration of Omnichannel Distribution in E-commerce

Outbound logistics has now become increasingly complicated by the transition to omnichannel retailing. It becomes a challenge to merge online and physical distribution channels for a smooth operational flow (Agrawal et al., 2020). Organizations must develop sound transportation networks and inventory positioning for true real-time order fulfillment across multiple channels, which is arguably where reliable IT systems come into play.

13.2.5.4 Technological Contributions

13.2.5.4.1 Autonomous Vehicles and Drones

There are two types of emerging technology in the present world, one of which is drones and the other, autonomous vehicles. They are two examples of the emerging technology that will revolutionize last-mile deliveries. Autonomous systems speed up delivery, reduce labor costs, and lower errors. Zhang et al. (2022) opine, "Drones can prove effective in saving operational costs for rural and remote areas where conventional means of distribution would be less viable."

13.2.5.4.2 Blockchain for Transparency and Trust

The application of blockchain technology in outbound logistics has become to bolster trust and transparency further. Blockchain ensures accountability and reduces fraudulent activities through immutability of the records regarding transactions and shipment status. On the other hand, distribution models can simplify reverse logistics through better traceability and promise to adhere to regulations (Kouhizadeh et al., 2020).

13.2.6 Integration of Inbound and Outbound Logistics

Integrated logistics then helps organizations really cross the gaps created by internal inbound and outbound processes. Just generally, integrated logistics makes resources utilization maximally and is responsive to market needs by bridging gaps, for example, in distribution, inventory management, and procurement processes (Christopher, 2016). Most of the research done to improve the alignment between operations on supply and demand usually looks at roles and responsibilities through reducing lead times and operational redundancies (Chopra & Meindl, 2020). For example, Ivanov et al. (2019) argued that integrated businesses show better order accuracy and significantly reduced inventory holding costs. Thus, coordination of processes such as scheduling and production can lead to better resilience of the supply chain.

13.2.6.1 Collaborative Models

The two models mainly focused on were Collaborative Planning, Forecasting, and Replacement (CPFR) and vendor managed inventory (VMI). These models are based on suppliers and buyers working together for the maximization of inventory levels and the minimization of the stockout. This involves timely replenishment and lower holding costs for the buyers, enabling the suppliers to manage the inventory on the buyer's behalf (Sari, 2008). With CPFR, data sharing and collaborative forecasting improve the accuracy of demand forecasting and production planning (Agrawal et al., 2020).

Collaborative planning approaches are especially successful in situations when demand is extremely volatile, such that traditional logistics systems do not perform as promised. For example, research carried out by Yang et al. (2017) has shown the efficacy of CPFR in retail as it has improved supplier-buyer relationship and, thus, reduced stockouts to about 30%.

13.2.6.2 Sustainability Focus

There is an ever-increasing belief among people that sustainability is a necessity for integrating circular economy concepts into logistics. Circular economy strategies are aimed at reducing waste and maximizing resource efficiency by transforming the traditional linear supply chain concept into a closed-loop system by Geissdoerfer et al. (2017). This shift requires a strong tie between inbound logistics and outbound logistics to cater to the reverse flows of returned material as well as refurbish product distribution.

Energy-efficient storehouses and low-carbon transport are two main approaches of greening logistics, which are already being adopted by many businesses. The major hurdles in its adoption are cost and lack of infrastructural support, which are found very high initial investment (Govindan et al., 2021). As stated by Kumar et al. (2020a, 2020b), the claims of those organizations implementing integrated green logistics systems are improved brand value and greater compliance with environmental regulations.

13.2.7 Role of Resilience in Logistics

The significance of logistics resilience may be explained by the growing complexity and globalization of supply chains. Researches have emphasized that business continuity, minimize downtime, and sustain service levels ensure resilience in times of emergency. According to Chopra and Sodhi (2021), claim that as businesses with robust logistics systems are better equipped to handle volatility, resilience has a direct impact on long-term profitability and customer satisfaction.

The COVID-19 pandemic clearly highpoints how resilience in logistics was put to the test globally. Companies with a more resilient logistics network, such as those having diversified supplier bases and flexible transportation systems, find better adjustment in supply chain disruptions in the time of the pandemic. Pharmaceutical companies, for example, easily adjusted their logistics network to support the distribution of vaccines by including cold chain systems and utilizing technology for real-time monitoring (Sharma et al., 2021). For example, Wieland and Durach (2021) described how players like Amazon actually ensured operational stability at the face of the COVID-19 pandemic through investments in inventory buffers, dynamic routing systems, and supplier diversification, which not only ensured the continuity but also allowed reduction in lead times even in extreme disruption.

One of the best practices regarding resilience is scenario planning, risk mapping, and developing redundancies at nodal points in supply chains. Insights drawn from

Pettit et al. (2019) show that collaboration between stakeholders gives rise to resilient networks. In sectors like automotive and consumer goods, models in collaboration for risk sharing and joint contingency planning have been effective.

13.2.7.1 Emerging Trends

Resilience emerging trends thus call for the use of AI and predictive analytics to complement logistics management. AI-based systems monitor volumes and find possible risks and send warnings early, giving room for a manager to prepare ahead of time (Ivanov & Dolgui, 2021). Predictive analytics with integrating algorithms of machine learning, work to improve forecasting accuracy, enabling effective demand planning, and stock inventory management during disruptions.

Blockchain technology is found to significantly strengthen resilience in logistics through enhancing traceability and transparency. According to Queiroz et al. (2020), a study revealed that blockchain mitigates the risks associated with counterfeit products and supply chain fraud, thereby strengthening trust and reliability. Also, sustainability-focused resilience strategies like circular logistics and green transportation are picking up. Since they are environmental risk reducers in line with global sustainability goals, but enhance system robustness to disruptions induced by climate change (Govindan et al., 2021).

13.3 KEY CHALLENGES

Effective logistics management is critical to ensuring supply chain efficiency and resilience. However, its processing is often hampered by a range of challenges. Challenges in logistics can be put into three main categories: inbound, outbound, and integrated logistics processes, together with technological and sustainability dimensions.

13.3.1 Challenges in Inbound Logistics

Inbound logistics—activities include supplier selection, procurement, transportation, and inventory management—involves all those activities that ensure the seamless flow of raw materials and components into the production process. Ironically, such a critical portion of supply chain management is plagued with problems that can seriously affect the functional efficiency and add to costs.

13.3.1.1 Supplier Variability and Lead-Time Uncertainties

Primary inbound logistics challenges involve variability from suppliers. Suppliers may vary on delivery schedules, order volumes, and the quality of raw materials delivered. Such differences bring variability in lead times. Often, lead times are unpredictable due to differences within the suppliers' processes. The inefficiency or external disturbance, such as a labor strike or production delay, causes variation in suppliers' delivery systems. Such disruptions can result in stockouts or unnecessary inventory holding, which further puts a strain on operational efficiency (Chopra & Meindl, 2020). The uncertainty surrounding supplier performance requires companies to maintain buffer inventories or invest in alternate suppliers, which in turn adds to their costs.

13.3.1.2 Quality Problems with Raw Materials

The quality of raw materials is also an issue for inbound logistics to ensure. Low-quality inputs will produce defective products, high wastage, and increased costs of rework and checks. Quality consistency can only be ensured if suppliers are strongly involved, yet most companies suffer from a lack of visibility into their suppliers' production processes. Strong supplier evaluation and development programs can mitigate risks such as this (Christopher, 2016).

13.3.2 Challenges in Outbound Logistics

Rising Transportation Costs: Transportation is a key spending area in logistics. Increasing fuel prices and irregular freight rates, and inefficient routing make transportation costs worse (Rushton et al., 2017).

Inefficiencies of Last-Mile Delivery: Last-mile delivery is a key component of satisfaction for the customer that suffers from traffic congestion, limited delivery infrastructure, and high costs to deliver in an urban area, mainly during e-commerce logistics (Govindan et al., 2021).

Reverse Logistics Complexity: Return growth in e-commerce has added complexity to the reverse logistics operations because managing product returns and refurbishment, and recycling requires very efficient systems that are often not effectively deployed (Agrawal et al., 2020).

13.3.3 Problems in Technology Integration

Fragmented Systems: The diversity in IT systems between suppliers, manufacturers, and distributors creates barriers to smooth information exchange and decision-making (Ivanov & Dolgui, 2020).

High Implementation Costs: Advanced technologies like IoT, blockchain, and AI are more expensive in initial investment, thus not feasible for small to medium enterprises (Queiroz et al., 2020).

Cybersecurity Risks: Increased dependence on digital technologies will expose logistics networks to cyber threats, risking data integrity and operational continuity (Pettit et al., 2019).

13.3.4 Sustainability Challenges

Green Logistics Implementation: requires transitions to a new platform of green transportation and eco-friendly packaging. Such transitions have barriers associated with high costs and limited technological capabilities (Govindan et al., 2021).

Circular Economy Integration: Aligning logistics operations with circular economy principles, such as reusing and recycling materials, requires significant process reengineering and stakeholder collaboration (Chopra & Sodhi, 2021).

Regulatory Compliance: According to Wieland and Durach (2021), the different country-specific environmental regulations add complexity to the global logistics operation.

13.3.5 Difficulties in Integration

Coordination Gaps: Inefficiencies and higher expenses result from a lack of coordination and misalignment between the inbound and outward logistics tasks (Christopher, 2016).

Silos of Data: Decision-making is slowed down and visibility is diminished when data is not standardized between supply chain partners (Mentzer et al., 2001a, 2001b).

Organizational and Cultural Barriers: Every organization has a unique culture and stakeholders who are wary of change may postpone or completely oppose cooperative initiatives like CPFR or VMI (Wagner & Bode, 2014).

13.4 TECHNOLOGICAL ADVANCEMENT IN LOGISTICS

Logistics is undergoing a paradigm shift due to technological improvements, which are making it more sustainable, transparent, and efficient. By facilitating smooth communication and data-driven decision-making, they have completely changed the logistics industry. The Internet of Things (IoT) and artificial intelligence (AI) are two of the most revolutionary.

13.4.1 IoT for Real-Time Tracking of Goods and Assets

Real-time tracking capabilities made possible by IoT technologies have revolutionized logistics by improving commodities monitoring and control. Throughout the supply chain, IoT sensors that are linked to GPS and RFID offer information on temperature, location, and handling conditions to improve product quality. According to Ng et al. (2021), IoT decreases theft and spoiling losses while increasing supply chain efficiency and transparency. For example, IoT-enabled containers make it easier and less disruptive for the shipping industryto track and manage smart containers predictively (Verdouw et al., 2016).

13.4.2 AI for Demand Forecasting and Logistics Optimization

Demand forecasts and streamlining delivery processes are revolutionizing AI-powered logistic systems. According to Waller and Fawcett (2013), opined by lowering inventory costs, predictive analytics powered by machine learning algorithms analyzes vast volumes of market and historical data to precisely predict demand levels. Additionally, UPS' ORION system uses AI-driven optimization tools to identify delivery routes efficiently. It has demonstrated significant savings in fuel and reduced emissions. Additionally, these solutions increase the agility of logistics systems by making them more sensitive to changing market conditions.

13.4.3 Automation and Robotics

Automation and robotics, which are at the forefront of logistics technologies, address manpower constraints and efficiency requirements. Leading the way in last-mile delivery innovation are drones and self-driving cars. The most expensive part of logistics, last-mile delivery, has benefited from drones and self-driving cars. According to research by Boysen et al. (2021), "drones could solve the problem of congesting cities" by dropping delivery materials directly to the consumer, thereby cutting through the time and cost entailed by deliveries. Some real-life applications of drones include Amazon's Prime Air initiative and Zipline's medical supply deliveries in Africa. In contrast, autonomous vehicles have sensors and AI systems that ensure safe and efficient navigation (Goodchild & Toy, 2018).

13.4.4 Warehouse Automation in Improving Inbound Logistics

Warehouse operations have made tremendous progress since the emergence of robotics. AGVs and robotic picking systems have reduced human errors and increased speed in sorting, storing, and retrieving goods. Kshetri (2018) further adds that with warehouse automation, better space utilization and scalability are also achieved. Hybrid automation, where both humans and robots work together, enables a company to approach operational excellence. Amazon's robotic capabilities combine with human workers to exhibit such potential.

13.4.5 Blockchain in Supply Chain Management

Blockchain technology addresses important concerns related to transparency, security, and trust in logistics by taking advantage of its decentralized and immutable architecture. Blockchain provides safe, tamper-proof tracking of goods throughout the supply chain, according to Saberi et al. (2019). This ensures that all stakeholders get onto the same sheet of paper, thus minimizing conflicts and lawsuits. Demonstrating one of its key strengths, blockchain will apply to Walmart's traceability of food items, such that it reduces response time to potential crises from days to seconds (Kamath, 2018). Further, smart contracts are considered to be an integral part of blockchain. They automate compliance checks and reduce administrative overheads (Perboli et al., 2018). Inter-operability issues and implementation costs remain a challenge for blockchain adoption, and it still necessitates industry-wide standards. Addressing these barriers is truly essential for realizing its full potential.

13.4.6 Sustainability Innovations

Sustainability is an emerging focus in logistics driven by concerns of the environment and regulatory forces. Green logistics deals with carbon footprint attenuation through technological and operational innovations. Even electric and hydrogen-based transportation is getting a lot of attention in lieu of fossil-based forms of transportation (Jabbour et al., 2019). Electric trucks introduced by Tesla and hydrogen-based trucks by Toyota are a few examples to be considered. Being an effective tool, as discussed by McKinnon (2018), route optimization helps to reduce the fuel consumption as it provides delivery routes.

Using solar-powered warehouses and wind-powered transportation systems are few trends that integrating renewable energy into logistical operations. Renewable energy can lower operating expenses and match logistics procedures with the objectives of sustainable development. Businesses are adopting the circular economy and biodegradable packaging that bolsters their dedication to environmental sustainability.

13.5 EMERGING TRENDS AND FUTURE DIRECTIONS IN LOGISTICS

13.5.1 Sustainability in Logistics

The efforts are being made to reduce carbon emissions and mitigate environmental damage globally and it has led to a growing acceptance of sustainability in logistics.

13.5.1.1 Carbon-Neutral Supply Chain Growth and Green Logistics Practices

Logistics industry is adopting carbon-neutral efforts because of stricter environmental restrictions and consumer demand for sustainability. Businesses are investing in vehicles that run on renewable energy to reduce emissions. For instance, trucks run on hydrogen or electricity, and they are also optimizing their routes. Jabbour et al. (2019) emphasized use of biodegradable packaging and reverse logistics systems that recycle products are examples of common green logistics techniques. Any organizations may achieve a carbon-neutral supply chain like those utilized by the DHL's GoGreen initiative. Integration of renewable energy in logistics infrastructure, including solar-powered warehouses and electric charging stations, is going to grow as these solutions will align with one of the United Nations Sustainable Development Goals, SDG 13.

13.5.2 Omnichannel Integration

With the increasing growth of e-commerce, the rise of omnichannel logistics is truly apparent, where companies implement an integrated model for serving customers across various channels. These customers benefit through a seamless experience across channels, namely online, in-store, and hybrid models. This trend consequently needs agile and flexible logistics systems that can accommodate heterogeneous fulfillment methods such as home delivery, curbside pickup, and in-store returns. Chopra (2016) emphasizes the need for logistics to support customer expectations in e-commerce and calls for agile distribution networks. The rising demand for fast delivery services has resulted in cities being dotted with fulfillment centers as well as flexible warehousing strategies such as micro-fulfillment centers. The intersection of channels and AI-driven demand forecasting is similarly critical in ensuring product availability without overstocking and associated costs.

13.5.3 Decentralized Logistics Networks

The decentralized logistics networks focus on localized operations and distributed systems to reshape the traditional supply chains. COVID-19 pandemic has caused disruption in the supply chain, and it has accelerated the shift toward decentralized logistics. In such a model, businesses distribute production, storage, and distribution facilities across multiple locations to circumvent dependence on single points of failure. Decentralized logistics networks increase resilience and responsiveness because companies can respond rapidly to disruptions (Perboli et al., 2018). Recently, additive manufacturing, known as 3D printing, has created the possibility of localized production closer to demand points. This type of decentralized system further contributes to reducing lead times, cost of transportation, and carbon footprint; therefore, most businesses would like to see such decentralization in systems for efficiency and sustainability.

13.5.4 The Role of Artificial Intelligence and Predictive Analytics

The increase in data access in logistics serves to unlock more advanced applications of AI and predictive analytics. AI and predictive analytics arm logistics managers with tools to proactively make informed decisions. Predictive models analyze historical data, market trends, and real-time information with a view to anticipating such disruptions as supply shortages, weather events, or geopolitical tensions. Consequently, route adjustments become dynamic and inventory reallocations occur, minimizing the impact of unforeseen events. As Waller and Fawcett note (2013), AI optimizes operational efficiency, knowing patterns that could only take massive manual effort.

13.6 CONCLUSION

The backbone of the modern supply chain, logistics represents efficiency, connectivity, and innovation in an increasingly complex and competitive global environment. All kinds of hurdles the logistics sector has had to face—from growing organizations to increasing customer expectations—have led to disruptive supply, increased technology needs, and increased sustainability demands. However, these challenges are among the causes for radical innovations that have changed the sector by providing immediate solutions while preparing it for more complex challenges in the future. The integration of sustainability is now a global issue that revolves around logistics and is one of the major issues in this transition.

A commitment to sustainability is growing within the industry, and with that new trend, carbon-neutral supply chains are being developed. Some organizations have taken on various green logistics practices, including adopting renewable energy vehicles, optimizing transportation routes to reduce emissions, and making items from biodegradable materials. Reverse logistics systems are increasingly prominent and are likely to reflect the shifting direction of the industry from linear to circular economy principles. Sustainable logistics is already a mandate because organizations must balance competitiveness through profitability and environmental responsibility.

Technological advances are transforming logistics processes into greener and stronger systems for developing sustainability. With artificial intelligence and predictive analytics at work, businesses can predict interruptions and manage inventory, as well as optimize supply routes. This highly data-oriented work may actually be a much more straightforward approach to environmental sustainability that contributes to waste reduction and efficiency of operations. The import of a middleware phenomenological approach of blockchain in supply chains would increase trust as being part of a sustainable practice and thus would be verifiable.

The increased trend toward decentralized networks in logistics. Here, various nodes each take care of a portion of the tasks with a view to achieving increased robustness and efficiency. This measure would reduce the costs for energy consumption in long-distance transportation, as transportation and delivery could be more localized through local manufacturing. Coupled with this paradigm shift, omnichannel integration has also transformed the business of logistics, as it synergizes all channels to render the services to meet wider customer needs. In fact, new fulfillment methods, such as micro-fulfillment centers and last-mile options using drones or electric cars, will increasingly support such transformations.

The dilemma now with these changes comes with the implication of continuous investment in research and innovation, because there are costs of capital for sustainable technology, shortage of qualified workers for handling highly automated systems, and unpredictably disarrays of the global supply chains. Eagerly awaited and of paramount importance is cooperation between industry, governments, and academic institutions to overcome these challenges and create a conducive atmosphere within which logistics may grow in a sustainable and inclusive manner, thus directing future research toward developing industry-specific logistics solutions, standardized sustainability indicators, and hybrid models combining machine skills and human expertise.

Therefore, as far as the current global change is concerned, logistics is the first to bring together the two sides of this move to make it more operationally sustainable. This perfect adaptability and innovativeness in the new process of transition will determine the successful way forward for logistics in today's complex environment. Advancements embraced into technology and prioritized service sustainability can take the business place to the next level and set it up for construction resilience toward future relevance and influence.

REFERENCES

Agrawal, S., Singh, R. K., & Murtaza, Q. (2020). "Prioritizing Challenges for Lean Implementation in Indian SMEs." *Production Planning & Control.*

Bowersox, D. J., & Smykay, E. W. (1969). *Physical Distribution Management: Logistics Problems of the Firm.* The Macmillan Company.

Boysen, N., Fedtke, S., & Schwerdfeger, S. (2021). "Last-Mile Delivery Concepts: A Survey from an Operational Research Perspective." *OR Spectrum*, 43(1), 1–58.

Chopra, S. (2016). *Supply Chain Management: Strategy, Planning, and Operation.* Pearson Education.

Chopra, S., & Meindl, P. (2020). *Supply Chain Management: Strategy, Planning, and Operation.* Pearson.

Chopra, S., & Sodhi, M. S. (2021). "Managing Risk to Avoid Supply-Chain Breakdown." *MIT Sloan Management Review*, 62(3), 52–60.
Christopher, M. (1998). *Logistics and Supply Chain Management: Strategies for Reducing Cost and Improving Service*. Financial Times/Prentice Hall.
Christopher, M. (2016). *Logistics and Supply Chain Management*. Pearson.
Dubey, R., Gunasekaran, A., & Papadopoulos, T. (2017a). "Disaster Resilience in Supply Chains: The Role of Big Data and Predictive Analytics." *Computers in Industry*, 101, 116–127.
Dubey, R., Gunasekaran, A., & Papadopoulos, T. (2017b). "Green Supply Chain Management: Theoretical Framework and Further Research Directions." *Benchmarking: An International Journal*, 24(1), 184–218.
Frohlich, M. T., & Westbrook, R. (2001). "Arcs of Integration: An International Study of Supply Chain Strategies." *Journal of Operations Management*, 19(2), 185–200.
Geissdoerfer, M., Savaget, P., Bocken, N. M., & Hultink, E. J. (2017). "The Circular Economy – A New Sustainability Paradigm?" *Journal of Cleaner Production*, 143, 757–768.
Goodchild, A., & Toy, J. (2018). "Delivery by Drone: An Evaluation of Unmanned Aerial Vehicle Technology in Reducing CO^2 Emissions in the Delivery Service Industry." *Transportation Research Part D: Transport and Environment*, 61, 58–67.
Govindan, K., Mina, H., & Alavi, B. (2018). "A Decision Support System for Demand Management in Sustainable Supply Chains." *Journal of Cleaner Production*, 184, 115–129.
Govindan, K., Rajendran, S., Sarkis, J., & Murugesan, P. (2021). "Multicriteria Decision Making Approaches for Green Supply Chain Management." *Transportation Research Part E: Logistics and Transportation Review*, 135, 102362.
Govindan, K., Soleimani, H., & Kannan, D. (2015). "Reverse Logistics and Closed-Loop Supply Chain: A Comprehensive Review to Explore the Future." *European Journal of Operational Research*, 240(3), 603–626.
Gunasekaran, A., Patel, C., & McGaughey, R. E. (2004). "A Framework for Supply Chain Performance Measurement." *International Journal of Production Economics*, 87(3), 333–347.
Ivanov, D., & Dolgui, A. (2020). "A Digital Supply Chain Twin for Managing the Disruption Risks and Resilience in the Era of Industry 4.0." *Transportation Research Part E: Logistics and Transportation Review*, 136, 101933.
Ivanov, D., & Dolgui, A. (2021). "Post-COVID-19 Supply Chain Recovery: The Role of Digital Twins, AI, and Blockchain." *International Journal of Physical Distribution & Logistics Management*, 51(2), 567–584.
Ivanov, D., Dolgui, A., & Sokolov, B. (2019). "The Impact of Digital Technology and Industry 4.0 on the Ripple Effect and Supply Chain Risk Analytics." *International Journal of Production Research*, 57(3), 829–846.
Jabbour, C. J. C., Jabbour, A. B. L. S., Sarkis, J., & Filho, M. G. (2019). "Unlocking the Circular Economy: Sustainability and Innovation in Logistics and Supply Chains." *Business Strategy and the Environment*, 28(2), 309–320.
Kamath, R. (2018). "Food Traceability on Blockchain: Walmart's Pork and Mango Pilots with IBM." *Journal of the British Blockchain Association*, 1(1), 1–12.
Kouhizadeh, M., Saberi, S., & Sarkis, J. (2020). "Blockchain Technology and the Sustainable Supply Chain: Theoretically Exploring Adoption Barriers." *International Journal of Production Economics*, 231, 107831.
Kouvelis, P., Chambers, C., & Wang, H. (2020). "Supply Chain Management Research and the Realities of the Pandemic: An Overview." *Production and Operations Management*, 29(4), 771–777.
Kshetri, N. (2018). "Blockchain and Electronic Healthcare Records [Cybertrust]." *Computer*, 51(12), 59–63.

Kumar, G., Singh, R. K., & Shankar, R. (2020a). "Role of Flexibility in Linking Operations Strategy and Supply Chain Performance." *Benchmarking: An International Journal*, 27(5), 1777–1805.

Kumar, S., Teichman, S., & Timpernagel, T. (2020b). "A Green Supply Chain is a Requirement for Profitability." *International Journal of Production Research*, 58(1), 1–15.

Lee, H. L., Padmanabhan, V., & Whang, S. (1997). "Information Distortion in a Supply Chain: The Bullwhip Effect." *Management Science*, 43(4), 546–558.

Marchet, G., Perego, A., Perotti, S., & Mangiaracina, R. (2021). "Logistics and Supply Chain Management in the Omnichannel Era." *Transportation Research Part E: Logistics and Transportation Review*, 150, 102268.

McKinnon, A. (2018). *Decarbonizing logistics: Distributing goods in a low carbon world.* Kogan Page Publishers.

Mehrjerdi, Y. Z. (2021). "Collaboration in Supply Chain Networks: A Comprehensive Framework." *Supply Chain Management: An International Journal*, 26(4), 389–406.

Mentzer, J. T., DeWitt, W., Keebler, J. S., Min, S., Nix, N. W., Smith, C. D., & Zacharia, Z. G. (2001a). "Defining Supply Chain Management." *Journal of Business Logistics*, 22(2), 1–25.

Mentzer, J. T., Flint, D. J., & Hult, G. T. M. (2001b). "Logistics Service Quality as a Segment-Customized Process." *Journal of Marketing*, 65(4), 82–104.

Ng, D. T. K., Leung, J. K. L., Chu, S. K. W., & Qiao, M. S. (2021). "Conceptualizing AI Literacy: An Exploratory Review." *Computers and Education: Artificial Intelligence*, 2, 100041.

Perboli, G., Rosano, M., & Saint-Guillain, M. (2018). "Disrupting the Logistics Industry with Decentralized Networks." *Transportation Research Part E: Logistics and Transportation Review*, 118, 465–487.

Pettit, T. J., Fiksel, J., & Croxton, K. L. (2019). "Ensuring Supply Chain Resilience: Development and Implementation of an Assessment Tool." *Journal of Business Logistics*, 40(1), 1–22.

Queiroz, M. M., Telles, R., & Bonilla, S. H. (2020). "Blockchain and Supply Chain Management Integration: A Systematic Review of the Literature." *Supply Chain Management: An International Journal*, 25(2), 241–254.

Rajeev, A., Pati, R. K., Padhi, S. S., & Govindan, K. (2019). "Evolution of Sustainability in Supply Chain Management: A Literature Review." *Journal of Cleaner Production*, 207, 1–9.

Rushton, A., Croucher, P., & Baker, P. (2017). *The Handbook of Logistics and Distribution Management*. Kogan Page.

Saberi, S., Kouhizadeh, M., Sarkis, J., & Shen, L. (2019). "Blockchain Technology and Its Relationships to Sustainable Supply Chain Management." *International Journal of Production Research*, 57(7), 2117–2135.

Saenz, M. J., Revilla, E., & Knoppen, D. (2020). "Resilient Supply Chain Design: A Case Study Approach." *International Journal of Production Research*, 58(3), 817–832.

Sari, K. (2008). "On the Benefits of CPFR and VMI: A Comparative Simulation Study." *International Journal of Production Economics*, 113(2), 575–586.

Shapiro, R. D. (1984). "Get Leverage from Logistics." *Harvard Business Review*, 62(3), 119–126.

Sharma, M., Joshi, S., & Luthra, S. (2021). "Interrelation of Resilience and Sustainability for Supply Chain Effectiveness in COVID-19 Disruptions." *Journal of Cleaner Production*, 131, 857–875.

Sheffi, Y. (2015). *The Power of Resilience: How the Best Companies Manage the Unexpected.* MIT Press.

Stevens, G. C. (1989). "Integrating the Supply Chain." *International Journal of Physical Distribution & Logistics Management*, 19(8), 3–8.

Verdouw, C. N., Wolfert, J., Beulens, A. J. M., & Rialland, A. (2016). "Virtualization of Food Supply Chains with the Internet of Things." *Journal of Food Engineering*, 176, 128–136.

Wagner, S. M., & Bode, C. (2014). "Supplier Relationship-Specific Investments and the Role of Safeguards for Supplier Innovation Sharing." *International Journal of Operations & Production Management*, 34(9), 1124–1149.

Waller, M. A., & Fawcett, S. E. (2013). "Data Science, Predictive Analytics, and Big Data: A Revolution that Will Transform Supply Chain Design and Management." *Journal of Business Logistics*, 34(2), 77–84

Wang, G., Gunasekaran, A., Ngai, E. W., & Papadopoulos, T. (2019). "Big Data Analytics in Logistics and Supply Chain Management." *International Journal of Production Economics*, 203, 1–12.

Wang, Y., Zhang, J., & Lee, C. K. (2022). "AI and Machine Learning in Logistics and Supply Chain: Trends and Challenges." *Journal of Business Logistics*, 43(2), 156–172.

Wieland, A., & Durach, C. F. (2021). "Two Perspectives on Supply Chain Resilience." *Journal of Business Logistics*, 42(1), 56–63.

Yang, J., Pan, S., & Ballot, E. (2017). "Innovative Retail Logistics Fulfillment by Cloud-based Solutions." *International Journal of Production Research*, 55(4), 1177–1194.

Zhang, A., Yu, K., & Shen, L. (2021). "The Impact of IoT on Supply Chain Performance: A Review of Empirical Evidence." *Journal of Business Logistics*, 42(1), 50–67.

Zhang, A., Yu, K., & Shen, L. (2022). "The Impact of Drone Delivery Systems on Last-Mile Logistics: A Case Study." *Transportation Research Part E: Logistics and Transportation Review*, 157, 102503.

14 Unchaining Humanity: Pioneering Solutions to Eradicate Exploitation in Global Supply Chains

Jasmine Mariappan, Chitra Krishnan, and Supriya Lamba Sahdev

14.1 INTRODUCTION

In the complex tapestry of global trade, the dark threads of exploitation are often those that wove into the global supply chains of industries that stretch across the continents. Child labor, forced labor, and human trafficking are nothing else than a huge threat against numerous people all over the world (International Labour Organization, 2017). It may shed light on how widespread such severe yet inhumane practices are in global supply chains, and it may also address these from ethical, legal, and humanitarian standpoints. It also presents a hierarchical structure of the chapters, indicating that they are part of a multifaceted framework for attaining a clear understanding of and finding a solution to these complicated issues.

Human rights violations reach their most extensive form when child labor enforcement combines with forced work alongside human trafficking activities. The same human rights violations occur today in agricultural work as well as manufacturing and construction together with services industries (World Health Organization, 2020). The evolution of these exploitative practices is shocking, and figures imply that more than a hundred children and adults are victims of labor or sex exploitation among manufacturing supply chains worldwide. These workers face the toughest and harsh job settings frequently in the unhealthy environment They cannot receive an education, be fairly compensated, etc., and be free from exploitation. This is not limited to those directly impacted by the injustice, rather, it contaminates the reputations of the companies involved while further upholding passive voices of both poverty and injustice.

Ensuring child labor, forced trade, and human trafficking in any global supply system is more than the ethical responsibility and legal conformity; it is the responsibility of companies to honor the principles of dignity, justice, and respect for human rights (Bales, 2016). Ethically, we cannot condone the continuation of such a practice with the excuse that it is a fair competitor. Children deprived from education and voiceless people depict the absence of the value for human worth and human dignity. Secondly, this is legally supported in international conventions and national

DOI: 10.1201/9781003477242-17

regulations, that provide the basis of fighting these issues by obliging governments and businesses to protect human rights and bear accountability for the violations. Nevertheless, the success of legal ones depends on solid follow-up enforcement efforts and on transnational organized crime.

Further, from the humanitarian perspective, the urge to eradicate a child, exploitation forced labor, and human trafficking is primarily owing to respect and sanctity associated with human rights and welfare that a vulnerable section of society requires (Assembly, 1948). It requires active and professional leaders who are going to dismantle part of the system of exploitation of the vulnerable groups and assist the victims, as well as find a way of solving part of the causes of the problem, such as poverty, widespread illiteracy, and social injustice. This chapter will seek to provide an analysis of the various issues of child labor, forced labor, and human trafficking within the supply chains around the globe. It will explore not only all aspects of the issue but also will deal with such questions as legal frameworks, corporations' roles, collaboration, and innovation. Each part of the chapter (e.g., section) is expected to provide the audience with some essential information, analysis, and recommendations ensuring the overall issue addresses which may involve research, case studies, and real-life facts of the matter, but it is ultimately aimed at making the dialogue continued and the involving action more activated on combating exploitation of labor in the global supply chains. Let us all stick to this agitator while believing in ourselves to eradicate injustice, publicize human rights, and take the world into one in which human beings live in an environment where they are all treated as equals.

14.2 UNDERSTANDING THE PROBLEM

Today, child labor, forced labor, and human trafficking retain the highest status of human rights violations that originate from global supply chains, whereas the international actors have tried to fight against them. In this part, the definitions and distinctions between the types of exploitation are provided, as well as the multilateral factors contributing to their persistence are explored, then the most effective case studies based on reality are presented to illustrate the harm caused to the individuals and communities.

14.2.1 Definitions and Distinctions

The ILO defines child labor as a work that denies children a proper chance to grow up, play, learn, and be children, besides negative lifelong consequences like physical or emotional trauma. It incorporated activities that were physical, mental, social, and moral risks and could seriously disregard the growth, learning, security, and health of children, which include hazardous work and work that can interfere with a child's education and health.

Forced Labor: This is a set of circumstances where individuals are forced or the ones lying to them to continue working, and cannot leave because of threats, violence, or other intimidation methods (Frey & Fletcher, 2015). It is distinguished by a lack of agreed-upon obligation to participate and features conditions of exploitation such as mandatory overtime, denial of wages, and physical or sexual oppression.

Human Trafficking: Involves the recruitment, transportation, transfer, harboring or receipt, of persons for the purpose of exploitation through force, fraud, or deception. This can be shown by different modes, including, for example, the slave trade, sexual harassment, organ transplanting, and forced unions. Many victims of human trafficking undergo extreme situations of sexual violence and hardship, such as working as wage slaves, enslavement, and being violently abused.

14.2.2 Contributing Factors

Economic Factors: Poverty, no schooling, and not enough jobs are what force many children into working long hours and even slavery. In regions where poverty is pervasive and social safety nets are inadequate, individuals and families become much exposed to exploitative behaviors by some employers and text traders as they look for simple ways to get out of poverty and get necessities.

Social Factors: Discrimination, social exclusion, and gender inequality, which itself is the cause of extreme vulnerability, are the reasons why women, children, migrants, and indigenous people (marginalized groups) are particularly prone to manipulation. Customary social norms and social-specific menus can be grounds for the reproduction of cycles of oppression, and the victims are unable to seek asylum, or the victims are blocked from receiving assistance.

Regulatory Factors: Poorly tailored or shady labor laws combined with ineffective enforcement and corruption often create perfect catalysts for more abuse to take place uncontested. Sometimes it happens that the total absence of regulation in a certain field or even inconsistency among numerous authorities is the case when slackers can breach the law and continue unscrupulous operations off the radar (Klymak, 2023).

14.2.3 Examples

Child Labor in Cocoa Production: In West Africa, particularly in countries like Ivory Coast and Ghana, child labor is prevalent in cocoa farming (Allain, 2013). While some apply harmful chemicals and are unable to attend school, children as young as five years old are diverted to work and cut down trees. This results in unsustainable poverty circles and kids' deprivation of their ability to get an education and live a promising life.

Forced Labor in Garment Factories: In nations such as Bangladesh and Cambodia, global brand names buy from factory supplier sources of clothes. Yet, a bulk of the workers, from women to guest workers, go through unjust working circumstances of long hours, low wages, and an unsafe working environment. They are often controlled and threatened for production goals completion; so the situation of forced labor is a glaring example in garment industries.

Human Trafficking in the Fishing Industry: Southeast Asia is considered a notorious site of modern-day slavery and slave labor in the commercial fishing industry (Latham-Sprinkle et al., 2019). Workers from countries such as Myanmar and Cambodia arrive in these countries using false dreams of getting well-paid jobs and end up aboard fishing vessels for months or even years, which is also accompanied by physical cruelty and debt slavery.

These are several real-life cases that shed light on the devil of the horrifying consequences for individuals and communities, such as child labor, forced labor, and human trafficking, proving the necessity of concerted action against these difficult matters in global supply chains.

14.3 LEGAL AND REGULATORY FRAMEWORKS

Efforts to combat child labor, forced labor, and human trafficking rely on a robust legal and regulatory framework at both the international and national levels. This section provides an analysis of key international conventions and treaties, examines relevant national laws and regulations, and evaluates the effectiveness of existing legal frameworks, highlighting challenges in implementation and enforcement.

14.3.1 International Conventions and Treaties

ILO Conventions: The International Labour Organization (ILO) has set up different conventions concerning the child labor put in force (No. 182). Facilitation of the movement of goods, people, and services (No. 182), as well as the migration and trafficking of humans (No. 29), to mention but a few. These conventions, which stipulate the least standards to be met for the elimination of these (ILO, 2017) forms of exploitation and serve as a guide for member state actions to enact the legislations and policies that go well with these standards.

UN Protocols: The UN has come up with a catalog of arrangements to be built in order to suppress, fight, and bill human exploitation, especially victims of females and kids. The protocol of Palermo, which is an addition to the UN Convention against Transnational Organized Crime is a comprehensive framework for addressing human trafficking; these include prevention, creation of a protective environment for the victim, as well as persecution of the culprits.

Regional Agreements: The European Union has included applicable directives and regulations to efficiently counter human trafficking as well as safeguard the rights of victims. These treaties serve functions similar to international arrangements and help members states combat the transduction of trafficking networks through joint aid.

14.3.2 National Laws and Regulations

Legislative Measures: Several countries are taking steps by promulgating child labor laws, forced labor laws, and human trafficking legislations. These laws usually make it illegal to do these things, put in a system for recognizing and prosecuting perpetrators, and give victims the protection and power they need.

Labor Standards: Minimum employment standards built into national labor laws are designed to cover fair wages, working hours, safety at work, and other job-related issues. Such observance of these norms is imperative as they will aid in purging this exploitation and safeguarding the rights of workers in many occupations.

14.3.3 Effectiveness and Challenges

Implementation: Nevertheless, the rights of children against child labor, forced labor, and trafficking are protected by diverse international and national legal frameworks, but the effectiveness of such measures depends on the adoption of effective measures at the national level. This will have to be accompanied by adequate resources, training, and synchronization between the government departments, law enforcement units, and civilian agencies.

Enforcement: The challenges in the enforcement of the law include, among others, limitation to capacity and resources, corruption, and the absence of political will. Frequently, the police agencies operate typical in that they prioritize other criminal activities than labor exploitation so they become the protectors rather than the defenders.

Cross-border Cooperation: Global issues of child labor, forced labor, and human trafficking should be solved with respect to their transnational character through the cooperation and interaction of all countries of departure, transit, and destination. On the other hand, disparities tied to legal systems, jurisdiction, and diplomatic sensitivities can impede the further expansion of efficient cooperation.

Legal and regulatory frameworks that focus on combating child labor, forced labor, and human trafficking are of great importance in the struggle for their abolishment. Although in terms of international and national level, considerable advancements have been made, issues of non-conformance to agreements and implementation of plans are still to be dealt with. However, the above challenges can only be sufficiently dealt with if governments, civil societies, and the private sector are consistent in their effort to put the respect of human rights and the dignity of everyone.

14.4 CORPORATE RESPONSIBILITY

Businesses play a pivotal role in combating child labor, forced labor, and human trafficking within their supply chains. This section examines the responsibilities of businesses, explores effective corporate due diligence practices, and presents case studies illustrating successful approaches implemented by companies in various industries.

14.4.1 Role of Businesses

Ethical Imperative: Businesses have their responsibility to make sure that their activities are not sustaining the labor practices which imply exploitation of workers. Holding human rights together and ethical labor standards apart from society expectations is a key toward sustainable business practices (Akbari & McClelland, 2020).

Supply Chain Influence: As key players in global value chains, companies are endowed with immense capability to influence not only their suppliers and subcontractors but also those of their supplies. Through the use of this influence to enforce labor standards and ethical practices, companies can repair things and protect the environment from exploitation (Congdon Fors, 2024).

Risk Mitigation: Eliminating labor exploitation is not only a moral issue but it is a risk mitigation issue as well. If a company fails to identify and curb the problems of child labor, forced labor, and human trafficking, it can suffer from a damaged reputation, legal liabilities, and financial losses.

14.4.2 Best Practices for Corporate Due Diligence

Risk Assessment: Businesses must carry out inclusive risk assessments to determine whether there are any labor rights violations in their supply chain. The place covered in the analysis is the geographical location, industry sector, relationships with subcontractors, examples from history, etc (Latham-Sprinkle et al., 2019).

Monitoring and Auditing: In terms of efficient compliance, monitoring and auditing labor standards are essential, together with the detection of exploitation cases. In-site inspections of workers, interviews, document reviews, and self-assessments of suppliers are effective tools for supply chain activities monitoring (Aristova et al., 2024).

Remediation and Accountability: Defining a well-structured remediation policy and accountability strategy is crucial when a labor violation is found. This entails introducing amendments, assisting the workers affected, and holding suppliers to account through contractual agreements or such methods as termination of such business connections.

14.4.3 Examples

Patagonia: Patagonia, an outdoor clothing retailer, has developed features in its supply chains to prohibit labor exploitation. Through its Social & Environmental Responsibility program, Patagonia carries out regular audits, invests in supplier training, and collaborates with other partners to address systemic problems (Miller, 2023).

Nestlé: Nestlé has been able to take care of child labor in the cocoa supply chain with the Cocoa Plan initiative. By working directly with farmers to increase productivity and promote sustainable farming practices, Nestlé has had a significant effect in reducing the number of cases of child labor (Bubicz et al., 2019).

They illustrate that diverse methods were used by the companies to fight child labor, forced labor, and human trafficking in the supply chains. This shows that corporate responsibility is a key factor driving positive social impact and sustainable business practices.

14.5 COLLABORATIVE INITIATIVES

Governments, businesses, civil society organizations, and international bodies are indispensable in partnership opportunities created to efficiently deal with complex and systemic issues of child labor, forced labor, and human trafficking. In this block, we will discuss multi-stakeholder collaboration, clarify its importance, explain diverse partnerships, and use illustrative cases from collaborative initiatives and their effectiveness in eliminating exploitation in the supply chains.

14.5.1 Importance of Multi-Stakeholder Collaboration

Comprehensive Approach: To overcome child labor, human trafficking, and forced labor, we need a united and comprehensive approach. Not every actor or institution can tackle these problems on their own. Collaborative projects unify partners from different sectors possessing varied expertise and assets that enable not only the creation of an integrated, successful plan but also the actualization of an important impact.

Shared Responsibility: Child labor, forced labor, and human trafficking constitute the main complex challenge that results in cross-border movement and weaves itself into multiple sectors of the economy. Through cooperation among governments, enterprises, non-governmental organizations, and international unions, the stakeholders can share knowledge, support efforts, and coordinate their actions to help put an end to exploitation as well as to promote sustainable development.

Sustainability and Scale: Through collective efforts and actions, it is possible to attain more sustainability than by just engaging in stand-alone projects. Through a joint effort, aligning objectives, and channeling collective action, stakeholders can perk up the efficacy of the efforts, assist more afflicted individuals and communities, amplify the change at the systemic level, which results in lasting reforms, and give respect to human beings and their dignity.

14.5.2 Overview of Partnerships

Public-Private Partnerships: Governments and businesses sometimes collaborate through public-private partnerships (PPPs) to identify and address labor exploitation in supply chains. Those collaborations do exactly that: they merge the efforts of different sectors, set up regulatory structures, define industry standards, and provide aid and support to the victims of exploitation.

Multi-Stakeholder Platforms: There are multi-stakeholder sites that enable governments, businesses, NGOs, and other influential parties to hold dialogue, information exchange, and to collaborate on labor rights problems. These platforms are platforms to let stakeholders share their best practices, indicate common problems, and develop joint action plans to solve the issue.

International Initiatives: International organizations such as the International Labour Organization (ILO), the United Nations Office on Drugs and Crime (UNODC), and the International Organization for Migration (IOM) are a key part in implementing cross-country cooperation among governments, businesses, and civil society organizations which promotes efforts against child labor, forced labor, and trafficking of human beings.

14.5.3 Examples of Collaborative Initiatives

The Alliance 8.7: The Alliance 8.7 is a world's stakeholders and states working toward achieving the SDG Target 8.7 that aims to mitigate child labor, forced labor, and human trafficking by the end of the year 2030. The Alliance comprises governments, businesses, civil society organizations, and international bodies, including the United Nations, to cooperate in sharing data and resources and tracking the progress toward the set objective of eliminating extreme poverty across the globe by the year 2030 (Labour, 1994).

The Dhaka Principles for Migration with Dignity: The Dhaka principles are a group of guidelines evolved through the process of multi-stakeholders, including government, business, trade unions, and civil society organizations, in place of ethical recruitment practices to safeguard the rights of migrated workers. Through this means, stakeholders subscribe to such principles with an obligation to respect labor standards, maintain open recruitment processes, and steer clear of exploitation and abuse of these migrant workers (Kamruzzaman et al., 2022).

The Responsible Business Alliance (RBA): Previously, the Electronic Industry Citizenship Coalition (EICC), the RBA is a global platform of electronics companies that practitioners to a higher level of ethical standards in global supply chains. Cooperation between the RBA and such as governments, NGOs, and enterprises is vital for establishing common rules, methods, and tools for improving labor conditions, environmental friendliness, and ethical sourcing throughout the electronics industry (Schmutzler et al., 2013).

14.5.4 Impact of Collaborative Initiatives

Policy Development and Advocacy: Collaborative projects are the tools that build the agendas of the policies, bring legislative reforms, and enlighten the general society on the significance of labor rights and humanity. Through the involvement of stakeholders and getting their voices heard, such initiatives can eventually impact the decision-makers and consequently lead to systemic change that goes beyond national and international contexts.

Capacity Building and Empowerment: Collaborative efforts permit the conduction of capacity building, training, and empowerment for all stakeholders, including governments, private enterprises, civil society, and workers themselves. Through the process of improving knowledge, skills, and resources, these actions give individuals the ability to identify and adequately address the exploitation of labor.

Measurable Impact: Collaborative initiatives have seen success by documenting workable outcomes to eliminate child labor, forced labor, and human trafficking in supply chains. Through the establishment of focused interventions, close surveillance of results, and results evaluation, stakeholders can monitor their impact and make changes where necessary, hence coming up with effective and sustainable remedies.

14.6 INNOVATION AND TECHNOLOGY IN SUPPLY CHAIN ETHICS

When it comes to supply chain ethics, technology is pivotal in terms of detecting, preventing, and dealing with the occurrence of labor abuses. This part examines the disruptive technology in this context and describes the application of blockchain, artificial intelligence (AI), and other emerging technologies for improving transparency, accountability, and ethics in supply chains. Besides that, it offers case studies of innovative strategies addressing the problem of labor exploitation.

14.6.1 Role of Technology in Detecting and Preventing Labor Exploitation

Technology serves as a critical enabler in identifying and mitigating labor exploitation within supply chains:

Data Analytics and Monitoring: The modern data analytics tools allow for continuous control of supply operations and the identification of misdemeanors and signs of labor exploitation. Technological tools that aid in combating money laundering work by analyzing vast amounts of data from different sources like transaction records, sensor data, and satellite images to detect suspicious patterns and activities for deeper investigations (Ogbuke et al., 2022).

Digital Platforms and Reporting Systems: Digital platforms act as the central data system for gathering, processing, and transmitting information regarding working conditions at the various stages of the supply chain. Through these platforms, key stakeholders can publish grievances, upload compliance reports, and monitor success in a real-time framework, thereby enhancing transparency and accountability throughout the supply chain (Ebinger & Omondi, 2020).

14.6.2 Use of Blockchain, Artificial Intelligence, and Emerging Technologies

Blockchain Technology: Blockchain technology promises immutable registers that are permanently tracked and transparent to everyone involved in the supply chain. A blockchain-based intervention can be a valuable tool for stakeholders to build secure, auditable records of labor practices, which enable transparency and traceability while simultaneously mitigating the risk of fraud and exploitation (Waqar et al., 2024).

Artificial Intelligence (AI): AI-empowered tools drive supply chain transparency and elucidation by optimizing processes, analyzing enormous volumes of data, and recognizing the patterns suggestive of labor exploitation. AI algorithms can find anomalies, predict dangers, and enhance the decision-making process to be in advance to respond to violations of workers' rights (Toorajipour et al., 2021).

14.6.3 Examples of Innovative Solutions

14.6.3.1 IBM Food Trust

14.6.3.1.1 Technology Overview

IBM Food Trust is based on blockchain and, as a result, creates a system of incorruptible records that track all the activities in the food chain. It means that all stakeholders including farmers, processors, distributors, and retailers together get real-time information about food items right from the farm gate to the dinner table of the consumer.

Furthermore, it enables and enhances secure and clear data transfer among the stakeholders. In this case, the shared information includes data that is necessary for tracing food safety issues, including the origin of the products, the conditions of their processing, and the conditions under which they are transported. Thus, through creating an atmosphere where participants are able to share information freely, IBM Food Trust strengthens the accountability and trust within the chain, resulting in a safer and more ethical food supply chain.

14.6.3.1.2 Features and Functionality

IBM Food Trust has full traceability, which means users can track a certain food product across the whole supply chain. This capability allows stakeholders to visualize where products were raised, how they were processed, and routes they underwent to reach users. They include transparency because getting the details of the products affected is essential when doing food recalls, thus promoting food safety. It also employs smart contracts, which are contracts that operate on predefined conditions and enforce the suppliers' compliance with ethical labor practices. These contracts can pay or perform some other action on the verification of compliance, thus minimizing the risk of labor exploitation by making suppliers accountable for standards that they agreed to meet.

Further, it is worth mentioning that IBM Food Trust improves consumers' interaction using connected platforms that enable clients, for instance, to begin scanning the QR codes on products. It presents the background of the food, the process that was used in producing it, and details about sustainable practices that the consumers can use to make informed decisions. Overall, through the principles of transparency and accountability, IBM Food Trust supports a more responsible food system.

14.6.3.1.3 Accountability and Ethical Practices

- Supplier Accountability: By making data accessible, suppliers are incentivized to adhere to ethical practices and labor standards. Companies that fail to comply risk losing customers and market access, as consumers increasingly demand transparency.
- Labor Compliance Monitoring: The system can incorporate labor compliance data, allowing companies to verify the ethical treatment of workers throughout the supply chain. Regular audits and real-time reporting can help identify and address labor issues proactively.

14.6.3.1.4 Impact on Labor Exploitation

IBM Food Trust increases suppliers' responsibility for their share of the food chain by providing crucial information to all sides. This increases the levels of transparency in the industry and puts pressure on the suppliers to abstain from practices such as child labor, which may lead to customer boycotts and the subsequent loss of business. Consumers today also expect that suppliers be ethical in sourcing their products and fair to their employees.

In addition, the structure also supports labor compliance monitoring including the necessary data to check on the ethical treatment of workers as furnished by companies in the supply chain. Auditing and monitoring of labor affairs should be continuous and real-time, in order to respond to emerging challenges. Besides, it serves as a useful tool to avoid non-compliance with labor standards and to create a fairer context for workers, which would generate more sustainable and efficient food industries.

14.6.3.1.5 Broader Implications

Thus, the success of IBM Food Trust opens the prospect of the further application of blockchain technological solutions in various industries, including textiles and electronics industries, where cases of labor exploitation were mentioned. As a case study revealing how blockchain can be leveraged to improve issues of transparency and accountability in the food chain, the IBM Food Trust can be applied to similar industries struggling with ethical practice and labor issues.

In addition, since the platform entails collaboration between multiple and qualified supply chain entities, it is advantageous as well. It is a community purpose where people share the best of practice, knowledge and resources, and pledge for ethical business with sustainable practice. So, we see that the nature of collaboration between companies is to promote change dynamics in the global apparel industry by making ethical sourcing and responsible labor standards the new vision. Such continuity not only enhances each affiliated business, but also helps to promote a fair and balanced global economy.

IBM Food Trust exemplifies how innovative technology can revolutionize supply chains, promoting transparency, accountability, and ethical practices. By harnessing the power of blockchain, the platform not only improves traceability but also empowers consumers and workers alike, contributing to a more ethical global economy. As technology continues to evolve, solutions like IBM Food Trust will play a critical role in addressing labor exploitation and ensuring that the production of goods is equitable and responsible worldwide.

14.7 FUTURE DIRECTIONS

The process of identifying and reducing the remaining issues of child labor, forced labor, and human trafficking in global supply chains is paramount for the development of ethical practices and the preservation of vulnerable workers. Besides, the ability to study trends and risks that are coming up is also important to create strategies that are effective against these problems. The next part examines how, despite the existing challenges, new trends are taking place and offers future guidance concerning the role of governments, businesses, and other stakeholders.

14.7.1 Emerging Trends and Risks

Technology and Digitalization: Technology effectuates the realization of heightened supply chain transparency and accountability, and new risks are introduced through the use of digital platforms for exploitative labor practices, the proliferation of online trafficking networks, and the exploitation of workers who have a vulnerable position in the gig economy.

Climate Change and Environmental Degradation: Climate change and environmental degradation are factors that make the situation more difficult and create conditions for more people to make a living through exploitation in supply chains. Dumping of people from their homesteads, loss of their means of livelihood, as well as the occurrence of natural disasters, can cause the people involved to be exploited, while conflicts and displacement occasioned by extractive industries and land grabbing can result in the subsequent conflicts and displacement.

Pandemics and Global Crises: Through pandemics and global crises as an example of the COVID-19 pandemic, we can understand that global supply chains are highly interconnected, and their workers are at risk of shocks and disruptions caused by such crises. Economic states of depression, border restrictions, and restrictions on mobility can aggravate already existing problems and create new obstacles for zeroing in on labor exploitation.

14.7.2 Recommendations for Future Action

Strengthen Legal Frameworks: The national authorities should toughen up the legislation and improve the current mechanism for the labor standards implementation; also, the government bodies should take prompt action against those who exploit workers. It involves the ratification and implementation of the connected treaties and standards, the reinforcement of the labor inspection systems, and the punishment of violations.

Promote Responsible Business Practices: Businesses should have a conscience and work to safeguard the fundamental rights of workers as well as to eliminate labor abuse within their supply chain. They are required to set up impactful due diligence systems, collaborate with suppliers and subcontractors to monitor all the labor standards, and be transparent on the measurable results of their initiatives in this regard.

Enhance Collaboration and Information Sharing: The participation of national governments, businesses, civil society organizations, and the role of international agencies all play a major role in the suppression of labor exploitation. Stakeholders should share experiences, best approaches, and lessons learned, and work together to come up with programs, initiatives, and solutions to promote ethical sourcing, protect workers' rights, and address systemic issues.

Invest in Education and Livelihoods: Assisting in education, skills development, and alternative livelihoods through the provision of education and skills, and other alternative support mechanisms is crucial in empowering workers and their communities to reduce vulnerability to exploitation. It is of utmost importance that governments, businesses, and development institutions target their investments in education, skills development, and economic opportunities to groups and communities who are the first victims of exploitation at workplaces, with an emphasis on the marginalized ones.

Address Root Causes: Moreover, the resolution of labor exploitation requires eradicating poverty, inequality, and social exclusion. Governments are advised to roll out social protection inclusions, pursue equitable development regimes, and eliminate the bottom-layer structures that have long been the cause of exploitation.

Developing lasting solutions to those unsolved issues and new risks is an ongoing task for public authorities, companies, and NGOs. This can be achieved by strengthening legal frameworks, developing responsible business practices, enhancing collaboration and cooperation, supporting skills development, and improving livelihoods, as well as addressing underlying causes, which unite all stakeholders in protecting labor rights and supporting dignified lives for all individuals.

14.8 POLICY RECOMMENDATIONS AND FUTURE DIRECTIONS

14.8.1 Strengthening Legal Frameworks

14.8.1.1 Recommendations for Improving Laws and Regulations

The implementation of protection against exploitation in international supply chain management requires improvement and synchronization of legal measures locally and in the global domain. Here are key recommendations:

Harmonize Standards: Standardize global codes of conduct regarding labor relations and apply them globally. Cooperate with such international actors as the International Labour Organisation (ILO) in formulating and adopting seventh, clear meaningful global standards.

Update Legislation: Work, update, and modernize laws and regulations related to labor and employment that are related to exploitation in detail. This comprises updating existing legal provisions to accommodate new forms of exploitation as well as checking whether the laws are well-suited for a fast-evolving world economy.

Strengthen Worker Protection: Strengthen the labor laws and set higher standards for wages, working conditions as well as organizations' rights for workers. Establish procedures that would help in relaying the protection details and ensure that they are adhered to.

Improving Supply Chain Transparency: It is crucial for ensuring ethical production practices and protecting vulnerable groups. By requiring producers to release comprehensive information about their supply chains, including sourcing and manufacturing conditions, companies can be held accountable for exploitative practices, ultimately empowering vulnerable workers and communities.

14.8.1.2 Enhancing Enforcement Mechanisms

Effective enforcement is crucial for the success of any legal framework. Recommendations include:

Increase Funding and Resources: Allocate more resources to enforcement agencies to ensure they can investigate and address violations. This includes funding for training, technology, and personnel.

Develop Specialized Units: Create specialized units within enforcement agencies focused on labor exploitation and human trafficking. These units should have the expertise and tools necessary to address complex supply chain issues.

Promote Whistle-blower Protections: Implement strong protections for whistle-blowers who report exploitation. Encourage reporting by ensuring that whistle-blowers are protected from retaliation and have access to legal and financial support.

Enhance International Cooperation: Foster greater collaboration between countries to address transnational exploitation. This includes sharing information, coordinating investigations, and harmonizing enforcement efforts.

14.8.2 Promoting Ethical Practices

14.8.2.1 Encouraging Widespread Adoption of Ethical Sourcing

To promote ethical practices within global supply chains, several strategies can be employed:

Incentivize Ethical Sourcing: Provide incentives for companies that adopt ethical sourcing practices. This could include tax breaks, certifications, and public recognition for businesses that demonstrate commitment to ethical standards.

Support Certification Programs: Encourage the development and adoption of certification programs that verify ethical practices. These programs should be transparent, rigorous, and widely recognized to ensure their credibility.

Educate and Train Businesses: Offer training and resources to businesses on ethical sourcing practices. This includes providing guidance on how to implement and maintain ethical practices within their supply chains.

14.8.2.2 Supporting Corporate Transparency

Transparency is a principle that has to be followed as a part of proper ethical conduct. Recommendations for enhancing corporate transparency include:

Mandate Public Reporting: Make it obligatory for companies to disclose information about their operations in their supply chain, and the measures that they are willing to take to eradicate exploitation. This will help the consumers in making the right decisions as well as pressurize companies into achieving the right standard.

Facilitate Stakeholder Engagement: Recommend firms to consult with its stakeholders, such as employees, non-governmental organizations, and residents of the supply chain affected areas, about their supply chain management processes. This engagement will assist in finding out about and perhaps solving possible problems.

Promote Best Practices Sharing: Develop channels where companies can post and discuss cases that demonstrate proper ethical sourcing. This collaborative approach will go a long way to disseminating effective work strategies as well as enhancing transparency.

14.8.3 Leveraging Technology

14.8.3.1 Investing in and Supporting Technological Innovations

It is very clear that technological interventions can go a long way in addressing issues of exploitation. Key recommendations include:

Support Research and Development: Discharge more resources into research and development of technologies in supply chain visibility and tracking. This includes backing innovations such as in the area of blockchain, artificial intelligence, and IoT.

Provide Funding for Technology Adoption: Provide cash incentives to encourage firms to accept and integrate new technologies. This could be in the form of grants, subsidies, or relatively low interest charges on technology acquisition.

Foster Innovation Hubs: Create innovation hotspots or linkages between technology industries, research organizations, and companies to enable the creation of better ways to track and prevent exploitation.

14.8.3.2 Addressing Limitations and Challenges

On one hand, there is incredible potential in using technology. On the other hand, there are problems in doing it. Recommendations for addressing these challenges include:

Ensure Inclusivity: Buy technological solutions to make them easily and reasonably available to firms in the market even those of SMEs. Do not adopt solutions that allow the large players in the supply chain to lock out the small supply chain firms.

Address Privacy Concerns: Ensure that procedures aimed at monitoring the data have adequate measures toward the privacy of the data collected. Make sure that technology is not abused and conform to the different privacy standards.

Evaluate Effectiveness: Introduction of technological solutions should be monitored on a regular basis in attempts to determine their efficiency and make necessary changes. This concerns the observation of effectiveness brought by technology on minimization of exploitation and management of horns arising from use of technology.

14.8.4 Fostering Collaboration

14.8.4.1 Building Stronger Partnerships Between Stakeholders

This implies that there must be synergy among different stakeholders in dealing with exploitation. Recommendations for building stronger partnerships include:

Facilitate Multi-Stakeholder Dialogues: Encourage and facilitate annual meetings between governments, corporations, civil society, and other partners to review difficulties, lessons, and future action.

Create Collaborative Platforms: Develop forums or channels wherein the key stakeholders may actively engage for activities concerning the ethics in supply chain and exploitation. These platforms should facilitate exchange of information and work together in addressing emerging challenges.

Encourage Joint Initiatives: Proceed to give aid and financial backing to collaborative projects between distinct stakeholders that will tackle certain forms of exploitation. Joint initiatives can benefit from the synergies of the involved stakeholders and have a greater reach.

14.8.4.2 Expanding Multi-Stakeholder Initiatives

To broaden the scope and effectiveness of multi-stakeholder initiatives, consider the following:

Increase Participation: Encourage greater participation from diverse stakeholders, including businesses of all sizes, governments, and civil society organizations. Broad participation ensures that initiatives are more comprehensive and representative.

Promote Global Standards: Work toward establishing global standards for multi-stakeholder initiatives to ensure consistency and effectiveness across different regions and sectors.

Measure and Report Impact: Implement mechanisms to measure and report the impact of multi-stakeholder initiatives. Regular reporting will help assess progress, identify areas for improvement, and demonstrate the value of collaborative efforts.

14.9 CONCLUSION

Altogether, this chapter reviewed and discussed multifaceted issues associated with child labor, forced labor, and human trafficking in the global supply chain environment. This way, some important findings and insights about specific dimensions, challenges, and strategies have been revealed.

To begin with, there can be no doubt that even after significant efforts have been made, child labor, forced labor, and human trafficking are still pervasive in global supply chains. Complex supply chain structures, poverty, weak institutional frameworks, and structural discriminations continually enshrine these abusive acts, which call for collective action at multiple levels.

In addition, awareness of the part of governments, corporations, civil society organizations, and international agencies is crucial in addressing these issues. Multi-stakeholder transparency and responsibility are important in ensuring ethical actions, monitoring and promoting labor rights, and driving change in the supply networks. Furthermore, the appearance of new patterns and risks, including technological advancement, climate change, and global crises, poses challenges and opportunities to deal with labor exploitation. Thus, if these difficulties are to be handled in the future, it becomes important to accustom itself to these changes and adopt measures that were not undertaken before.

In light of the outcomes highlighted in this chapter, one can draw the conclusion that our work is still in progress. There is a need for an imperative that will support activities that seek to eliminate child labor, forced labor, and human trafficking in global supply chains. Policy makers need to strengthen legal frameworks, increase the intensity of regulation enforcement, and focus spending on social protection. Companies should ensure ethical conduct, work hard in conducting vigorous investigations, as well as embrace the well-being of their employees in every operational process. CSOs are expected to advocate for policy changes, increase awareness, and assist affected populations. People should go out in search of information, support moral businesses, and demand accountability from companies.

Despite the challenges, there are prospects for a new period where all employees will be honored and respected. Together we can create a world where workers' rights are considered and protected, and where no one is exploited to the point when they can no longer work and achieve success in life. It is my hope that we all work toward this positive vision of justice, equitable opportunities, and human dignity for all at each link in the global supply chain.

REFERENCES

Akbari, M., & McClelland, R. (2020). Corporate social responsibility and corporate citizenship in sustainable supply chain: A structured literature review. *Benchmarking: An International Journal*, *27*(6), 1799–1841.

Allain, J. (2013). 2000 protocol to prevent, suppress and punish trafficking in persons, especially women and children. Supplementing the United Nations convention against transnational organized crime. In *Slavery in International Law* (pp. 410–421). Brill Nijhoff.

Aristova, E., Higham, C., Higham, I., & Setzer, J. (2024). Corporate climate change responsibilities under the OECD guidelines for multinational enterprises. *International & Comparative Law Quarterly*, *73*(2), 505–525.

Assembly, U. G. (1948). Universal declaration of human rights. *UN General Assembly*, *302*(2), 14–25.

Bales, K. (2016). *Blood and Earth: Modern Slavery, Ecocide, and the Secret to Saving the World*. Random House.

Bubicz, M. E., Barbosa-Póvoa, A. P. F. D., & Carvalho, A. (2019). Incorporating social aspects in sustainable supply chains: Trends and future directions. *Journal of Cleaner Production*, *237*, 117500.

Congdon Fors, H. (2024). *Child Labour Background, Challenges, and the Role of Research in Achieving Sustainable Development Goal 8.7 (No. 840)*. University of Gothenburg, Department of Economics.

Ebinger, F., & Omondi, B. (2020). Leveraging digital approaches for transparency in sustainable supply chains: A conceptual paper. *Sustainability*, *12*(15), 6129.

Frey, D. F., & Fletcher, C. (2015). Protocol to ILO convention no. 29: A step forward for International Labour Standards. *Human Rights Brief*. https://doi.org/10.2139/ssrn.2612932

International Labour Organization. (2017). Global estimates of child labour: Results and trends, 2012–2016. *International Labour Organization; International Institute for Labour Studies*.

Kamruzzaman, P., Williams, K., Wardak, A., Kabir, M. E., & Ayobi, Y. (2022). Exploring dignity in the context of displacement–evidence from Rohingyas in Bangladesh and IDPs in Afghanistan. *Third World Quarterly*, *43*(8), 1854–1874.

Klymak, M. (2023). The trade effects of information provision about forced and child labor. *World Development*, *167*, 106217.

Labour, P. P. (1994). *A Practical Guide*. World Health Organization.

Latham-Sprinkle, J., David, F., Bryant, K., & Larsen, J. (2019). *Migrants and Their Vulnerability to Human Trafficking, Modern Slavery and Forced Labour*. International Organization for Migration. http://hdl.handle.net/1854/LU-8636089

Miller, S. C. (2023). Sustainable trailblazing: A comprehensive analysis of Patagonia's corporate social responsibility initiatives and their ethical implications. *Belhaven University, Nov, 11*.

Ogbuke, N. J., Yusuf, Y. Y., Dharma, K., & Mercangoz, B. A. (2022). Big data supply chain analytics: Ethical, privacy and security challenges posed to business, industries and society. *Production Planning & Control*, *33*(2–3), 123–137.

Schmutzler, J., Gutiérrez, R., Reficco, E., & Márquez, P. (2013). Evolution of an alliance portfolio to develop an inclusive business. In *Social Partnerships and Responsible Business* (pp. 143–163). Routledge.
Toorajipour, R., Sohrabpour, V., Nazarpour, A., Oghazi, P., & Fischl, M. (2021). Artificial intelligence in supply chain management: A systematic. *Journal of Business Research*, *122*, 502–517.
Waqar, A., Khan, A. M., & Othman, I. (2024). Blockchain empowerment in construction supply chains: Enhancing efficiency and sustainability for an infrastructure development. *Journal of Infrastructure Intelligence and Resilience*, *3*(1), 100065.
World Health Organization. (2020). International standards for the treatment of drug use disorders: Revised edition incorporating results of field-testing. Geneva: World Health Organization and United Nations Office on Drugs and Crime; 2020. License: CC BYNC-SA 3.0 IGO

Index

Pages in *italics* refer to figures and pages in **bold** refer to tables.

For Product Safety Concerns and Information please contact our EU
representative GPSR@taylorandfrancis.com
Taylor & Francis Verlag GmbH, Kaufingerstraße 24, 80331 München, Germany

www.ingramcontent.com/pod-product-compliance
Ingram Content Group UK Ltd.
Pitfield, Milton Keynes, MK11 3LW, UK
UKHW022219190926
475702UK00003B/443
* 9 7 8 1 0 3 2 7 6 0 4 8 3 *